THE ART AND ARCHAEOLOGY OF FLORIDA'S WETLANDS

THE ART AND ARCHAEOLOGY OF FLORIDA'S WETLANDS

by

BARBARA A. PURDY

CRC Press
Boca Raton Ann Arbor Boston London

Library of Congress Cataloging-in-Publication Data

Purdy, Barbara A.
The art and archaeology of Florida's wetlands / by Barbara A. Purdy
p. cm.
Includes bibliographical references and index.
ISBN 0-8493-8808-2
1. Indians of North America—Florida—Wood carving. 2. Indians of North America—Florida—Antiquities. 3. Wetlands—Florida. 4. Florida—Antiquities. I. Title.
E78.F6P85 1991
975.9′01—dc 20 91-15430
CIP

Developer: Telford Press

Direct all inquiries to CRC Press, Inc., 2000 Corporate Blvd., N.W., Boca Raton, Florida 33431.

International Standard Book Number 0-8493-8808-2

Library of Congress Card Number 91-15430
Printed in the United States

To my grandchildren

Someday, you may have to make decisions about the future of the past. Please choose wisely and cautiously.

CONTENTS

PREFACE

THE INDIANS OF Florida created sophisticated sculptures and utilitarian items in wood (Frontispiece). These artifacts are as ancient as those recovered in Switzerland and as elaborate as many carved by the Indians on the Northwest Coast of North America. For generations, textbooks and popular literature have illustrated and described the Swiss Lake Dwellers and Northwest Coast art. People flock to museums in both areas to view representative examples of these unique specimens. When people visit the Florida Museum of Natural History, they too can see displays of Northwest Coast art but they see very little of the woodworking accomplishments of Florida Indians. The reasons for this situation are historic. The Indian populations of Florida were reduced drastically early in the contact period shortly after the discovery of the New World by Columbus in AD 1492. By AD 1700 the Indian cultures of Florida were "wiped out" and the only knowledge of their woodworking skills comes from scant references in Spanish and French accounts. One of the De Soto narratives, for example, reported that a wooden fowl with gilded eyes was perched on the top of a Tocobaga Indian temple. The intruders often destroyed the "pagan" idols. Northwest Coast cultures, on the other hand, were relatively undisturbed until the latter part of the 18th century. By that time, explorers and the governments that sponsored them were interested in describing all aspects of native lifeways and collecting samples of material culture that could be exhibited. Also, Indian artisans still produce traditional styles in the area.

My primary objective in writing this book was to bring together in one place a record of organic objects that usually perish and the archaeological wet sites where they were found. Until now this information has been scattered in often obscure documents and articles thus diminishing its importance. Perhaps someday the objects themselves can be brought together in one location and given the attention they deserve. Even though our ancestors may not have been Indians, we should take pride in their creations. After all, the Indians contributed to the state's heritage for more than 10,000 years. Florida loses also because her citizens and potential visitors are deprived of the opportunity to learn about these superb prehistoric wooden sculptures. People travel to other states and countries to see antiquities.

Florida does not contain stone, such as marble, that is suitable for creating sculptures. Wood is the only other naturally occurring medium that can be

OWL TOTEM

used to produce large effigies. Since wood perishes rapidly in climates like Florida, its survival depends upon fortuitous circumstances. It must become entombed in an anaerobic environment, usually an organic soil, that prevents degradation by bacteria that require oxygen. Ironically, since the objects in such deposits are invisible, most of them have been discovered during ambitious dredging, draining, and development projects.

Another major motive for writing this book was to bring the cultural and environmental heritage of Florida's wetlands to the attention of developers and state government agencies. There is much concern with protecting the modern-day flora, fauna, and water quality of wetlands but very little attention has been paid to the thousands of years of heritage they contain. Before permits are given to develop wetlands, archaeologists who are trained in this area of expertise should be allowed to test deposits. There is no way to transplant this heritage to another location. Once disturbed, it is gone forever. Large wood carvings, canoes, and human skeletons might be noticed and sometimes reported by dragline operators but certainly it would be impossible for them to spot the multitude of small artifacts and dietary items that are present also.

Students of archaeology who are accustomed to interpreting the accomplishments of Florida's Indians from stone, shell, and pottery remains will be amazed at how interesting and informative the past can become with the addition of the great variety of categories preserved at wet sites. The Preceramic Archaic Period, for example, generated very little research effort because it was known primarily by 5000 years of stone artifacts until sites such as Bay West, Little Salt Spring, Republic Groves, and Windover were discovered. We are now aware of what people ate, the environments they preferred, the decorative and utilitarian items they made out of wood and bone, their weaving skills, what they looked like, and what killed them 4000–9000 years ago.

It was my good fortune that, while preparing to write various chapters and sections of this book, there had been a great deal of activity associated with Florida's wetlands: concern on the part of environmentalists about wetland preservation, interest by the government in exploiting peat areas for energy production, and discoveries during development projects of wet sites containing human remains and artifacts that were subsequently investigated by archaeologists. I acknowledge with deep appreciation the assistance with or review of various sections of the book by David N. Dickel, Glen H. Doran, James S. Dunbar, John A. Gifford, Marion S. Gilliland, David W. Grattan, George M. Griffin, Mitchell E. Hope, Shirley Royal, William R. Royal, William H. Sears, Bruce D. Smith, S. David Webb, and Ryan Wheeler.

I am grateful to numerous individuals for permitting me to use their unpublished reports, and to individuals and institutions for providing me with photographs. In many cases, photographs were reproduced specifically for me to be included in this volume. I owe special thanks to John L. Knaub and his coworkers in the Office of Instructional Resources at the University of Florida for their patience and cooperation in preparing photographs and illustrations.

My initial interest in wet-site archaeology was stimulated by Richard D. Daugherty in 1966 when I was a crew member at the Ozette Village Site on the Olympic Peninsula in the state of Washington. Other individuals who have been particularly influential in shaping my thoughts about this important topic are John M. Coles, Frank Hamilton Cushing, George F. MacDonald, Ray M. McGee, Lee A. Newsom, and Elizabeth S. Wing. All of these folks deserve credit for the contribution this book may make, but I accept full responsibility for its deficiencies.

The University of Florida granted me sabbatical leave for two terms during 1989–1990 which furnished time free from teaching and other responsibilities that was essential for research and writing.

Barbara A. Purdy
Gainesville, Florida
June 1990

INTRODUCTION

ARCHAEOLOGISTS IN the United States are trained as anthropologists whose discipline focuses on a holistic view of culture. Holism is "the theory that whole entities, as fundamental and determining components of reality, *have an existence other than as the mere sum of their parts* [emphasis mine] (Random House College Dictionary 1984:631). Culture has been defined variously depending upon the orientation of its author but a good general definition of culture is the way of life of a group based on symboling. The word *group* infers that individual actions are not important (in an analytical sense) unless they are part of a norm that can be incorporated into a model. For the same reason, idiosyncratic objects are not significant, no matter how interesting they are, because social scientists cannot determine what they represent to the culture. The word *symbol* restricts the definition of culture to humans because *Homo sapiens* is the only animal that creates symbols and transmits them from generation to generation. These terms, holism and culture, signify that the primary goal of anthropology is to comprehend a people's way of life as it is and has been influenced by its history, environment, institutions, ideals, and much more. Any culture, therefore, is extremely complex and cannot be easily dismembered or reassembled. For example, an anthropologist may concentrate on the economic institution of a specific society but, in interpreting the data, he or she will attempt to place the results of the study in juxtaposition to all other known factors.

Herein lies the problem for archaeologists who wish to use a holistic approach to the behavior of people who lived in the past. The archaeologist, as anthropologist, is constantly frustrated by how difficult it is to understand cultures, especially those for which no written documents exist and only a few fragments survive in quantities large enough to be studied statistically. The past not only is a foreign country but it usually lacks more than 90% of its material culture. Consequently, we find ourselves explaining lifeways based on stone implements ("oh, they were hunters") or dishes ("oh, they were horticulturists"). The above statement is too critical and too simple, of course, but most archaeologists would agree that there is a great deal of speculation associated with their interpretations. Stone tools and dishes may have been of minor importance to the people who used them but, when nothing else remains, they are assigned great importance by the archaeologists who analyze them.

Archaeologists are interested in learning how, why, when, and under what circumstances people produced tools or ornaments and carried out activities associated with them. They also want to know how the tools and activities fit into the total functioning of the culture and what prompted them to change through time.

Water-saturated archaeological sites yield a large portion of environmental information and material culture that is usually missing from upland areas. Wet sites are found in permanently saturated deposits that entomb and preserve organic objects that seldom survive elsewhere. This description distinguishes wet sites from shipwrecks and from inundated terrestrial sites where degradation preceded submergence (Purdy 1988:xi). The ideal site is where all biological materials are preserved. Because of an acid pH, bone often does not survive at some wet sites where plants are found in superb waterlogged condition. Bones, on the other hand, are sometimes recovered from terrestrial sites whereas botanical species never survive in upland areas unless they have been charred. In this book I have limited my discussion to sites where waterlogged botanical materials were recovered. Most of these sites contained bone also.

Florida has an abundance of wet sites (Fig. 1) and the unique categories of materials recovered from them furnish excellent opportunities to broaden the data base and provide a more holistic view of the past. By assigning a number to each surviving category and by assuming that each category was of equal importance in the culture, a formula can be used to demonstrate how many relationships are available for study:

$$\frac{k(k-1)}{2} = X$$

where k = number of categories and X = number of relationships. Thus, if only stone implements are available for study, there are no relationships:

$$\frac{1(1-1)}{2} = 0$$

The three categories of material culture that usually survive in Florida (stone, shell, and ceramics) yield three relationships:

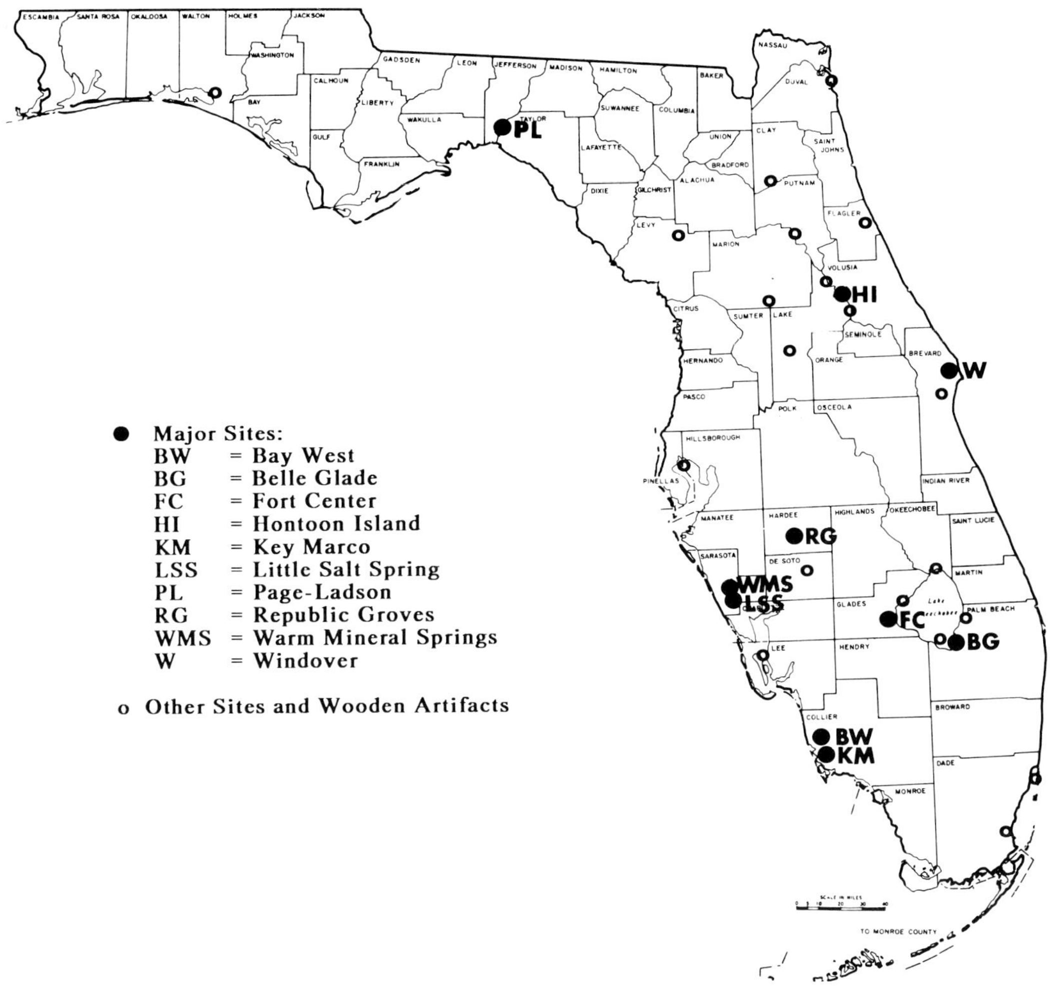

FIGURE 1 Locations of sites mentioned in the text (for canoe locations see Fig. 108).

$$\frac{3(3-1)}{2} = 3$$

but the seven major constituents of material culture from a wet site such as Hontoon Island (stone, shell, ceramics, bone, wood, noncultivated plants, and cultivated plants) can be expressed as

$$\frac{7(7-1)}{2} = 21$$

The four organic categories (bone, wood, noncultivated plants, and cultivated plants), when added to the three nonperishable categories, provide magnitudes of associations that usually do not exist (Fig. 2). Obviously, these added relationships furnish greater possibilities for interpretation. The increase in constituents from wet sites, then, are not just single additional pieces of material

FIGURE 2 Specimens representing seven major categories of material items recovered at Hontoon Island (photo by the author).

culture; they provide an exponential potential for increasing our understanding of the past. This topic is considered in greater depth in Purdy (1988:325–326).

Not all wet areas in Florida will contain the types of deposits essential for the preservation of fragile organic materials. The environmental circumstances essential to their survival are discussed in Chapter 2. I have incorporated into a model the major constituents that must be present in order to expect the kind of preservation desired.

In Chapter 3, I discuss the contents of all of Florida's wet sites that I was able to document. The excavation techniques employed at these sites are given when they are available. The information these sites disclose is truly astounding. The preservation of dozens of species of wood, for example, makes it possible to understand more fully what the environment was like in the past and why the aborigines valued the properties of particular woods over others to perform certain tasks. The adze marks on the wood can be matched to shell or stone adzes permitting a study of woodworking techniques and even motor skills. The added categories make it possible to determine the relative importance of plants to animals, cultivated plants to noncultivated plants, aquatic to terrestrial species, and much more. The enlarged inventory from wet sites also furnishes data about continuities and discontinuities in cultural practices that are usually discernible only from stone, shell, and ceramic remains. Abrupt changes in Indian cultures that followed European contact, for example, were documented in water-saturated strata at Hontoon Island. These changes would not have been noted at all if excavations had been restricted to the terrestrial portion of the site where the deposits were trampled or were compressed when the organic fraction decomposed. The wooden sculptures recovered from wet sites suggest that there may have been cultural divisions within the state that are not readily perceived from utilitarian items and subsistence practices. Cushing (1897) recognized the individuality of the wood carvings at Key Marco, and Schwehm (1983) describes the different ways in which animals were portrayed in various regions of Florida.

Each wet site provides more diverse information than does an upland site and each wet site is different with regard to what it reveals about past environments and cultures. Elaborate wood carvings, grave goods, human skeletons, dietary items, and much more have given us expanded views of people and their behavior that is not possible under any other conditions except from dry tombs and caves. It is logical that information gained from wet site excavations can be used, with caution, to make reasonable extrapolations about objects and activities that were formerly present on nearby upland sites. The contents of wet sites can be compared globally to each other and to organic materials preserved under extremely dry conditions (Purdy 1988:328–333). Most important of all, we have not yet turned the first corner in utilizing the full potential of the added information recovered at archaeological wet sites.

Florida canoes are discussed in Chapter 4. Aboriginal canoes are Florida's stepchild. They are big, difficult to transport, preserve, store, and exhibit. Yet, the information furnished by canoes provides important insights about adaptations to the aquatic resources that supplied the Indians' major subsistence items and probably led to the cultural elaboration and population increase that began more than 5000 years ago. Florida has the oldest watercraft in the Western Hemisphere and the largest number recorded anywhere in the world. Perhaps it is this latter fact that has diminished their importance. In states where only one or two canoes have been found, unlimited resources have been allocated to study and preserve them.

No investigator should make plans to excavate an archaeological wet site without first making proper arrangements for the preservation and analysis of the vast quantities of diverse materials that can be expected. A conservator should be part of the research team and should be ready to process objects as they are removed from the water-saturated deposit. At the very least, preparation should include containers, waterproof markers, preservatives, and a recording system. The archaeologist is faced with a collection of items for which there is no comparable material. Adequate funding should be made available to ensure that the organic components of the site are properly analyzed. Study and preservation of wet-site materials are a problem because most archaeologists and granting agencies have not worked at wet sites and do not realize how rapidly the botanical specimens will disintegrate or how much time is needed to analyze the diverse environmental and cultural components when every seed, nut, berry, and wood scrap has survived. A rough calculation of what is involved in analyzing artifacts and ecofacts from wet sites can be arrived at by estimating how long it would take to sort, record, describe, and interpret typical materials found at terrestrial sites, multiply this figure by the number of elements at wet sites, then add time for preservation to prevent degradation and for flotation processing. These matters are taken up in Chapter 5.

The broad geographic distribution, the diverse environmental situations, and the great time span of the artifacts recovered from Florida's archaeological wet sites suggest that a great wealth of wooden artifacts and other botanical specimens may still be undiscovered. With the recovery of this heritage comes a major responsibility to preserve it for the future. We have a great deal yet to learn about treating wooden artifacts. One major problem is a matter of linguistics. We speak of waterlogged wood as being *preserved* yet the wood is obviously not in pristine condition. One need only compare the changes that take place in *old* waterlogged wood if it is permitted to dry with a piece of *recent* wood exposed to the same conditions. In most cases, the wood is not preserved at all. We can still see it because its macroscopic, physical dimensions have remained pretty much intact in the oxygen-free environment in which it was entombed. This only means that organisms needing free oxygen have not been able to degrade the wood. Other mechanisms are operating, however, because certain constituents are replaced by 87–93% water giving the impression that the piece is structurally sound as long as it remains waterlogged. Preservation treatment should not be hurried if one is anticipating that these unique specimens will last for thousands of years.

Sampling strategies are very important for wet-site excavations. If an attempt were made to save, study, and store all material from wet sites, it would take years to analyze and warehouses to store. It is the responsibility of the archaeologist to develop sampling methods most appropriate for a specific site and for the problems under investigation. One of the requirements of science is to have data upon which conclusions are drawn available for other workers who may wish to verify for themselves how results were obtained. I believe the data should include not only a compilation of records on paper, but also the actual objects, curated and stored in accessible locations. Obviously, the quantity of the material must be manageable.

In 1979, I wrote to all of the County Extension Directors, the Soil Conservation Offices, and the peat producers in the state of Florida asking them for information about organic deposits and artifacts that had been found in them. The response from the Extension Directors and the Soil Conservation Service was excellent but none of the peat producers replied to my inquiry. I was able to identify a number of soil types that are good candidates for predicting that organic materials might be preserved in them (Purdy 1981). One point mentioned by several individuals was that certain U.S. Geologic Survey maps are out of date. These include those for Bradford (1913), Duval (1921), Escambia (1906), Flagler (1918), Gadsen (1903), Hernando (1914), Hillsborough (1916), Indian River (1913), Lake (1923), Leon (1905), Orange (1919), Pinellas (1913), Polk (1927), Putnam (1914), and St. Johns (1917) Counties. Some of these may have been revised since 1979. It is interesting to contemplate why a soils map would become outdated. Under normal, natural processes it is unlikely that this would occur in 50 years, 500 years, or longer. Some of the soil maps that have been revised reflect drainage of about 80 years and it may not be apparent from republished maps if organic soils were more abundant earlier. The Dade County map was republished in 1947 but it probably does not provide a true picture of the soil types today because of the extensive development that has taken place around Miami since 1947. In areas where drainage has occurred, subsidence is estimated to average 2.5 cm/year. At this rate, it takes only about 6 years to erase the accumulated history of 100 years of "rapid" peat development based on 16 cm/century. Future studies should include a history of peat mining and drainage projects and a search for predrainage period maps. Individuals with long-time knowledge of local events should be consulted before an accurate picture of predrainage/dredging conditions can be depicted.

In 1980, I conducted a statewide field survey along with George F. MacDonald (Director, Museum of Man, Ottawa, Canada) to assess the kinds of deposits in which wooden artifacts and other botanical remains of archaeological interest might be preserved (Purdy 1981,1982). Most of the sites no longer exist where wood carvings were recovered in the past. Instead, powdery, subsiding soils have been created by drainage canals or housing developments repose upon dredged land. In 1980 also, I wrote to institutions throughout the state to inquire if they had wooden artifacts in their collections. By this method, I was able to obtain information about some of the specimens described in the last section of Chapter 3.

The usually perishable organic materials from wet sites provide the only broad opportunity to discover relationships between resources available and resources utilized by prehistoric groups of people. Despite the superb preservation that exists in Florida, however, not all materials of interest survive. Hair and flesh of humans and other animals, for example, have not been found.

References

Cushing, Frank Hamilton 1897 Exploration of Ancient Key Dwellers' Remains on the Gulf Coast of Florida. *Proceedings of the American Philosophical Society* 25(153):329–448.

Purdy, Barbara A. 1981 Cultural Resources at Wetland Sites in Florida: Investigations of Sites and Preservation of Artifacts. Final report, Survey and Planning Grant, Heritage Conservation and Recreation Service. On file, Division of Historical Resources, Tallahassee.

1982 Survey, Recovery, and Treatment of Wooden Artifacts in Florida. *Proceedings of the ICOM Waterlogged Wood Working Group Conference*, Ottawa, 1981, pp. 159–169.

1988 Archaeological Wet Sites: Untapped Archives of Prehistoric Documents. In *Wet Site Archaeology*, pp. 325–335. Edited by Barbara A. Purdy. The Telford Press, West Caldwell, New Jersey.

Random House College Dictionary 1984 Revised edition. Random House, Inc., New York.

Schwehm, Alice Gates 1983 The Carved Wood Effigies of Fort Center: A Glimpse of South Florida's Prehistoric Art. M.A. thesis, Department of Fine Arts, University of Florida, Gainesville.

WET SITE ENVIRONMENTS

Introduction

Archaeological wet sites are almost always discovered accidentally during development activities, because they are entombed in a matrix that renders them invisible. As a result of this situation, much significant information is lost even under the best of circumstances, and some sites probably have been totally destroyed without record.

The geographic extent, developmental history, and composition of most of Florida's organic soils are fairly well known because of economic interests in them for agricultural purposes and as potential energy sources (e.g., Davis 1946; Gleason 1974, 1984; Griffin et al. 1982). By comparing this information to reports of archaeological wet sites, it should be possible to advise developers and government agencies about the probability of archaeological wet sites occurring in specific areas and suggest ways to protect them or to extract representative samples of their contents before they are demolished. This chapter, therefore, focuses on conditions in the natural environment that may favor preservation of artifacts manufactured of organic materials and a model is proposed that can be used to predict the locations of archaeological wet sites. My interest in this topic has evolved over a period of 20 years while endeavoring to find out (1) what constitutes a wet site, (2) why floral and faunal materials survive at some waterlogged sites, while only floral materials survive at others, (3) why organic artifacts are preserved in a superb state at some sites but are extremely degraded elsewhere, (4) what conditions permitted human brains, but none of the other organs, to survive at several sites in Florida, and (5) why Florida does not have "bog bodies" like those found in northern Europe.

From Cushing's description, quoted below, one gets the idea that odor, texture, depth below surface, and the underlying deposits may have been important factors contributing to the survival of hundreds of masks, figurines, and utilitarian items made of wood and bone that were recovered from the famous Key Marco site in Collier County, Florida.

> . . . the surface deposit consisted of a stratum of spongy black or dark brown muck, permeated by both rotting and living rootlets . . . eighteen or twenty inches thick throughout the middle. Below this was a somewhat thicker stratum of brownish gray peaty marl, soft, tremuous, exceedingly foul-smelling, and rich in the best preserved relics we discovered. Underneath it, throughout the middle of the court was a less well-defined layer of the less peaty marl, intermixed with shells and other debris, and also with abundant ancient remains—which indeed we continued to encounter even in the underlying, comparatively firm shell and clay-marl bottom (Cushing 1897:358).

Depletion of oxygen by rotting organic material in the organic soils prohibits most bacterial growth, thus preserving the residual remains. The term "odor" suggests hydrogen sulfide, a poisonous gas when breathed in large quantity, which can only exist when oxygen has been depleted (Griffin, personal communication 1989).

In order to produce a covering rule that can predict where archaeological wet sites occur and to assess how these sites will be impacted if they are disturbed, it is necessary to examine the importance of variables such as bedrock, climate, vegetation, chemistry, and drainage (oxidation) that have been involved in the formation, deformation, and loss of Florida's organic soils.

The Organic Soils of Florida

Composition of Underlying Deposits

The topography, chemical composition, and physical characteristics of bedrock have influenced the overlying sediments in several ways (Gleason et al. 1974:288). All of Florida is underlain by limestones. In some locations, particularly South Florida, organic soils lie directly on limestones or on freshwater marls that formed either in the late Pleistocene or early Holocene (Fig. 3). There are numerous incidences, in fact, where there is interbedding of marls with peat or muck deposits (Davis 1946:162–164; Gleason et al. 1974:298). Because of the high calcium carbonate content that exists under these conditions, the organic soils are alkaline with a pH above 5.5, especially at the bottom of the sequences. Marls, also called calcitic muds, are thought to form when blue-green algal periphyton precipitate calcium carbonate (Gleason and Spackman 1974:146). Calcitic muds develop only in the presence of limestone because a source of calcium and carbonate ions is required. Marls occur when there is a relatively long dry period alternating with a wet period. They are located in areas which dry out shortly after the termination of the rainy season. The severity of a short hydroperiod coupled with a long seasonal drought favors calcitic mud deposition (Gleason et al. 1974:298). These freshwater

FIGURE 3A Interbedding of marl and peat (A) from Brooks 1974:277, (B) from Gleason et al. 1974:325).

deposits are important to our study because they cover over 1 million acres of land in South Florida (Gleason and Spackman 1974:146) and because the presence of calcitic muds underlying or interbedded with peat deposits in the Everglades and elsewhere is indicative of climatic conditions different from those conducive to the continuous accumulation of peat (Gleason et al. 1974:291). Calcitic muds, for example, have been identified in Warm Mineral Springs associated with human skeletal remains, animal bones, and artifacts (Clausen et al. 1975:197). The implication is that alternating wet and dry periods, and considerably more aridity than present, must have prevailed at the site from around 10,000 years ago to 8500 years ago. Little Salt Spring has a similar deposition history (Clausen et al. 1979:609).

In other areas of Florida, the organic soils are distinctly acid with a pH as low as 3.5. The underlying sediments of these deposits are often quartz sands derived from Upper Miocene or Pleistocene origins. Davis (1946:207) illustrates and describes a former mining operation at Leesburg where the peat deposit rests directly on sand. Stricklin Peat Co. near Grandin is also a good example of this type of organic soil. Davis provides a possible explanation for the development of peat on top of pervious sand or clay deposits.

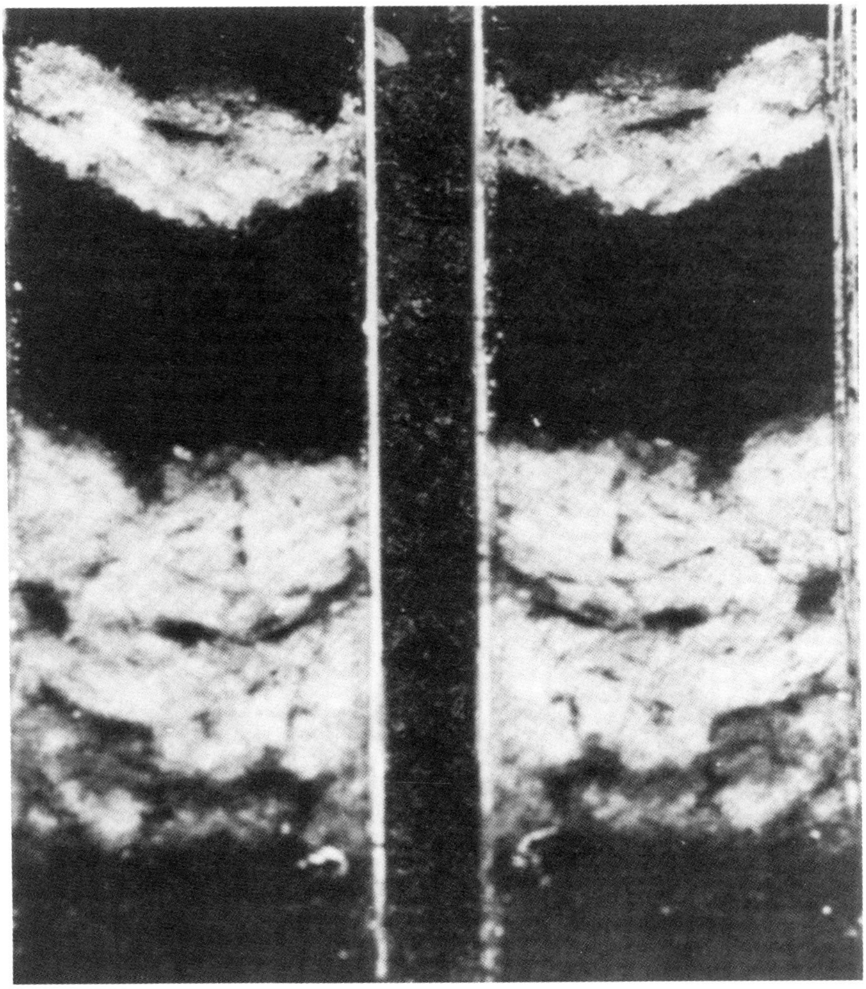

FIGURE 3B

> The limestones underlying sands and clays are far below the lake bottoms in some places but the solution channels in this rock are effective drainage passages as long as the sands and clays above them remain pervious. With some gradual changes in the sands and clays these frequently become less pervious causing a reduction in underground drainage. This condition affects the water in the basins causing a slow to rapid rise (1946:182).

As a result of other variables operating independently of the underlying sediments, pH can fluctuate considerably within a region or even within a single deposit depending upon the location or depth that a sample is tested. Since pH is a simple attribute to measure, and an important one, it is recommended that it be assessed for every site.

To summarize, bedrock conditions affect the nature of the overlying organic soils. A major influence is the hydrogen-ion concentration expressed by pH which indicates acidity or alkalinity. It is essential to consider pH in constructing a model to predict locations of archaeological wet sites. For example, bone will not survive in an acid environment because calcium carbonate, a main constituent of bone, dissolves in the presence of acid. Wooden artifacts, however, survive in superb condition in acid peat soils. Bedrock also affects drainage to a certain extent; however, most limestones, marls, clays, and sands tend to be more or less porous and the development of organic soils above them probably depends to a great extent upon a rise in the water table that decreases drainage through solution channels in the limestone bedrock.

Climate

Glaciers that formed during the last ice age had reached their maximum extent about 17,000 years ago. They buried northern Europe under a mile-thick sheet of ice and, in North America, they blanketed all of Canada, the Great Lakes, and reached into New England and the Midwest. Overall, the glaciers covered some 11 million square miles of land that is now ice free. They began to melt some 14,000 years ago; however, at 11,000 years ago, when they had retreated north of the Great Lakes, meltwater surged down the St. Lawrence River and jammed a kind of conveyor belt that was bringing warm salty water to the North Atlantic from the Gulf of Mexico. When the conveyor belt was stopped, it triggered a cold interlude (the Younger Dryas) that lasted for 1 thousand years. The above model is only one of many designed to explain the sudden relapse into deep cold after glaciers had been in full retreat for several thousand years (the Two Creeks Interstade). This theory has been supported with evidence from sediment cores taken in the Gulf of Mexico (Kunzig 1989:74–79). Freshwater calcitic muds from the southeastern end of Lake Okeechobee, dated at >12,000 BP, were deposited during the Two Creeks Interstade (Gleason et al. 1974:300,310).

The world's environments have been adjusting to postglacial conditions for much of the time since the Ice Ages terminated at least 10,000 years ago. A date of 10,000 years is a good place to begin a discussion of Florida's climate because human presence in Florida has not been securely documented prior to that time although limited evidence exists for an earlier occupation (Clausen 1979; Purdy 1981).

While there is not a general consensus about the level of the sea surrounding the Florida peninsula 10,000 years ago (Fairbridge 1974:223–232; Robbin 1984:437–458), there is agreement that it was many meters lower than present and that Florida had more land area than today. From extensive investigations by William A. Watts (see Watts and Hansen 1988 for a recent summary article) and others, it is clear also that xerophytic vegetation existed at that time and very little surface water was present. Fairbridge explains the glacial scene as follows:

> In the normal course of events, water that is evaporated from the ocean surface moves inland as cloud to precipitate rain or snow and is returned eventually to the sea by way

> of rivers and by seepage. But in an ice age the water heat budget gets out of balance. Snow is converted to glacier ice and steadily accumulates while the level of the ocean drops resulting in less ocean surface to provide for evaporation. As the sea surface becomes cooler there is a reduction in the rate of evaporation; thus, there is less cloud and decreased precipitation. Without rain the subtropics essentially dry up and great deserts develop and expand. The entire globe gets colder and colder as well as more and more arid. Desert sands blew . . . and built giant . . . dunes in western and central Florida (Fairbridge 1974:223).

There are many archaeological sites that provide support for the climatic and geologic evidence that Florida was desert like prior to 10,000 years ago. These sites occur in springs, beneath peat deposits, and along the coasts now buried under many meters of water particularly where old river beds lead out to the Gulf. These sites will be discussed in detail in the next chapter.

With the melting of the glaciers, the seas rose, and increased moisture resulted in more available fresh surface water and higher water levels in Florida's aquifers. Drainage was retarded by longshore drifting of sands, especially along the Atlantic Coast, and ponding began in existing depressions, especially in the Everglades (Griffin, personal communication 1989).

Vegetation

The vegetation patterns seen in Florida's wetlands today are extremely young (<6000 years old) and have been continually adjusting to sea level rise throughout their development (Gleason and Spackman 1974:167).

The kinds of plants that decompose partially to form major constituents of organic soils vary as a result of local conditions of geology, climate, and topography. Davis (1946:81–84) identifies types of fibrous, woody, and plastic peats and mucks including sawgrass, mangrove, and prairie and pond. Investigations have shown that several generations of plants form the organic soil deposits of Florida and, in places, one distinct type of vegetation has replaced another through time (Davis 1946:16–17). Beneath coastal mangrove swamps, for example, is found a transgressional sequence of terrestrial peats, overlain by freshwater calcitic muds, mangrove peats, and marine muds reflecting a postglacial Holocene rise in sea level (Wanless 1974:190; Davis 1946:21–22). In other cases, marsh type vegetation is replaced by a terrestrial flora as basins fill in or, as in the Everglades, sawgrass replaces water lily. Sawgrass is the dominant marsh vegetation in the Everglades today probably because it has low nutrient requirements (Steward 1974:37).

This succession from one plant type to another signals a change in climate, depth, and moisture requirements. At a number of archaeological sites these factors have been observed as important to the survival of artifactual materials and to the identification of resources available to prehistoric residents (Doran and Dickel 1988; Clausen et al. 1975; Clausen et al. 1979; Brown and Cohen 1985).

Chemical Composition

Peat accumulates and matures when the addition of new plant debris exceeds its rate of decomposition. A change to arid conditions allows air to enter the soil and aerobic bacteria flourish to oxidize the peat (Griffin, et al. 1982:24). It is instructive to review the mechanisms involved in the decomposition of wetland vegetation as it forms peat because the same processes affect cultural items manufactured of organic materials that become entombed in peat deposits.

Plants undergo a degree of disintegration to form peat but complete decomposition is prevented because the plant remains are submerged in oxygen-depleted waters where the dissolved oxygen saturation approaches zero, aerobic bacteria are inhibited, and bacterial oxidation of the plant remains is minimized. The activities of fungi and burrowing animals are also retarded. Decomposition is further inhibited by the build-up of partial decay products such as methane, alcohol, hydrogen sulfide, and organic acids (Moore and Bellamy 1974 in Griffin et al. 1982:4). CO_2 is produced by anaerobic bacteria using oxygen supplied by the slow decomposition of organic matter within the soil but this process is not rapid (Goode 1982:51).

The sulfate ion appears to play a major role in the development and maintenance of many organic deposits. As saturated or buried soils become increasingly anaerobic due to the lack of atmospheric oxygen, the sulfate ion undergoes a series of transformations that reduces it to sulfide. This reduction is carried out by microorganisms, primarily bacteria, which exist in anaerobic and poorly oxidized environments. Sulfide ions combine with hydrogen ions to form hydrogen sulfide (Goode 1982:2–3) which accounts for the "exceedingly foul" odor noted by Cushing (1897:358) and all of us who have excavated archaeological wet sites. The presence of this odor is probably a significant clue to the possibility of well-preserved organic matter. In contrast, when aerobic conditions are established, oxidation causes the sulfide to change back to sulfate or elementary sulfur. In the absence of sulfide in oxidizing levels, aerobic bacteria, fungi, insects, and a lowered moisture content are effective in destroying the texture and structure of the peat deposit. (It is a well-known fact also that mass mortalities of fish occur when phosphate levels are high and large quantities of hydrogen sulfides are released during drought conditions [Wood and Maynard 1974:126].) The optimum moisture content for anaerobic conditions is 80% and optimum oxidizing conditions exist at 50% moisture (Goode 1982:48–49). In addition, once peat is dried below 40% moisture, it is not readily wetted again (Davis 1946:46). This property of resistance to readsorption or reabsorption of water by dried peat is present also in wood and other floral remains of interest to archaeologists.

Another feature of sulfide in an anaerobic environment is its reaction with reduced iron minerals to form iron sulfides (pyrite) (Goode 1982:3,56). Pyrite is stable only in anaerobic conditions and may have an effect on how well organic materials respond to preservation treatments. This situation is discussed in Chapter 5.

The constituents of living plants are numerous and are subjected to microbiological decomposition immediately after death. Some of these constituents do not become part of the peat but most of them contribute to peat formation. A comparison of the inorganic minerals (expressed as ash content) and organic compounds of peats with those of the peat-forming plants from which they were derived illustrates the degree of degradation that occurs. Some of the inorganic components of the peat such as calcium oxylate (weddellite) were present in the living plant tissue (Griffin et al. 1984) but others were brought into the deposit in free water (Sawyer and Griffin 1988). Organic substances include lignins, celluloses, hemicelluloses, proteins, fats, resins, waxes, and gums (Davis 1946:61). In the formation of peat, celluloses, and hemicelluloses are broken down rapidly. Living plants are composed of about 50% cellulose and hemicellulose while peat contains only about 6% (Davis 1946:72). Other constituents, such as lignins, waxes, and proteins, that have a higher percentage of carbon increase proportionately. The lignins are usually the most important of the carbohydrate substances retained in the peat.

In general, peats contain a many-fold increase in ash, proteins, and lignin content over the plants from which they were derived.

Peats vs. Mucks

In Davis' monumental treatise on the *Peat Deposits of Florida,* he laments the fact that while many European countries use peat for fuel, this has not been done on a large commercial scale in the United States. Florida, he says, has the third largest amount of peat in the country and the organic nature of peat with a high carbon content indicates that it should prove useful as a fuel for heat and power.

> *The fact remains that the expansion of industries in Florida greatly depends upon the development of cheaper power developed from cheaper fuels. Consequently, if peat could be mined, processed, and burned efficiently on a large scale this abundant resource of the State might provide the needed cheaper power for industrial expansion* [italics in the original] (Davis 1946:221–222).

Davis' investigations were supported heavily by the United States Bureau of Mines.

It is interesting to examine this issue as it has evolved since Davis' time. Two major factions exist: (1) the environmentalists, on the one hand, who view with alarm the destruction of wetland environments and (2) others who wish to exploit wetland environments for economic purposes. The latter pursuit became intensified with the fossil fuel crisis in the early 1970s which prompted studies of Florida's present-day peat resources. One of these studies was commissioned by the U.S. Department of Energy (Griffin et al. 1982) and another by Coastal Petroleum Company (Williams Brothers Environmental Services 1980).

These publications, in addition to two volumes of the *Environments of South Florida* (Gleason 1974; 1984), Goode (1982), and Davis (1946), furnish excellent descriptions of various classifications of organic soils, some of which have a direct bearing on the model we are attempting to build in order to predict if materials of archaeological significance are entombed in them.

Peats and mucks are both considered organic soils. Soil scientists distinguish bestween peats and mucks based on the percentage of organic matter present. Peats contain more than 75% organic matter while mucks are composed of 25–75% organic matter in the dry state. Mineral soils contain <25% organic matter. Organic carbon content is inversely related to the ash (mineral) content. This is because ash represents the inorganic portion of the soil and, therefore, as the quantity of one of these variables increases, the other decreases. There is a striking relationship between moisture and organic carbon. Since organic matter is composed of a high percentage of organic carbon, when the organic matter decomposes as a function of oxidation, CO_2 is produced, and the quantity of organic carbon is decreased (Davis 1946:15, 25–29,38–40,65; Goode 1982:14; Griffin et al. 1982:3).

Of particular interest to our study is the fact that mucks tend to be nonfibrous (<5%) whereas peats are usually fibrous. Most researchers agree that the presence of muck indicates that the deposit has undergone a degree of biochemical oxidation or it had a sedimentary origin where the materials forming the muck were largely transported to the place of deposition (allochthonous). One example of this latter situation from an archaeological site is the colloidal (rubber) "peat" observed at the Windover site which is thought to have resulted from a saltwater intrusion 8000–9000 years ago coupled with high algal/bacterial activity (Doran and Dickel 1988:283). Fibrous peats are those that are formed in situ (authochthonous) where little or none of the organic constituents have been transported (Davis 1946:15) or rendered amorphous by oxidation. The same processes that render the plant constituents of an organic soil deposit unidentifiable when moisture content is reduced below a critical level also destroy wooden artifacts that have been entombed in the deposit.

In Chapter 3, I will describe and illustrate some wooden artifacts that were found at sites that no longer exist in the Everglades and other areas of Florida because they were destroyed as a result of drainage and subsequent oxidation of the organic deposit.

The Model

Theoretically, artifacts manufactured of perishable materials need only standing water to survive. But water seldom stands still over long periods of time. Archaeological sites that preserve these kinds of artifacts, therefore, are usually found in permanently saturated organic soils. These soils are generally fibrous because nonfibrous soils suggest that oxidation has occurred. An exception to this rule would be if artifacts or human bones were intruded into a muck deposit that had already stabilized. The odor of hydrogen sulfide should be expected because it indicates that anaerobic conditions prevail which retard decomposition. Both floral and faunal material will survive under the above conditions if the hydrogen ion concentration (pH) is alkaline, but bone will perish if acid conditions exist. Bone should be well preserved in lower parts of peat deposits that overlie limestone or marl and should not be expected where peat deposits overlie sand. A major exception to this latter situation occurs if shell is incorporated into the matrix (as in shell middens), thus increasing the level of calcium carbonate. While the type of vegetation making up the peat deposit is indicative of environmental conditions, vegetation itself is not a prime factor in predicting preservation. Depth below surface may be a factor, with better preservation sometimes occurring in deeper levels that have been subjected to less oxidation during dry periods.

There does not appear to be any clear explanation at present why brain material, but no other soft tissue, has been preserved at several waterlogged archaeological sites in Florida. It is possible that brains are easy to see because they are contained in the crania whereas other organs may have become part of the soil mass. Angel (1970) proposed that when brain material persisted after skin and flesh had decayed, it was "because of a chemical change involving denaturation of protein, persistence of fat in myelin and cholesterol." There are reasons why some 2000-year-old bodies found in bogs in northern Europe retain flesh and stomach contents but no analogous situation exists in Florida. The requirements are

1. Deposition of the body must be in water deep enough to prevent attack by organisms, such as maggots, rodents, and foxes, and still enough to be oxygen deficient to inhibit decay by bacteria.
2. The bog pools must contain water with tannic acid of sufficient strength to begin to preserve the outer layers of the body by tanning. Of interest is the fact that many of the bodies do not have bones because of the extreme acid conditions.
3. The temperature of the water in the bog pool must have been below 4°C when the body was deposited. This normal refrigerator temperature prevented decay and inhibited rot (Coles 1988a:223–224).

Water temperatures in Florida have never satisfied the requirement of number 3.

Summary Comments

When vegetation decomposes to form peat, a major percentage of the celluloses and hemicelluloses are lost leaving lignins as the primary carbohydrate component. As long as the peat remains waterlogged further degradation is slowed; however, as soon as the moisture content is decreased below a critical level, air enters the soil and allows aerobic bacteria to oxidize the remaining organic compounds. This effectively upsets the critical balance that has been reached and maintained under waterlogged conditions. A weight loss of >80% occurs and the structure of most peats is destroyed rapidly when water is withdrawn; this accounts for the subsidence observed with the drainage of wetlands (Fig. 4). A weight loss of 87–93% accompanied by extreme shrinkage and contortion has been measured for artifactual woods from archaeological wet sites that have been allowed to air dry (see Chapter 5).

In Griffin's study of the peat resources of the northern Everglades, he concludes that

> Although agricultural usage of the soil is not usually thought of as consumptive, the drainage-induced oxidation of the peaty soils used for sugar cane production is evidently consuming peat at a high rate. For example, during the 35 years between the 1946 estimates of Davis and those of the present 1981 survey, the tonnage of peat in the Palm Beach-Hendry County area has been reduced on the order of 741 million tons; that figure amounts to peat consumption on the order of magnitude of 21 million tons per year from those two counties. That rate is 137 times the total 1979 production in Florida for all other use (Griffin et al. 1982:22).

Conditions in the natural environment established the organic soil deposits of Florida over a 4000- to 6000-year period. It has taken human manipulation of the environment approximately 100 years to use up these soils through drainage, agriculture, and the nursery business. Some countries regret the destruction of their organic soils. In 1987 the Dutch, whose record of drainage of wetlands is second to none in Western Europe, purchased several Irish bogs and gave them to the Irish nation. They said that they had allowed their own unique wetlands to perish and they did not want it to happen to the Irish bogs (Coles 1988b:72). Only a small percentage of Ireland's wetlands remain.

One would expect that the preservation of most floral and faunal material including human skeletons and wooden artifacts would date only to about 5000–6000 years ago when moist conditions in Florida became widespread. This is not always the case, however, since many spectacular archaeological wet sites in Florida are older than 6000 years. These include the sites at Little Salt Spring, Warm Mineral Springs, Windover, Republic Groves, and Bay West. The survival of these older sites may be the result of people congregating around available water supplies located in sinks or deep depressions at a time when surface water in Florida was still scarce.

A

B

FIGURE 4 (A) Illustration of desiccation of strata that occurs when deposits are drained (photo by the author 1986). (B,C) The concrete monument at the everglades experiment station showing subsidence losses (B) between 1924 and 1974 (from Stephens 1974:359), and (C) between 1974 and 1989 (photo by the author 1989). In 1924 the marker was flush with the ground surface; by 1974, 5.5 feet of subsidence had occurred; by 1989, 6.25 feet of subsidence had occurred.

FIGURE 4C

References

Angel, S. Lawrence 1970 Letter to Mitchell E. Hope, June 30. Dr. Lawrence worked in the Division of Physical Anthropology, Smithsonian Institution, Washington, D.C.

Brown, Janice G. and Arthur D. Cohen 1985 Palynologic and Petrographic Analyses of Peat Deposits, Little Salt Spring. *National Geographic Research* 1(1):21–31.

Clausen, Carl J., H.K. Brooks, and Al B. Wesolowsky 1975 The Early Many Site at Warm Mineral Springs, Florida. *Journal of Field Archaeology* 2(3):191–213.

Clausen, Carl J., A.D. Cohen, Cesare Emiliani, J.A. Holman, and J.J. Stipp 1979 Little Salt Spring, Florida: A Unique Underwater Site. *Science* 203(4381):609–614.

Coles, John M. 1988a An Assembly of Death: Bog Bodies of Northern and Western Europe. In *Wet Site Archaeology*, pp. 219–235. Edited by Barbara A. Purdy. The Telford Press, West Caldwell, New Jersey.

1988b The Peat Hag. *Council for British Archaeology*, Report No. 38. Cambridge.

Cushing, Frank Hamilton 1897 A Preliminary Report on the Exploration of the Ancient Key Dweller Remains on the Gulf Coast of Florida. *Proceedings of the American Philosophy Society* 35(153):329–448.

Davis, John H., Jr. 1946 The Peat Deposits of Florida. *Florida Geological Survey, Geological Bulletin* No. 30, Tallahassee.

Doran, Glen H. and David N. Dickel 1988 Multidisciplinary Investigations at the Windover Site. In *Wet Site Archaeology*, pp. 263–289. Edited by Barbara A. Purdy. The Telford Press, West Caldwell, New Jersey.

Fairbridge, Rhodes W. 1974 The Holocene Sea-Level Record in South Florida. In *Environments of South Florida: Present and Past*, pp. 223–232. Edited by Patrick J. Gleason. *Miami Geological Society, Memoir* 2. Miami.

Gleason, Patrick J., Editor 1974 *Environments of South Florida: Present and Past. Miami Geological Society, Memoir* No. 2, Miami.

1984 *Environments of South Florida: Present and Past II. Miami Geological Society,* Coral Gables.

Gleason, Patrick J. and William Spackman, Jr. 1974 Calcareous Periphyton and Water Chemistry in the Everglades. In *Environments of South Florida: Present and Past,* pp. 146–181. Edited by Patrick J. Gleason. *Miami Geological Society, Memoir* No. 2, Miami.

Gleason, Patrick J., Arthur D. Cohen, William G. Smith, H. Kelly Brooks, Peter A. Stone, Robert L. Goodrick, and William Spackman, Jr. 1974 The Environmental Significance of Holocene Sediments from the Everglades and Saline Tidal Plain. In *Environments of South Florida: Present and Past,* pp. 287–341. Edited by Patrick J. Gleason. *Miami Geological Society, Memoir* No. 2, Miami.

Goode, Richard W. III 1982 Vertical Distribution of Sulfide in the Organic Soils of the Northern Everglades, Palm Beach County, Florida. M.S. thesis, Department of Geology, University of Florida, Gainesville.

Griffin, George M., C.C. Wieland, L.P. Hood, R.W. Goode III, R.K. Sawyer, and D.F. McNeill 1982 Assessment of the Peat Resources of Florida, with a Detailed Survey of the Northern Everglades. United States Department of Energy Publication DDE/FE/05114-T1; DE84014060;190 pp. (Also published by State of Florida, Governor's Energy Office, Tallahassee.)

Griffin, G.M., R.K. Sawyer, and S.R. Melkote 1984 Weddellite Occurrence in Peats and Other Organic-Rich Sediments of Florida. *Journal of Sedimentary Petrology* 54:861–868.

Kunzig, Robert 1989 Ice Cycles. *Discover* 10(5):74–79.

Purdy, Barbara A. 1981 *Florida's Prehistoric Stone Technology.* University Presses of Florida, Gainesville.

Robbin, Daniel M. 1984 A New Holocene Sea-level Curve for the Upper Florida Keys and Florida Reef Tract. In *Environments of South Florida: Present and Past II.* Edited by Patrick J. Gleason. *Miami Geological Survey,* Coral Gables.

Sawyer, R.K., C.C. Wieland, and G.M. Griffin 1988 The Clay Mineralogy of Calcitic Seat Earth in the Northern Everglades of Florida. *Journal of Sedimentary Petrology* 58:81–88.

Steward, Kerry K. 1974 Physiological, Edaphic, and Environmental Characteristics of Everglades Sawgrass Communities. In *Environments of South Florida: Present and Past,* pp. 37–46. Edited by Patrick J. Gleason. *Miami Geological Society, Memoir* No. 2, Miami.

Wanless, Harold R. 1974 Mangrove Sedimentation in Geologic Perspective. In *Environments of South Florida: Present and Past,* pp. 190–200. Edited by Patrick J. Gleason. *Miami Geological Survey, Memoir* No. 2, Miami.

Watts, William A. and Barbara C.S. Hansen 1988 Environments of Florida in the Late Wisconsin and Holocene. In *Wet Site Archaeology,* pp. 307–323. Edited by Barbara A. Purdy. The Telford Press, West Caldwell, New Jersey.

Williams Brothers Environmental Services 1980 *Energy Production from Central Florida Peat Deposits.* Resource Sciences Center, Tulsa, Oklahoma.

Wood, Ferguson and Nancy G. Maynard 1974 Ecology of the Micro-Algae of the Florida Everglades. In *Environments of South Florida: Present and Past,* pp. 123–145. Edited by Patrick J. Gleason. *Miami Geological Survey, Memoir* No. 2, Miami.

FLORIDA'S ARCHAEOLOGICAL WET SITES

Introduction

The significance of archaeological wet sites is that they provide views of past cultures and peoples that were never known before. The nonperishable component of many archaeological wet sites is often so trivial that the sites would go unnoticed if the organic materials were not present. The intent of this chapter is to describe Florida's archaeological wet sites and assess how each has given us a broader knowledge of past environments and human utilization of those environments. The information is uneven, of course. Some sites have yielded vast quantities of materials that have been studied from all perceivable angles, and preserved. Others have not been studied or reported in detail. Since most of these sites would not be known at all were it not for the items made of organic materials, it is the organic component of each site that will be emphasized. The locations of Florida's wet sites are shown in Fig. 1. The story begins with Key Marco because it was the first wet site in Florida that was excavated and studied. For convenience, the remaining sites are listed in alphabetical order.

Key Marco

The Key Marco site, known also as the Court of the Pile Dwellers (Cushing 1897) and the Cushing site, is situated in Collier County at the northern edge of the Ten Thousand Islands (Fig. 1). The extreme biological diversity and abundance of this region have been amply documented (Gilliland 1975, 1989; Widmer 1988). Archaeologically, the site is related to the Glades Area. The Indians who lived in the region at the time of historic contact are known as the Calusa. The Calusa were a chiefdom in that they recognized a leader who had inherited his position and who received tribute from peoples of surrounding areas (Goggin and Sturtevant 1964; Marquardt 1988; Widmer 1988). The attainment of a chiefdom-level political structure has been attributed to the fact that the Calusa, because of their favorable ecological niche, were able to amass food surpluses permitting them to turn their attention to activities other than hunting, gathering, and foraging. There is no evidence that they were agriculturalists although two gourd varieties (*Lagenaria siceraria* and *Cucurbita pepo*) were found at Marco that were probably cultivated for use as containers, floats, rattles, and possibly for other purposes as well. They harvested the products of their maritime environment so successfully that there seemed little need to spend the time and effort necessary to raise crops that require land clearing, planting, and harvesting.

A great deal is known about the Calusa and their leader Carlos from Spanish accounts, especially those involving Pedro Menéndez de Avilés (Solis de Merás 1964) and from the narrative written by Fontaneda (1944), a young man who was held captive by the Calusa for 16 or 17 years in the mid-16th century. The Spanish described them as hostile, warlike, and heathen. Fontaneda said that the Indians could never be tamed and recommended that they be enslaved or exterminated. The culture of the Calusa, obviously, was a world apart from any familiar to the Europeans.

Cushing conducted a preliminary reconnaissance of the Key Marco site in 1895 and he and his crew excavated there for 2 months (February 26 to April 20) in the spring of 1896. Cushing presented his findings to the American Philosophical Society in a talk given on November 6, 1896 which resulted in a 120-page publication (Cushing 1897). A thorough investigation of the excavations and the artifacts, including their present locations and conditions, was conducted by Gilliland (1975) who later published an account of the controversies and personalities involved (Gilliland 1989).

Excavations and Stratigraphy

Frank Hamilton Cushing's account of his excavation techniques at Key Marco has very likely been read and criticized by all professional archaeologists in Florida, especially those who have never worked at a wet site. There were few methodological procedures in 1896 that Cushing could use as a model. He was certainly aware of the inundated sites in Switzerland because he continuously compared the situation at Key Marco to the Swiss Lake Dwellings; however, those sites had been explored primarily by pioneering underwater methods long before the development of SCUBA. The circumstances at Key Marco were very different. Cushing intended to "exploit the little triangular Court of the Pile Dwellers from border to border" (Cushing 1897:356). Fig. 5 (taken from Cushing 1897:422) shows that the court was laid out in a grid of 81 units, each of which was a 10-foot square except those that bordered the irregular outer limits of the site. Each unit was numbered and Cushing occasionally referred to certain areas when describing specific artifacts. He also discussed the location of previous excavations that were conducted by Collier

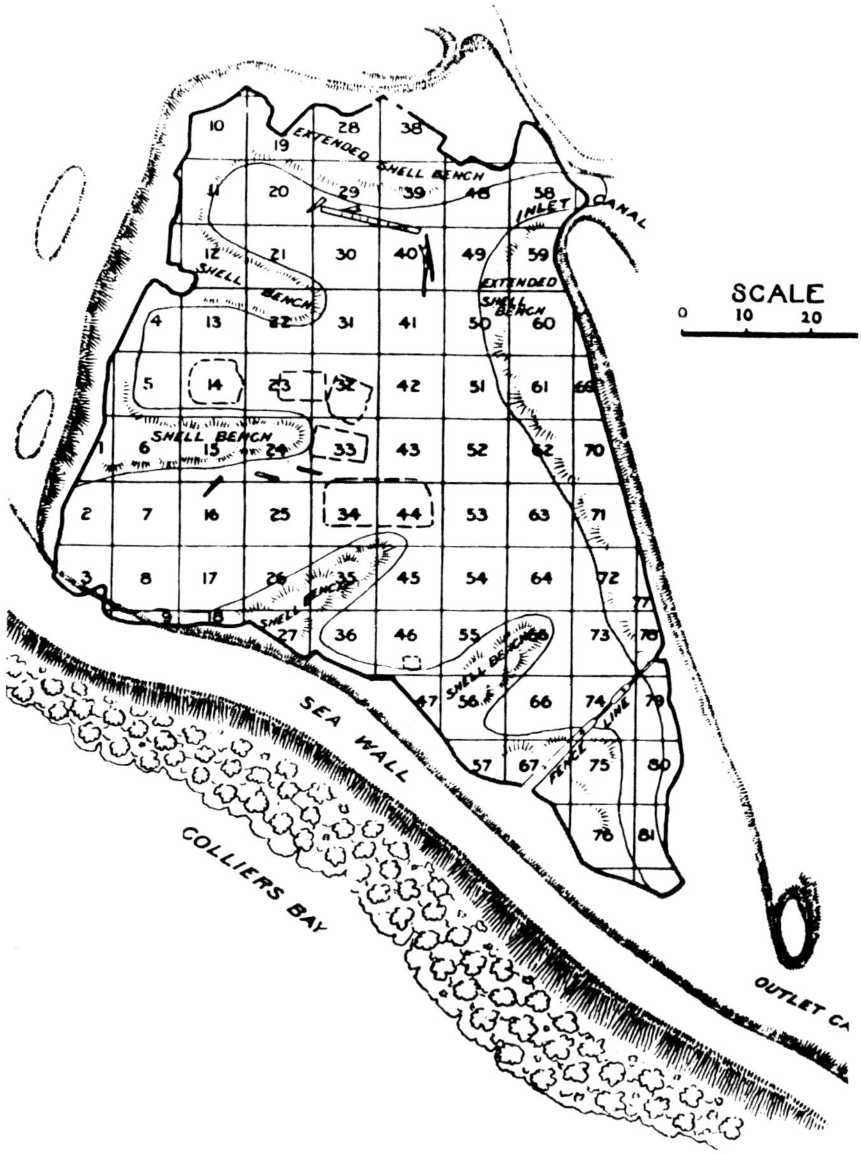

A

FIGURE 5 Court of the Pile Dwellers, Key Marco showing (A) Cushing's grid system (from Cushing 1897:422) and (B) flooded conditions of the partially excavated court (from Gilliland 1975:Pl. 6).

(the land owner), Wilkins, and Durnford (1895) and by himself in 1895 (Cushing 1897; Gilliland 1975, 1989). Cushing realized the problems he faced:

> The entire court was thickly overgrown with mangrove trees, underneath which also thickly grew, to a uniform height of six or eight inches, bright green aquatic weeds and mangrove shoots. Since the interior of this artificial and filled-up bayou was still not above the level of the surrounding tide-swept mangrove swamps through which the canals led, it lay almost continually under water, and its excavation looked at first to be almost

FIGURE 5B

impossible, and at best a most formidable undertaking. It would be necessary to cut away and uproot the mangroves and in some way to remove the water that filled to overflowing the excavations which had formerly been made, and thus covered the entire court. To begin, I had a few of the trees cleared away from the outer and southwesterly corner, and opposite my old excavation in section 34, 44, had a trench cut through the sea-wall to as great a depth as possible without letting water in from the bay outside. I then had a long trough of ship planks constructed and placed on stakes driven deep into the muck bed, so that one end rested over the excavation and the other, lower end, in the mouth of the sluice-way through the sea-wall. Then laying heavy planks over the boggy surface to furnish foothold for the men, I set them at work baling out the old excavation with buckets. It was at first like trying to bale out the sea itself, for water flowed in as fast as taken out; but after two or three hours of steady work, it began to lower, not only in the excavation, but over the entire court. On the following morning, however, there was almost as much water in the excavation thus enlarged, and elsewhere, as on the previous day; but it was much sooner disposed of by baling and by the banking up of the place last excavated, and I soon found that by thus proceeding each morning for a couple of hours more or less, the water could be kept sufficiently low to enable us, working in sections, or bins as it were (roughly corresponding to those shown in the plan), to excavate the entire place. Yet, even thus, much of our search in the lower depths had to be made merely by feeling with the fingers. I deem it unnecessary to give further details of our operations, save to say that three or four of us worked side by side in each section, digging inch by inch, and foot by foot, horizontally through the muck and rich lower strata, standing or crouching the while in puddles of mud and water (Cushing 1897:356–357).

Cushing's jerry-built system permitted the excavations at Key Marco to proceed in a fairly efficient manner, but it had an unfortunate consequence. In a letter written at the site to Dr. Pepper on March 14, 1896, Cushing says: "The draining of the basin of the water court has caused the peat and marl containing the specimens to sink somewhat over a foot in the middle of the basin, and the carrying off of so much water or rather brine will at last, after all these centuries expose the specimens more or less to the air and lead to their rapid decay if left thus for any length of time" (Gilliland 1989:83). As mentioned elsewhere in this volume, this is a problem noted by all of us who have excavated at wet sites. It did not seem to occur to Cushing to cover the artifacts temporarily until they could be recovered systematically. Instead, he quickly removed several masks and figureheads (Gilliland 1975:20, 1989:84). Cushing was also worried that what he left unexcavated would not be there for another season's work because his draining had made it easy for people to vandalize the site (Gilliland 1989:84). Sawyer gives a slightly different description of the "muck hole" at Marco and the initial reaction ("absolute revolt") of the crew (Gilliland (1989:72).

What Cushing could have done with pumps or well points! The only well-grounded reproof of Cushing's methods, however, is his lack of vertical control during his excavations. Even here he makes some observations about stratigraphy:

> As the northwesterly half of the court became cleared of its contents and the bottom was thus more and more revealed, we found that it was generally concave, or perhaps I may say, tray shaped; that is, comparatively shallow at the sides—not more than from eighteen inches to three feet deep—but throughout the middle and thence toward the mouths of the two canals, from four-and-a-half to five-and-a-half feet deep (Cushing 1897:357).

He distinguished the culturally sterile muck from the peaty marl containing the artifacts. This observation influenced his excavation strategy: "Thus despite the fact that after removing the surface muck from the sections, we dug only with little hand trowels and flexible pronged garden claws—and, as I have said before, with our fingers . . ." (Cushing 1897:359). Although he may not have been aware of it, his description of the culturally sterile organic stratum that had accumulated since the site was abandoned provides valuable insights about recent depositional history in the area which have climatic and environmental implications. He described five strata at the site:

> The surface deposit throughout the entire court consisted of a stratum of spongy black or dark brown muck, permeated by both rotting and living rootlets. It was . . . thin at the margins but eighteen or twenty inches thick throughout the middle [this was the culturally sterile zone]. Below this was a somewhat thicker stratum of brownish gray peaty marl, soft, tremulous, exceedingly foul-smelling, and rich in the best preserved relics we discovered. This stratum [the second zone] directly overlaid and surrounded

> the benches I have described. Finally underneath it, between the benches and throughout the middle of the court, was a less well-defined layer of less peaty marl, inter-mixed with shells and other *débris* and also with abundant ancient remains [the third zone]—which, indeed, we continued to encounter even in the under-lying, comparatively firm shell and clay-marl bottom [the fourth zone]. This, however, although nearly a foot and a half thick, we could not venture to excavate, since the slightest opening made through it into the sandy reef below [the fifth zone] let in a steady stream of water from the sea (Cushing 1897:358).

Cushing was describing a situation that I have observed at Hontoon Island and other wet sites in Florida, mentioned also by Davis (1946), that lateral movement of water through peat deposits is very slow but it is difficult to keep an area drained once the bottom is pierced. Later, in an article that appeared in the New York Journal, Cushing says:

> It was impracticable to penetrate this cement-like bottom . . . because it was our only protection against the waters of the sea. A small cut through it into the underlying shell near the sea wall . . . resulted in an instant flooding of the place and had to be packed and tamped that the work might go on; and this profoundly regretted, for every foot of even the underlying material contained charcoal, bones, and priceless ancient remains of wood, fibre, and other perishable material (Gilliland 1975:18).

Cushing also stated that the "various ancient artifacts we found. . . occurred at unequal depths . . ." (Cushing 1897:360). The feline figure, for example, "the gem of our art collections [was found] at a depth of not more than twenty inches, just between the overlying muck and the middle stratum of peat-marl" (Cushing 1897:429). A few objects are identified as from the "first relic bearing stratum or second relic bearing stratum" (Gilliland 1975:26, quoting Cushing).

Chronology

Cushing can be excused for not attempting to establish a chronology for the finds he recovered. In fact, he would undoubtedly have been censured if he had because in the late 19th century it was not popular to discuss the antiquity of the American Indians. The integrity of the entire profession of anthropology was at stake in the issue. Ales Hrdicka and William H. Holmes in particular, both prominent anthropologists of the time, had reacted strongly against "evidence" that was presented as "proof" that humans had lived in the Americas as long as they had in the Eastern Hemisphere. They insisted on irrefutable documentation whenever antiquity was mentioned. Since no chronometric techniques had been developed yet, the question of age was simply ignored. This is unfortunate because there were times when this firm stance may have worked against the scientific community as with the antiquity of the human remains found at Vero Beach and Melbourne in association with extinct megafauna. It is interesting to note, however, that these same gentlemen accepted, without reservation, assumed cultural relationships between the Caribbean, Central America, South America, and Florida, yet this topic is hotly debated today.

It is important to discuss the time range of occupation at the Key Marco site. In November 1974, I met with Willard F. Libby, developer of the radiocarbon dating technique. He was fascinated with the wooden artifacts from the Key Marco site and offered to date a few of them. I received permission from the Florida Museum of Natural History to take samples from six specimens, five of which were radiocarbon dated in Dr. Libby's laboratory in Los Angeles as follows:

Sample No. 1: float peg fragment, catalog number 40550, 1100-1300 BP (AD 675-875).

Sample No. 2: part of paddle, catalog number 40537, 1920±60 BP (AD 55±60).

Sample No. 3: part of post, no number, 1125±50 BP (AD 850±50).

Sample No. 5: piece of board painted with woodpecker, catalog number 40697, 1305±60 BP (AD 670±60).

Sample No. 6: mask fragment, catalog number 40710, 1275±50 BP (AD 700±50).

These dates were received from Libby in time to be listed in Gilliland's book (1975:257). When these results are calibrated to dendrochronology, the maximum variation amounts to only about 50 years (Stuiver and Becker 1986:863–910). Four of the five specimens fall between approximately AD 650 and 900. A disagreement arose about the results of these analyses. A prior assessment of the age of the site had placed it in the Glades III period because designs on some objects were purportedly similar to those of the Southern Cult of the Southeastern United States (a deduction challenged by Gilliland [1975:39]), a small number of sherds were of a pottery type known as Surfside, and a previous single radiocarbon analysis from the University of Michigan yielded a date of AD 1675 ± 100. The dissension arose because the results of these analyses, conducted by the world's authority on the method, did not conform to the conclusion drawn in the 1940s that the site was occupied in the late 15th and early 16th centuries. This conclusion appeared to be supported somewhat by the 17th century radiocarbon date. The argument was that the Key Marco collection may have been contaminated by petroleum-based pesticides applied at the Florida Museum of Natural History (Milanich 1978:682). An examination of recent literature describing experiments to determine the effects of contamination on organic specimens is revealing. A 1% addition of modern ^{14}C to a 50,000-year-old sample will result in an apparent age of about 35,000 years; a 5% addition of modern ^{14}C contamination will result in an apparent age of 25,000 years. Organic remains falling within the Holocene are not altered as dramatically as older materials by contamination with modern ^{14}C. For example, a 1% addition of modern ^{14}C to a sample with an actual age of 5000 years will result in an apparent age of about 4950 years; a 5% addition would result in about a 350-year decrease. The addition of 1% "dead carbon" (e.g., coal, lignite, etc.) to a modern sample is not severe; it increases the age by only 80 years (Gupta and Polach 1985:129–134). Other researchers have shown that the addition of certain reagent-grade chemicals to a sample of modern oak yielded a date >500 years in the future (Tankersley 1984). The sample pretreatment of ^{14}C laboratories is very efficient in removing most contaminating organics. In the vast majority of cases any remaining contamination in Holocene age samples will not be sufficient to alter the indicated age of the sample by an amount in excess of the statistical variance attached to the age estimate. It would be rare that samples, such as those from Key Marco, would yield dates 600–800 years older than anticipated. "It seems reasonable to suggest that specific geochemical evidence be offered when 'contamination' is used as an explanation for an anomalous ^{14}C value" (Gupta and Polach:1985).

Gilliland concedes that a few objects suggest a terminal date for the Key Marco site of around AD 1500 (the Glades IIIb Period) but she also points out that it is quite impossible to determine a beginning date for the occupation of the Cushing site at Key Marco. She mentions that the majority of pottery from the site is Glades Plain which is common to several time periods in the Glades area and says "the wealth and variety of artifacts throughout . . . point to occupation over a considerable period." She further emphasizes that absolutely no European materials were recovered (Gilliland 1975:37). Cushing occasionally referred to the depth in the deposit of some finds (e.g., 1897:36) and mentioned differences in color, odor, and composition of the matrices. It is not possible, however, to determine for sure from his descriptions if the stratigraphy he noted had any temporal significance but it is illogical to conclude that Key Marco was a single component site when every other site in the vicinity (see for example, Van Beck and Van Beck 1965) spans a time period of 1500 years or more. Cushing did not commit himself in print about the age of the site but he may have been truly prophetic when he told Mrs. Margaret McIlvane Collier that he believed the inhabitants of Key Marco had come from Central America about 1100 (now 1200) years ago (ca AD 800!). This comment was recorded in Mrs. Collier's diary on March 13, 1896 (Gilliland 1989:82).

Further support for the dating of the Key Marco site comes from Schwehm's (1983) thesis concerning south Florida's prehistoric art in which she describes similarities in art styles between Key Marco and Fort Center:

> Nevertheless, it seems apparent that the sculptures of both areas were intended for use within a religious framework but for different ceremonial purposes.
>
> Tools recovered from the Key Marco and Fort Center sites reveal that the Indians of both areas used similar shell and shark tooth tools to fashion their wood sculptures. Not only does there seem to be a link between the symbolism and technology of the two sites, but an outstanding similarity exists between the art styles of Key Marco and Fort Center in the sensitive naturalism with which the animal effigies of both cultures have been portrayed (Schwehm 1983:69–78).

Schwehm continues her investigation with a comparison of the Belle Glade carvings to those from Key Marco and Fort Center and concludes that the

Indians of the three sites shared affinities in art styles and all three fall within the same general time period (AD 500–1000). She concludes that the sites are linked by chronology, geography, technology, and possible shared religious beliefs. ". . . these three prehistoric Indian sites of south Florida also manifested a recognizable regional art style apparent in their carved wood animal effigies" (Schwehm 1983:81–85).

The dating of the Key Marco site is discussed fully and clearly also by Widmer (1988:89–93).

Flora and Fauna

Throughout Cushing's report he mentions the great diversity of flora and fauna he encountered in the deposits. Speaking of botanical remains he describes cane and bark matting, palmetto or flag leaves (some still green), various kinds of cordage, mangrove and button wood, and abundant refuse of meals including vegetable and fruit foods (Cushing 1897:350,363,368–369,388). Gourd remains were referred to above. Unfortunately, many of the plants were not identified and, in fact, Cushing probably did not save a representative sample of all of them. Wood anatomist Lee Ann Newsom (1989, ms), identified the following species of wood that were used to manufacture some of the artifacts now stored at the Florida Museum of Natural History:

Catalog No. A6310. Five tool handles found by Cushing associated with Busycon hammers. Two were made of pine (*Pinus* sp.) and three were made of poorly preserved hardwoods including *Eugenia* sp., *Nyssa sylvatica* (black gum), and *Forestierce segregata* (Florida privet).

Catalog No. A6307. Two tool handles, one with end shaped for socketed implement made of *Eugenia* sp.; the other made of pine.

Mask (see Fig. 6): *Swietenia mahogani* (mahogany), anatomical structure is very degraded, cells exhibit mechanical and chemical dissolution.

Catalog Nos. A5631, A5744, A5654B. All parts of boxes, made of *Taxodium distichum* (cypress).

Catalog No. A5497. Large bowl with scalloped edge, cut from the tree sideways (the radial section is on the upper edge of the rim) made of cypress.

Catalog No. A5480. Large tray, made of cypress.

Catalog No. A5496. Small tray, made of cypress.

Catalog No. A5487. Smallest bowl, made of cypress.

Catalog No. A5482. Larger of two small bowls with this number, pine.

Catalog No. A5524. Large mortar, made of cypress. Woodpecker (Fig. 7). Made of cypress.

Catalog No. A5519. Bottle-shaped pestle, complete collapse of cells.

Catalog No. A5517. Mushroom pestle, made of *Guaiacum* sp. (lignum vitae).

Catalog No. A5500. Double-ended pestle, made of a hardwood, *Bumelia* sp.

Catalog No. A5749. Float pegs, made of cypress.

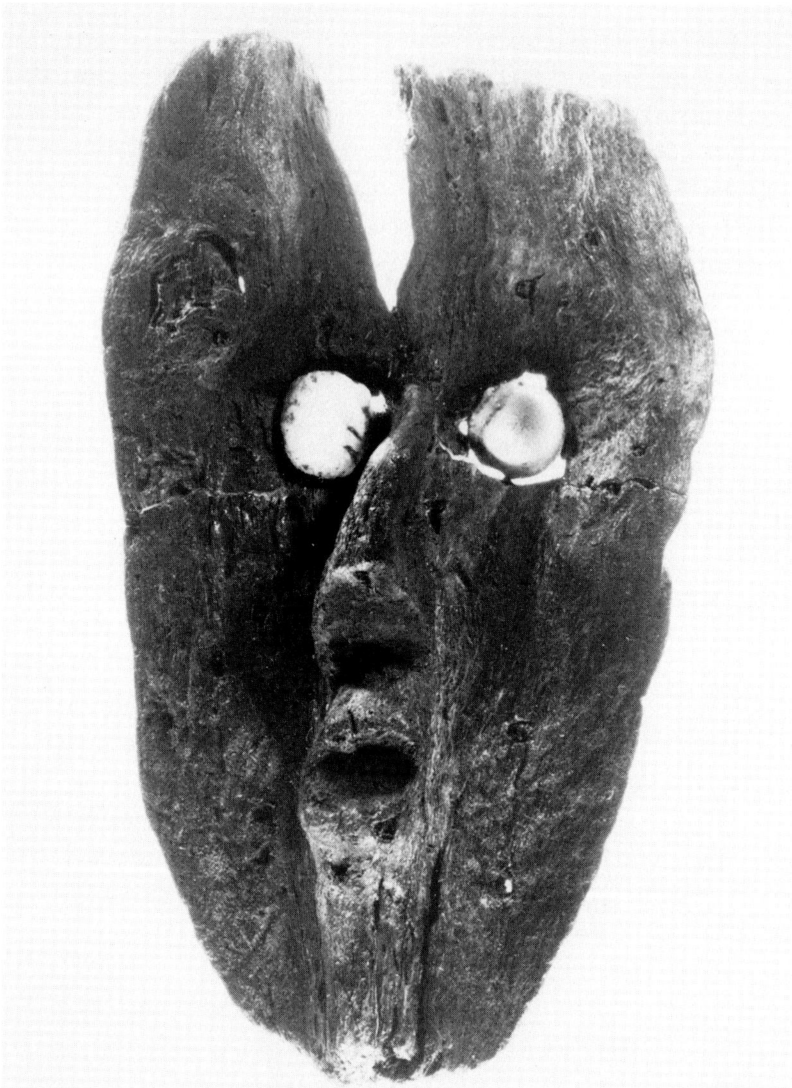

A

FIGURE 6 Mask with shell eyes made from mahogany. Cushing called it (A) Hawkbill Turtle Man and (B) front and back views 1990 (photos by the author).

The animal remains at the Key Marco site and other sites on Marco Island have been identified fairly thoroughly. The middens are composed of great quantities of species of marine shellfish such as conch, clam, oyster, and smaller varieties. Wing (1965) identified seven kinds of terrestrial mammals (opposum, rabbit, dog, bobcat, raccoon, gray fox, and deer), fourteen kinds of birds (loon, gannet, cormorant, great blue heron, American egret, snowy egret, white ibis, red-breasted merganser, turkey vulture, black vulture, long-billed curlew, herring gull, ring-billed gull, royal tern), seven species of amphibious or terrestrial turtles (mud turtle, box turtle, diamond-back terrapin, chicken turtle, gopher tortoise, green turtle, soft-shelled turtle), alligator,

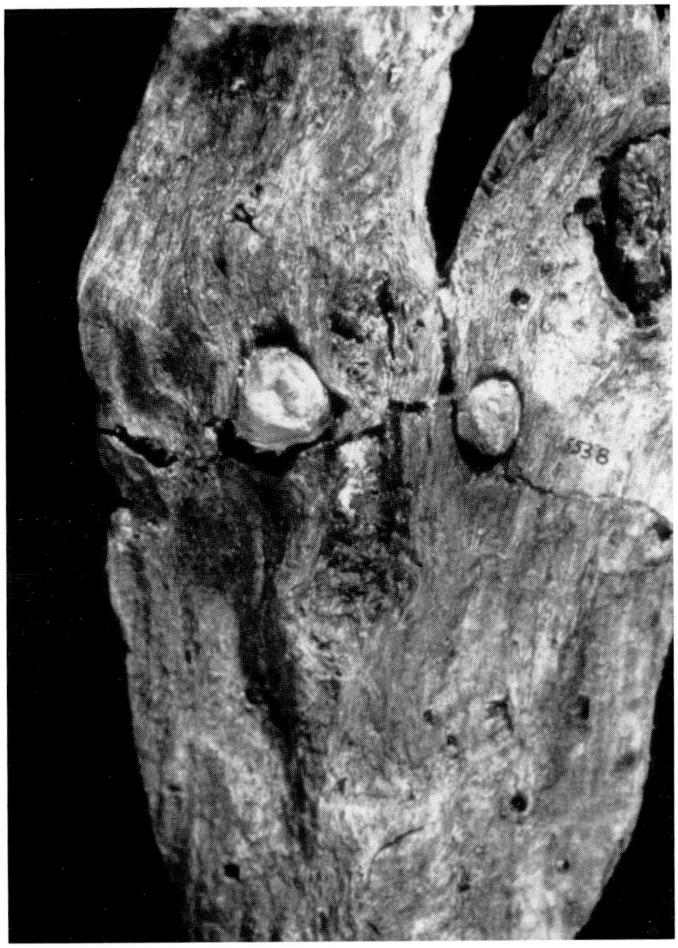

FIGURE 6B

whale, and twelve species of marine fishes (mako shark, requiem shark, tiger shark, saw fish, spotted eagle ray, sea catfish, snook, jew fish, jack, black drum, barracuda, mullet). Widmer (1988:238–243) furnishes an even longer list of fauna for the lower gulf coast in general. The flesh of these animals was used for food, their bones and teeth for weapons and tools, and their hides, skins, carapaces, and feathers for clothing, decoration, ornaments, and, perhaps, for shelter. The skin of the shark may have been used to smooth and polish wood and bone. Cushing mentions that many of the composite tools were "usually further secured with rawhide thongs—now mere jelly" (Cushing 1897:368) which confirms observations made at other archaeological wet sites in Florida that leather does not survive. Some manufactured items probably served as status symbols to establish class distinctions; others may have served as tribute.

A

FIGURE 7 Cypress board painted with the figure of a bird that Cushing called a jay or Kingfisher, but it was identified as a woodpecker by a Smithsonian ornithologist (Gilliland 1975:75). (A) Water color by Wills M. Sawyer, (B) field photo by Wills M. Sawyer, (C) 1974 photo by Kay Purington of the Florida Museum of Natural History showing fading of the figure and cracking of the board. Pictures taken by the author in 1990 are similar to those from 1974 but the board is more degraded.

FIGURE 7B

FIGURE 7C

Preservation

The preservation of the materials from the Key Marco site can be discussed under three categories: (1) within the deposits, (2) after they were removed, and (3) their present whereabouts and condition. Gilliland (1975, 1989) has researched these topics thoroughly by examining Cushing's (1897) published report, his field diary and letters, a catalog written by him in pencil on 1621 pieces of nickel tablet paper (brittle, crumbling, and yellow with age), through other documents and letters, and by locating and studying the remaining specimens.

The preservative qualities of the deposits at the Key Marco site fit very nicely into the model proposed in Chapter 2: permanently saturated peaty soils ensuring anaerobic ("foul-smelling") conditions that retard decomposition. The hydrogen ion concentration (pH) must have been alkaline because both flora and fauna survived. In addition, tidal action must have been quite gentle because it is doubtful that the organic component would have survived if the movement of water had been rapid or strong. Cushing says:

> The objects found by us in these deposits were in various conditions of preservation, from such as looked fresh and almost new, to such as could scarcely be traced through or distinguished from the briny peat mire in which they were embedded. . . .
>
> Articles of wood far outnumbered all others. I was astounded to soon find that many of these had been painted with black, white, gray-blue, and brownish-red pigments; and that while the wood itself was so decayed and soft that in many cases it was difficult to distinguish the fibre of even large objects of it, either by sight or by touch, from the muck and peat in which they were unequally distributed, but now more or less integrated; yet when discoverable in time to be cautiously uncovered and washed off by the splashing or trickling of water over them from a sponge, [they] . . .appeared not only almost perfect, but also deceptively well preserved so that I at first thought we might, with sufficient care, recover nearly all of them uninjured. This was especially true of such as had been decorated with the pigments; for owing to the presence in these pigments of a gum-like and comparatively insoluble sizing, the coatings of color were often relatively better preserved than the woody substance they covered, and enabled us the more readily to distinguish the outlines of these painted objects—when else some had been partially destroyed or altogether missed—and also enabled us to take them up on broad, flat shovels, and to more deliberately divest them of the muck and peat that so closely clung to them (Cushing 1897: 358–359).

Speaking of the wooden masks, Cushing said "we found fourteen or fifteen fairly well preserved specimens, besides numerous others which were so decayed that, although not lost to study, they could not be recovered" (Cushing 1897:388; Fig. 8).

Cushing mentions several objects that were "enfolded within decayed matting" or "wrapped about with strips of palmetto leaves that were still green in color, but which, of course, immediately decomposed on exposure to the air" (1897:385–388, 424). An example is the famous wolf head with detachable ears

A

FIGURE 8 (A and B) Field photographs by Wills M. Sawyer of fragile Key Marco masks that probably never left the field site (from Gilliland 1975:Plates 43 and 45).

and shoulder parts that were all wrapped separately in palmetto strips when found (pictured in Gilliland 1975:112, 1989:opposite 52). Other of these "packs" must have had ritual significance, like medicine bundles. One contained a box with a "horned crocodile" painted on the inner lid, nine adzes, a pair of painted shells, a knife with animistically carved handle, and other articles (Cushing 1897:385). Cushing also describes mats so thick they could have served as rugs. "It was obvious that the peoples who had inhabited the court understood well, not only platting, but weaving and basketry making too" but "except for masses of the peat upon which the remains lay and which, therefore, when dry showed traces of its beautifully and variously formed plies, naught of them could be preserved" (Cushing 1897:363).

Fortunately, Wells M. Sawyer, artist and photographer, who accompanied the expedition, photographed and painted many of the objects that did not survive (Gilliland 1989:126). Items that were too decayed to ship from the site were not included in Cushing's catalog (Gilliland 1975:26) so Sawyer's watercolors and photographs are the only evidence that some objects ever existed (Gilliland 1989:126). Sawyer left us a visual testimony that corroborates Cushing's often flowery descriptions. Why did Cushing have the foresight to take an artist and photographer with him to the site? Perhaps it was customary, or perhaps Cushing, after seeing what happened to the objects collected

Figure 8B

by Durnford and by himself in 1895, realized that many would not survive and insisted upon a picture record as well as a written account. Cushing was discouraged that

> fully twenty-five per cent of these ancient articles in wood and other vegetal material were destroyed in the search; and again, of those found and removed, not more than one-half retained their original forms unaltered for more than a few days. . . . Some of the things thus recovered could be preserved by very slow drying but it soon became evident that by far the greater number of them could not be kept intact. No matter how perfect they were at first, they warped, shrunk, split, and even checked across the grain, like old charcoal, or else were utterly disintegrated on being exposed to the light and air if only for a few hours.
>
> Unique to archaeology as these things were, it was distressing to feel that even by merely exposing and inspecting them, we were dooming so many of them to destruction, and to think that of such as we could temporarily recover only the half could be preserved as permanent examples of primitive art.
>
> I sought by every means at our disposal to remedy these difficulties, . . . (Cushing 1897:359).

Cushing also "exercised great caution in keeping related parts together . . ." (Cushing 1897:368), but some of these were separated when the collection was divided later on (Gilliland (1975:25).

He was able to "bring away, in measurably good condition, more than a thousand of these precious examples of prehistoric art in perishable materials (Fig. 9), not to mention many hundreds of examples in more durable substances such as shell, bone, and horn" (Cushing 1897:360).

Many preservative treatments were tried at the site but gave poor results. Some specimens were preserved by very slow dryng accomplished by packing them in dry sand for a time, then placing them on shelves under dry clothing for an additional period (Gilliland 1975:20). Mrs. Cushing had accompanied the expedition and she had "watched carefully in the cabin of the *Silver Spray* over the choicest objects found and happily succeeded in drying many a treasure . . ." (Gilliland 1989:72). Cushing shipped 11 barrels and 59 boxes of Key Marco specimens. Preparation of the specimens for shipment was a serious task (Gilliland 1989:88).

In Cushing's final comments before the American Philosophical Society, he said: "The specimens themselves are now sadly warped and shriveled. But happily some of them can be fairly restored by treatment with preservatives, and happily also, our photographs, drawings and paintings, and casts, made in the field, are almost equal for study to what the originals were when found" (Cushing 1897:415).

Cushing also "strengthened" the designs and forms of many of the mask and figureheads and it is not known for certain how extensively he altered them. "[I] *absolutely restored* the forms, proportions, and integrity of several of the most misshapen yet also the most important of these specimens." For example, he admits he took the deerhead to his hotel room to "complete the color scheme and show an artist friend" (Gilliland 1975:22,27). Yet Sawyer defended Cushing by saying "in almost every instance I saw these objects before they were cleaned and in all cases unless it is specifically mentioned the photographs represent the actual conditions after the muck and roots had been

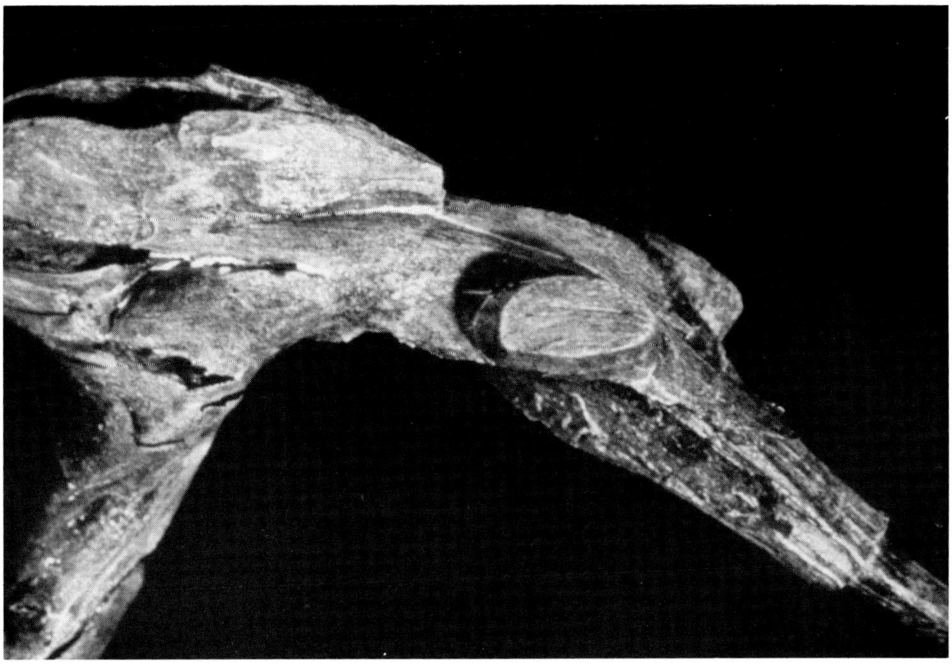

A

FIGURE 9 (A,B,C): Adze handle elaborately carved with large protruding eyes and perched atop the turn of the handle is a crouched rodent. A carved antler socket accompanied the specimen. (Gilliland 1975:137). (A and B) Both sides of carving, (C) shows entire specimen (photos by the author 1990), (D) feline figure (from Gilliland 1975:117, Plate 69).

removed. . . The restorations amount to so small a factor in the collection that it is scarcely worth devoting a paragraph to them" (Gilliland 1975:22).

It is fortunate that photographs, water colors, and casts were made for it is impossible now to recognize many of the artifacts in this remarkable collection and some of them are missing altogether (Gilliland 1975:22). Gilliland (1975, 1989) documents the agreement about the division of the Key Marco specimens between the United States National Museum, Washington, D.C. and the University Museum, Philadelphia. A controversy arose because the University Museum was assigned nearly all specimens "of the first order of excellence, . . . more than three-fourths of the relics and at least nine-tenths of the value of the collection" (Gilliland 1989:116;118). Some specimens were placed on exhibit and in storage at both museums and a number of objects were exchanged to the Museum of the American Indian (Heye Foundation) in New York. In the spring of 1964 a portion of the collection stored at the University Museum was shipped to the Florida Museum of Natural History in Gainesville for study by Gilliland (1975:5) most of which is still in storage there. In 1976, I secured from the United States National Museum and the University Museum a number of the Key Marco casts. Molds were made of the casts at the Florida Museum of Natural History and some of the casts reproduced from these molds are on display or in storage in Gainesville.

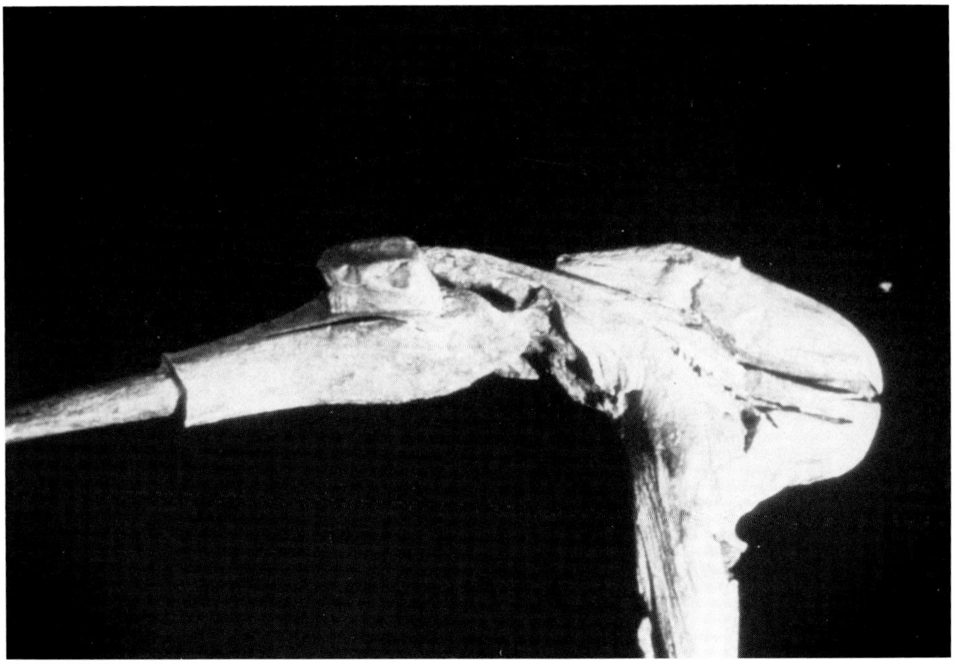

FIGURE 9B

Frank Hamilton Cushing

Frank Hamilton Cushing was a controversial figure (Fig. 10). He was considered a genius by some, an eccentric by others. He was accused of fabricating and passing off as genuine a jeweled toad he found during the time he spent with the Zuni. Later he was accused of painting the dancing figure on the inside of a shell found at Key Marco. Experimental reproduction was one of his greatest contributions to the infant science of anthropology. John Wesley Powell said that when Cushing was living at Zuni, there wasn't anything the Zuni could make that he couldn't make better. At Key Marco, Cushing said that the function of some of the implements he recovered was "clearly shown by the results of my experiments":

> Everywhere on the least finished surfaces of completed carvings, and on incipient works, not only in wood, but also in bone and horn, could be seen distinctive marks left by the finely serrated edges of these more than half-natural carving tools. As soon as we had discovered a few of them I secured fresh teeth and experimentally made knives and cutters of the various kinds I have described (Cushing 1897:369,371).

As mentioned above, it was no secret that Cushing *repainted* some of the masks and figureheads but he was simply trying to make them look like they did when he first saw them. Painted wooden specimens are very rare from prehistoric wet sites but they are not unknown. Coggins (1984) describes a

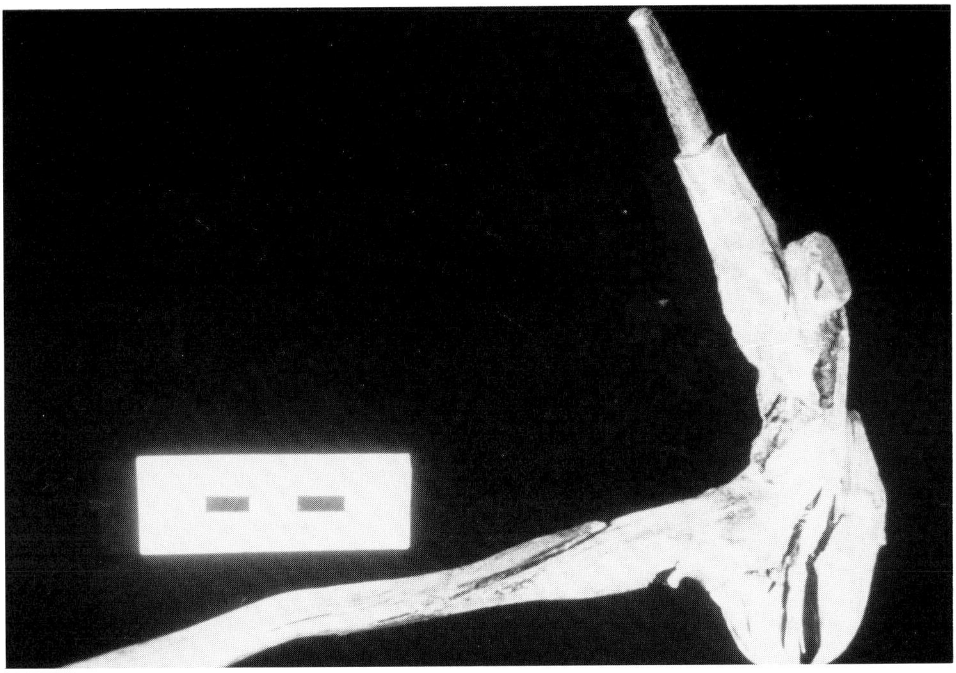

FIGURE 9C

number of wooden artifacts containing paint from the Cenote at Chichen Itsa, and Anderson (1985) recovered a painted paddle from a Mesolithic site in Denmark. After a careful reading of Gilliland's (1989) documentation of Cushing's activities, it is fairly clear that he was innocent of any intentional attempt to deceive.

Cushing has given us a gift that I believe has been overlooked largely because both he and his interpretations sometimes seemed bizarre. Cushing lived among the Zuni for a number of years in the 1880s when he was in his early 20s. America was playing cowboys and Indians at the time and many of the Southwestern groups, including the Zuni, still retained a large measure of their ritualism. Cushing took notes on ceremonies and sketched what he saw, he became thoroughly familiar with the Zuni language, he studied the Zuni inner life, he gained knowledge about their sacred rites, and he became a member of their most powerful secret society (Gilliland 1989:27,31,33,34,35). He published a 126-page paper on Zuni creation myths (Cushing 1896). Yet Gilliland (1989:3) and others do not place much importance on his consistent comparison of the Key Marco finds, particularly the masks and figureheads, to his experiences at Zuni. Gilliland (1975) recounts Cushing's names for the masks but she does not discuss his interpretation of them as representing beast gods.

In the section on "ancient artifacts from the Court of the Pile Dwellers," pages 362–387, Cushing describes in a very pragmatic manner what he recovered, including some rather elaborately carved tools. The reader feels comfort-

FIGURE 9D

able with what Cushing says and grateful to him for leaving such a lucid account. Seldom does he digress on these pages with speculations about ideology until page 387 when he talks about the "panther god," and from pages 388–394 when he discusses the masks and figureheads. About the panther god he says

> While the head and features—ears, eyes, nostrils and mouth—are most realistically treated, it is observable that not only the legs and feet, but also even the paws, which rest so stoutly upon the thighs or knees of the sitting or squatting figure, are cut off, unfinished; bereft, as it were, of their talons. And this, I would note, is quite in accordance with the spirit of primitive sacredotal art generally—in which it was ever sought to fashion the form of a God or Powerful Being in such wise that while its aspect or spirit might be startlingly shown forth, the powers associated with its living form might be so far curtailed, by the incompletion of some of its more harmful or destructive members, as to render its use for the ceremonial incarnation of the God at times, safe, no matter what his mood might chance, at such times, to be (Cushing 1897:397).

A

FIGURE 10 Frank Hamilton Cushing (A) at Age 22, (B) dressed as a Dakota Warrior, (C) fashioning a pottery jar (Smithsonian Institution portrait negative Nos. 53895, 8250, and 22-A, respectively).

Cushing tells us that the human masks were painted with practically the same face designs as occurred on the animal figures he recovered. He interprets this as the

> tendency peoples of this kind have, toward reproducing, in their face paintings or upon their masks, the characteristic marks or features of animal faces. . . . the Zuni Indians . . . observe that life is never manifest save in some sort of *form*, they argue that no form is without some sort of *life*, and since they further observe that each particular kind of life is manifest in some particular kind of form, they argue that form strictly conditions life—its powers and other characteristics. . . . To such a people, of course, form, semblance, aspect, is therefore all important; and they naturally think that by reproducing a given form or appearance which of itself gives rise to a certain effect, they may again and unerringly produce, or help to reproduce the same effect, with the form of their own making.

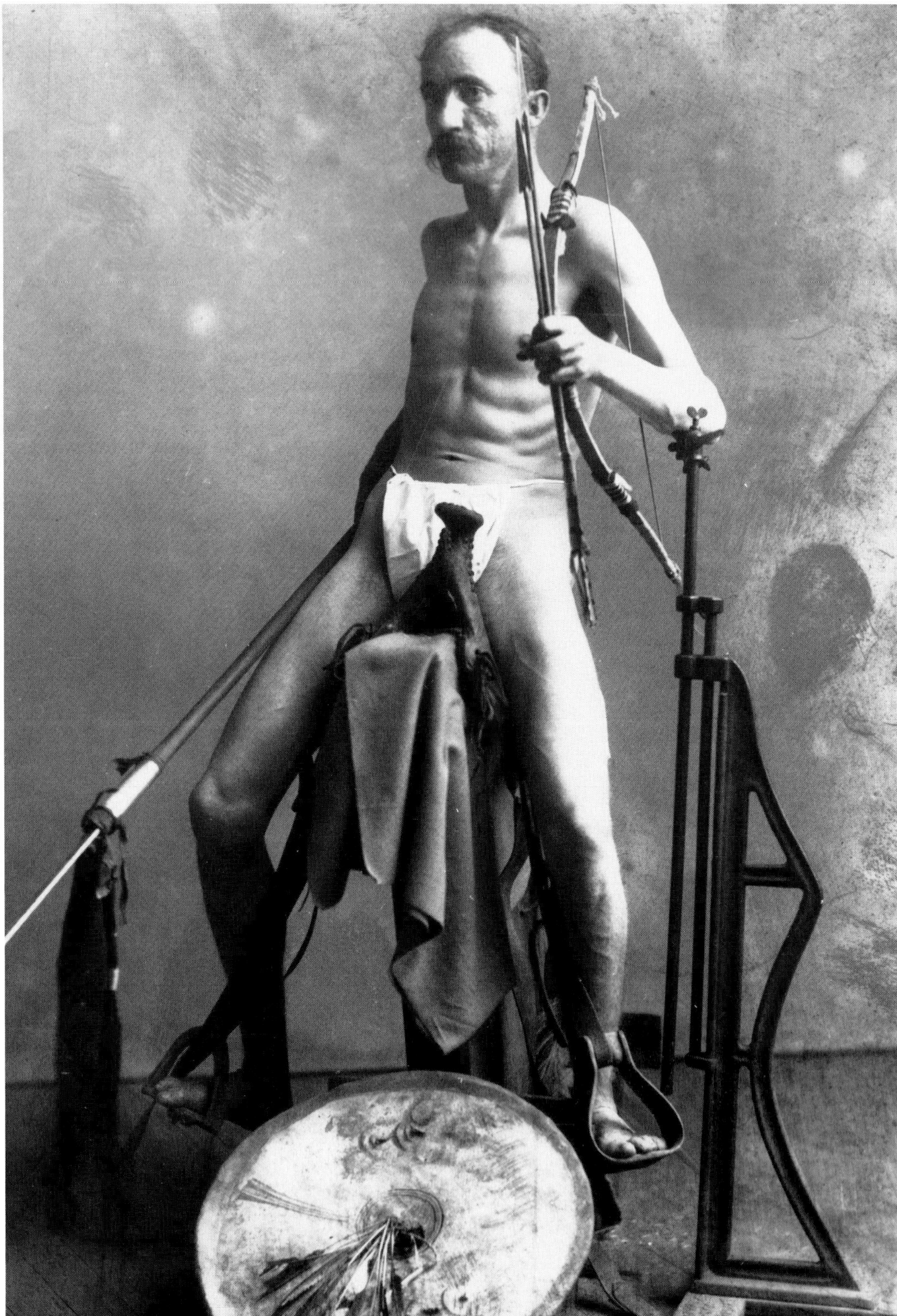

FIGURE 10B

FIGURE 10C

> . . . only according as his face changes so does his mood change, but also that his traits or his entire character may, for the time being be changed, by *wholly* altering, with paint or other marking, with mask or other disguise, the entire expression of his countenance or aspect. . . . by putting on a mask he can become . . . 'That which he thirsts to become.' . . . he will become that being or animal, or at least, but possessed by its spirit (Cushing 1897:388–400).

Cushing deduced that the Indians at Key Marco, as at Zuni, believed the "numberless best gods" depicted on implements, weapons, and amulets were alive. He also concluded that the Indians believed their canoes were animate objects like beasts of burden (Cushing 1897:413). There is no question that Cushing felt as comfortable with these insights about belief systems as he did with the practical descriptions of utilitarian objects.

Cushing's interpretations of the meaning of the masks and figureheads seem farfetched but he was one of the few people who ever gained an intimate (emic) knowledge of a North American Indian culture. The early explorers who observed native rituals were outsiders looking in (etic). They could not explain adequately activities that they could not even imagine as a possibility or those that were counterimages of what they considered their own high ideals of humanity (Gill 1982:4,8). For these same reasons we have had trouble accepting Cushing's analysis. Cushing had an inside view of native beliefs from his experiences at Zuni and he brought these to bear on the findings at Key Marco. It is now known that throughout North America masks were living things and that in masked performances the deities were present. In some Indian cultures, carvings were made to manifest the guardian spirit that was revealed to a man during a dream. The wearing of the face is not to cover or disguise the wearer but to present and animate the real presence of the spirit. Among the Hopi when a man dons a mask "he loses his identity and actually becomes what he is representing. . . . the mask is what is real" (Gill 1982:68–72). Cushing did not have academic training as an anthropologist but from childhood his life was devoted to anthropology. He was very well read and may have been influenced by people such as Edward Burnett Tyler and Heinrich Schielmann. Wardle (1951) seems to be the only author that accepted Cushing's interpretation of the place of the masks and "maskoids" among the people of the lower Gulf Coast of Florida.

Comments

Shortly after Cushing's work at Key Marco, several prominent archaeologists of the time, including C.B. Moore and Ales Hrdlicka, scrambled to locate a similar site in the area but to no avail. The Key Marco site is unique because it is still the only site on the Florida Gulf coast that has yielded such an abundance and diversity of usually perishable materials. Among the wooden objects alone there were trays, cups, bowls, tubs, mortars, pestles, toy canoes, paddles, stools, float pegs held together with cordage, ear buttons, plaques, amulets, boxes, clubs, atlatls, knife handles, floats, 3-dimensional anthropomorphic carvings, and masks (Fig. 11). The shell artifacts included cups, ladles, pendants, hammers, celts, picks, net weights, beads, and mask eyes. Bone included tools, weapons, beads, pins, antler, incised turtle carapaces, and shark teeth. Stone plummets and net weights were among the few stone remains recovered. Pigments and resins were also identified. Most utilitarian objects, especially those made of perishable materials, were found in such large quantities that Cushing was able to establish classes of artifacts (such as 622 float pegs examined by Gilliland). On the other hand, there was no such standardization possible with the decorative items. They were fairly abundant but there were no duplicates; each was unique.

The habitation at Key Marco ended as the result of a catastrophe. The site may have been abandoned because it was destroyed by fire (Cushing 1897:361).

There is evidence of fire on many of the artifacts and Gilliland says that both Cushing and Sawyer state that the fire was of major proportions and probably destroyed the superstructure (Gilliland 1975:33). It is not known whether the fire was natural or set by humans. The fact that no canoes were found indicates that the site was abandoned swiftly. Canoes were the primary means of transportation for these people (Gilliland 1989:19). The Key Marco site cannot be reevaluated. Marco Island now includes a full-scale city of homes and resort facilities (Gilliland 1975:27). The artifacts that still remain from the site need to be stabilized.

Key Marco had many of the attributes of other middens on the Gulf Coast, but it was more than just an ordinary garbage midden because so many usually perishable items had survived. When the site was excavated in the 1890s, it was the wood carvings that attracted the most attention. After nearly a century these spectacular artifacts have never been duplicated in quantity and diversity from any other archaeological site in Florida. The masks (Fig. 11), in fact, have no counterparts in prehistoric wooden art anywhere in North America; however, a clue to their probable existence elsewhere in pre-Columbian times is provided by greatly detailed 19th- and 20th-century ethnographic descriptions and illustrations of elaborate wooden masks carved by the Indians of the Northwest Coast. The use of masks during Calusa ceremonies was observed by Father Juan Rogel (Goggin and Sturtevant 1964; Lewis 1978). As far as is known, the early explorers, missionaries, and settlers took no examples of aboriginal material culture back to Europe from Florida. They usually destroyed Indian "idols" when they had a chance considering them pagan and they seldom bothered to describe clearly the mundane activities and tools used by the Indians. Even if some perishable items had made their way into museums of the time, they would have disintegrated by now through chemical and mechanical degradation. Thus, archaeological sites, especially those with a water-saturated component, furnish the only information available about the material culture and the environment of the prehistoric inhabitants of Florida. Cushing and his contemporaries recognized the uniqueness of the situation at Key Marco:

> Quite aside from all this, the large proportion of objects in perishable material, recovered by us, renders our collections from the keys unique in one respect at least; serves to illustrate how very little, after all, of the art of a Stone Age people (or in this case Shell Age people) is really represented by the remains that are commonly found on the camp sites and in the burial places of such peoples. Had my collections and observations been confined to the shell, bone, horn, pottery and other specimens in comparatively enduring materials found on the keys, the art that they represent would have seemed exceedingly crude . . . (Cushing 1897:410).

A discussion session held at the conclusion of Cushing's presentation before the American Philosophical Society was participated in by a number of prominent individuals. Professor F.W. Putnam remarked:

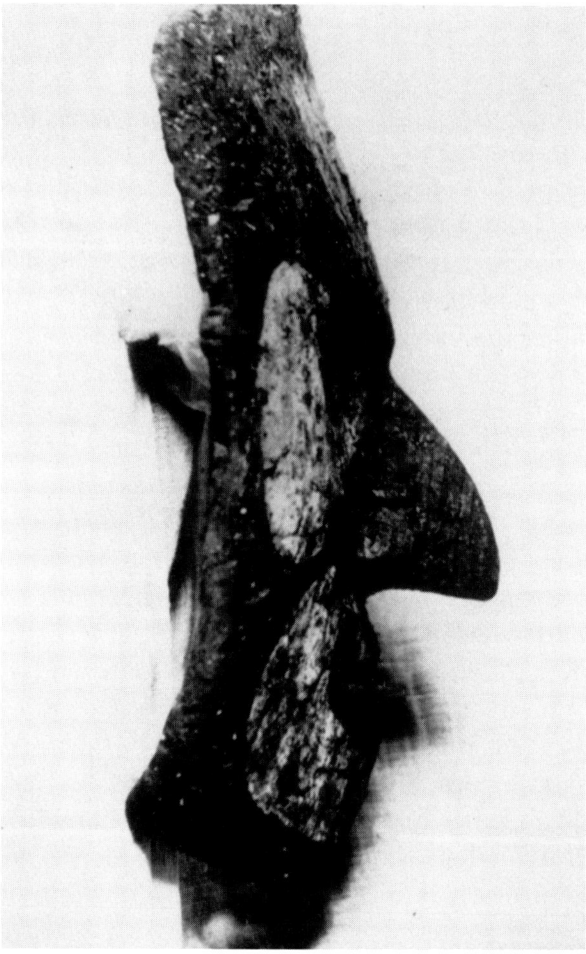

A

FIGURE 11 (A) Field photograph by Wills M. Sawyer of mask showing painted design in 1896 (from Gilliland 1975), (B) 1974 photo taken by Kay Purington of the Florida Museum of Natural History, (C) 1990 photo taken by the author.

> What I consider the most important point in Mr. Cushing's discoveries is that he was able to bring out of this muck deposit on the Florida Keys a large number of objects which by being buried in the muck were preserved; whereas the same objects if buried in a sand mound or lost in a shell heap would have perished. It is important to note that the objects in this collection, made of imperishable material, such as stone, bone, and shell, are of the same character as those already known from other parts of Florida. Thus it seems to me that Mr. Cushing's discovery instead of indicating a new culture, has thrown a powerful light upon, and greatly extended our knowledge of, the old culture of Florida (Cushing 1897:438).

Coles (1988) discusses the significance of wet sites and the broader perspectives they provide. He points out that archaeological wet sites having the

FIGURE 11B

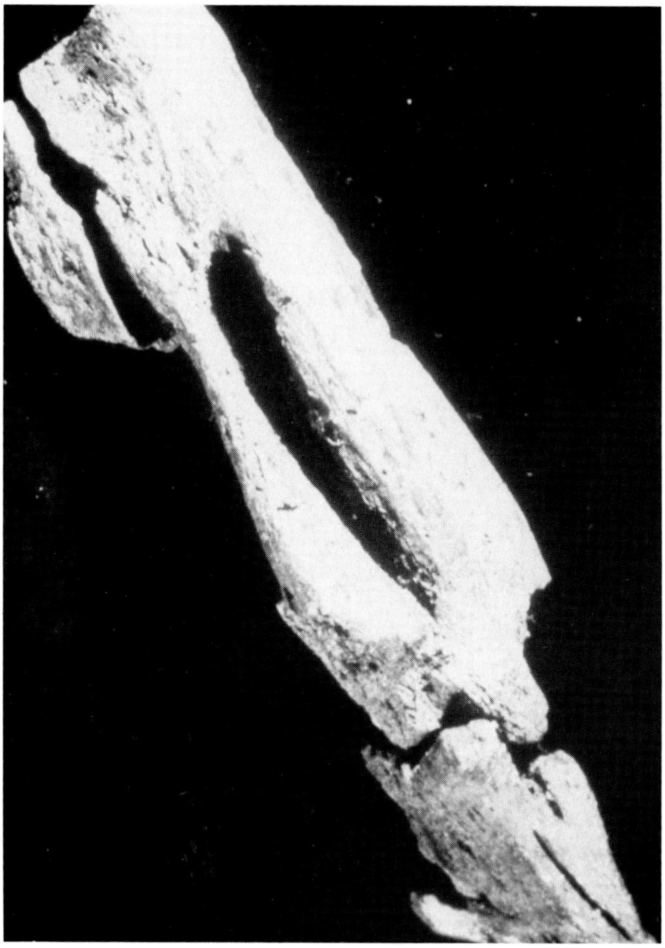

FIGURE 11C

potential to yield the greatest number of attributes should receive the highest priority with regard to funding, excavation, and protection. It is interesting to compare his list of attributes to those of the Key Marco site: (1) Key Marco was unique and of international significance, (2) the half-acre site was almost totally intact at the time it was excavated, (3) it contained a full range of artifactual materials from mundane utilitarian items to elaborate ceremonial carvings that were considered works of art and were of exhibit quality, and (4) it received considerable funding at the time. Today the Key Marco site does not exist, very few of the artifacts are on exhibit, the specimens have not been preserved, and intact preservation is not possible.

Our hats should be off to Frank Hamilton Cushing. He left us very detailed descriptions of the site and its artifacts accompanied by illustrations, photographs, and casts. Because of his integration into the Zuni culture he gave us rare insights into the probable function of some of the utilitarian and ceremonial objects. The difficulty that Gilliland had in working with Cushing's catalog bequeaths us with sound advice; that is, when excavating an archaeological site, keep good records because you and your crew will not always be around to clarify ambiguous entries.

Cushing did not complete his work with the Key Marco collection. He choked on a fishbone and died on April 10, 1900 when he was 42 years old. Other accounts say he died after an illness of a few days (Gilliland 1989:112). I think he died because he got tired of people criticizing him.

References

Allerton, D., G.M. Luer, and R.S. Carr 1984 Ceremonial Tablets and Related Objects from Florida. *The Florida Anthroplogist* 37:5–54.

Anderson, S.H. 1985 Tybrind Vig: A Preliminary Report on a Submerged Ertebolle Settlement on the West Coast of Fyn. *Journal of Danish Archaeology* 4:52–69.

Coggins, Clemency 1984 *Cenote of Sacrifice*. University of Texas Press, Austin.

Coles, John M. 1988 A Wetland Perspective. In *Wet Site Archaeology*, pp. 1–14. Edited by Barbara A. Purdy. The Telford Press, West Caldwell, N.J.

Cushing, Frank Hamilton 1896 Outline of Zuni Creation Myths. *Thirteenth Annual Report of the Bureau of American Anthropology*, pp. 321–447.

1897 Exploration of Ancient Key Dwellers' Remains on the Gulf Coast of Florida. *Proceedings of the American Philosophical Society* 25(153):329–448. Philadelphia.

Durnford, C.D. 1895 The Discovery of Aboriginal Rope and Wood Implements in the Mud in West Florida. *American Naturalist* 29:1032–1039.

Fontaneda, D°. d'E 1944 *Memoir of D°. d'Escalante Fontaneda Respecting Florida*. Trans. B. Smith 1854. Reprinted with revisions, University of Miami Press, Coral Gables, Florida.

Gilliland, Marion Spjut 1975 *The Material Culture of Key Marco, Florida*. The University Presses of Florida. Gainesville.

1989 *Key Marco's Buried Treasure*. The University Presses of Florida. Gainesville.

Goggin, John M. 1949 Cultural Occupation at Goodland Point, Florida. *The Florida Anthropologist* 2(3–4):65–91.

Goggin, John M. and William C. Sturtevant 1964 The Calusa, a Stratified, Non-Agricultural Society. In *Explorations in Cultural Anthropology: Essays Presented to George Peter Murdock*. Edited by Ward Goodenough. McGraw-Hill, N.Y.

Gupta, Sushil K. and Henry A. Polach 1985 Radiocarbon Dating Practices at ANU. Handbook. ANU Printing Services, Campora.

Lewis, C.M. 1978 The Calusa. In *Tacachale*, pp. 19–49. Edited by J.T. Milanich and S. Proctor. University Presses of Florida. Gainesville.

Marquardt, William H. 1988 Politics and Production Among the Calusa of South Florida. In *Hunters and Gatherers I. History, Evolution, and Social Change*. Edited by Tim Ingold, David Riches, and James Woodburn. St. Martin's Press, New York.

Milanich, Jerald T. 1978 The Temporal Placement of Cushing's Key Marco Site, Florida. *American Anthropologist* 80:682.

Newsom, Lee A. 1988 Key Marco Collection Wood Identification. Manuscript on file, Florida Museum of Natural History, University of Florida, Gainesville.

Schwehm, Alice Gates 1983 The Carved Wood Effigies of Fort Center: A Glimpse of South Florida's Prehistoric Art. M.A. thesis, Department of Fine Arts, University of Florida, Gainesville.

Sears, William H. 1956 The Turner River Site, Collier County, Florida. *The Florida Anthropologist* 9(2):47–60.

Solis de Merás 1964 *Pedro Menéndez de Avilés*. Trans. J.T. Connor. University Presses of Florida. Gainesville.

Stuiver, Minze and Bernd Becker 1986 High-Precision Decadal Calibration of the Radiocarbon Time Scale, AD 1950–2500 BC. *Radiocarbon* 28(2B):863–910.

Tankersley, Kenneth B. 1984 Personal Correspondence. Department of Anthropology, Indiana University, Bloomington.

Van Beck, John C. and Linda M. Van Beck 1965 The Marco Midden, Marco Island, Florida. *The Florida Anthropologist* 18(1):1–20.

Wardle, H. Newell 1951 The Pile Dwellers of Key Marco. *Archaeology* 4:181–186.

Widmer, Randolph J. 1988 *The Evolution of the Calusa*. The University of Alabama Press. Tuscaloosa.

Wing, Elizabeth S. 1965 Animal Bone Associated with Two Indian Sites on Marco Island. *The Florida Anthropologist* 18:21–28.

Bay West

Introduction

The melodramatic tale of the circumstances surrounding the recovery of materials from the Bay West site provides the best example possible of the weakness of legislation pertaining to development projects that modify wetland areas. Although stricter legislation has been enacted on paper in the years since this site was destroyed, the actual field situation remains virtually unchanged because there is little control over a project once a permit has been granted. A landowner becomes the possessor of the entombed heritage, and it is at his mercy.

In February 1980, dredging of a cypress pond at the Bay West Nursery was initiated to create a water source and disperse the fertile peat for use in the nursery business (Fig. 12). The sheriff's department was notified when human bones were observed but it was soon determined that the bones had nothing to do with a contemporary crime. It was not until March 7 that a member of the Southwest Florida Archaeological Society visited the site and realized its importance. An arrangement was made with the nursery owner whereby archaeological data could be retrieved without interfering with the ongoing dredging operation. Volunteers from the archaeological society worked until the first week in April salvaging material from the site (Beriault et al. 1981:39).

The Bay West site (8-CR-200) is located in Collier County a few miles northeast of Naples and approximately 9–10 km from the present coastline of the Gulf of Mexico (Fig. 1). The site and its significance are described in Beriault et al. (1981), Purdy and Hall (1981), Purdy (1981), and Purdy (1982). These accounts, as well as unpublished reports of the human skeletal and faunal materials, are summarized below.

A

FIGURE 12 (A and B) Before and after dredging the Cypress Pond to create a water source for the Bay West Nursery.

Methodology

Beriault et al. (1981:40) distinguish three ways that the material was salvaged at the Bay West site: (1) in an uncontrolled fashion from the spoil that had been deposited by the dredge, (2) by placing transects across the spoil piles which only served to make recovery less random and more systematic since the original provenance of the dredged peat was unknown, and (3) by collecting material as it was removed from the pond and dropped onto the spoil piles which made it possible to note some degree of association between the human skeletal elements, soil, and artifacts. The burials evidently came from approximately 1 m below the pond bottom; i.e., 1 m of peat had accumulated over the burials. Above the peat was about 2 m of water. Only material collected during the dredging operation was used for radiocarbon analysis.

An attempt was made to excavate a 2-m square along the west edge of the pond in an area that was to be destroyed the following day. This effort was conducted until nearly midnight using portable electric lights but water seepage prevented excavation to the depth of the burials and the square was abandoned. No attempt was made to retrieve material by diving because of poor visibility and the extensive number of trees and roots present in the pond (Beriault et al. 1981:40).

FIGURE 12B

Environment

The mortuary at the Bay West site was located in a 35-m diameter central pond depression of a cypress dome. These kinds of features are formed when the dissolution of limestone creates conditions favorable for the deposition of peat during wet periods. Marine shell underlying the peat is probably the major causative factor in promoting the essentially neutral pH values that assured the preservation of bone which would not have survived in an acid peat environment. Botanical specimens were preserved because of the low oxygen content.

Two core samples were taken and described by Meeder (Beriault et al. 1981:44). Core #1 from the west shore of the pond extended from 0–352 cm. The deposits from 63–76 cm consisted of a reddish brown fibrous peat and from 76–263 cm of a black amorphous peat. It is not stated which of these two zones is thought to correspond to the cultural stratum. The basal peat in this core collected at 253–263 cm was dated at 7550 ± 125 BP. Core #2 was taken from the east shore of the pond where the organic deposits apparently were much shallower. A black peat zone extended from 73–131 cm and, following a 2-cm lense of white quartz sand, a fibrous peat continued from 133–145 cm. Peat collected at 121–131 cm was dated at 5855 ± 120 BP. Beriault et al. (1981:43) believe the pond was expanding with time in a southeasterly direction.

On August 7, 1980, I submitted five soil samples which I had collected at the

Bay West site to the Soil Science Laboratories at the University of Florida. The samples were analyzed for pH, calcium (Ca), copper (Cu), iron (Fe), potassium (K), magnesium (Mg), manganese (Mn), zinc (Zn), and phosphorus (P). The pH ranged from 5.0–5.4 and all of the other values were higher for the sample with a pH of 5.4. In ppm: Ca, 5200–13,760; Mg, 104–224; P, 28.8–280+; K, 16–48; Zn, .44–8.48; Cu, .28–.96; Mn, 1.44–7.04.

Fauna

Beriault et al. (1981:47–48) mention that deer bone and antler were abundant in the peat from the pond. It did not appear worked and the authors concluded that it was present as a result of seasonal loss of antler or death of the animals by natural causes. There were exceptions. The upper cranial portion of an extremely large male deer was recovered from the peat in close proximity to a concentration of human bone and the authors suggest it may have been a deer headdress that was placed with a burial. In addition, two atlatl hooks made of deer antler and two deer leg bone points were recovered. They are discussed below with the artifacts.

Beriault et al. (1981:48) report that other animal remains were recovered during the salvage operation: sirens, snakes, birds, fish, turtles, and small mammals. A rear tooth from *Mammut gompotherium*, a Pliocene elephant, was collected near the pond but it probably had no association with native activities at the site.

Animal remains had been accidentally included with the human skeletal material that was sent to Dr. William R. Maples at the Florida Museum of Natural History for analysis. The animal bones were identified by Stephen Hale in the Zooarchaeology Laboratory in Gainesville. I am listing these below since this information has not been published and the great diversity of species represented is of interest.

Mammals

Didolphus virginianus (opossum)
Lynx rufus (bobcat)
Odocoileus virginianus (deer)
Rodentia (rodent)
Sylvilagus sp. (rabbit)
Mrocyon cinercoargenius (grey fox)

Fish

Osteichthyes (bony fish)
Amia calva (bowfin)
Lepisosteous (gar)
Micropterus salnoides (bass)

Birds

Anas sp. (duck)
Ardea herodias (great blue heron)
Casmerodius albas (egret)
Enlica americana (coot)
Grus canadensis (sandhill crane)
Meleagris gallopaoo (turkey)
Phalacrocorax auritus (anhinga)
Rallidae (rails, coot, gallinule
Lenaidura macroura (dove)

Reptiles and Amphibians	Shell
Alligator missippiensis (alligator)	Busycon contrarium (busycon)
Anura (frog)	Chione cancellata (clam)
Rana sp. (frog)	Dinocardium robustum (eastern cockle)
Amphiuma means (frog)	Macrocallista nimbosa (surf clam)
Siren lacertina (sirens)	Mercenaria campechirusis (quahog)
Crotalidae (snakes)	Planorbella sp. (water snail)
Cheloniidae (sea turtle)	Pomacea paludasa (apple snail)
Chelydra serpentina (snapping turtle)	Episula solidissma (clam)
Emydidae (snapping turtles)	Strombus sp. (conch)
Gopherus polyphemus (gopher tortoise)	Trachycardium sp. (bivalve)
Kinosternon sp. (mud turtles)	Mollusca pelecypoda (bivalve)
Oternotherous sp. (musk turtle)	
Terrapene carolina (box turtle)	

The most numerous bones in the sample by far were deer followed by mud turtles, snapping turtle, great blue heron, and clam (*Chione cancellata).*

Flora

Beriault et al. (1981) discuss the plant communities found today in the area of the pond and the upland at the Bay West site. They also identify several botanical species from the peat: *Lyonia* sp. (staggerbush/fetterbush), *Ilex* sp. (holly), *Myrica cerifera* (wax myrtle), *Sabal palmetto* (cabbage palm), *Serenoa repens* (saw palmetto), *Blechnum* sp. (fern), and bark of *Pinus ellioti* (slash pine).

Approximately 300 pieces of waterlogged wood from the Bay West site were sent to the Florida Museum of Natural History in May and June, 1980 to be preserved under my direction. The specimens were photographed, sketched, measured, and described prior to treatment with polyethylene glycol (PEG). Because of limited facilities, the wood was preserved in two batches between December 1980 and August 1981. The first batch was removed from PEG on March 15, 1981 and returned to the Collier County Museum in Naples on May 29, 1981 after we determined that treatment was successful. The second group was removed from PEG on August 1, 1981 and returned to the Collier County Museum in early 1982. In December 1982, the Director of the Collier County Museum returned all of the wood (except for the wooden artifacts and antler atlatl spurs) to Gainesville for permanent curation. These missing pieces are indicated in the notebook that describes each item from the site. The returned specimens were labeled according to the number assigned to them originally. As of February 1990, the wood from the Bay West site is in the Florida Museum of Natural History, Gainesville and is still in good condition.

The majority of the wood is pine including all of the poles or posts. Compression wood of pine was used to produce the ladle (Fig. 13). The pine

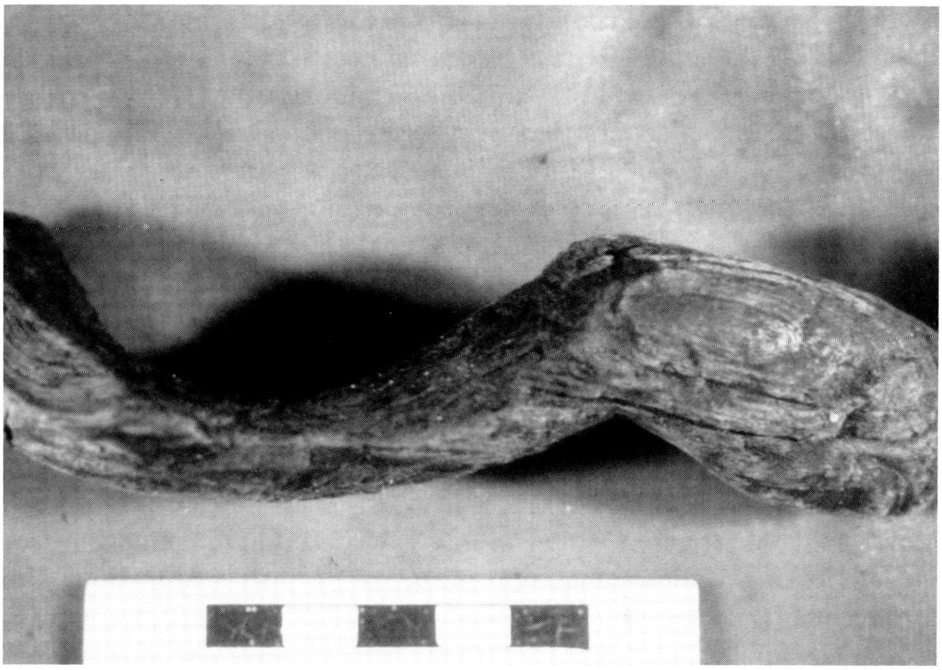

FIGURE 13 Ladle made from compression wood of pine. Specimen has shark teeth finishing marks on the handle (not visible). Photo by the author.

exhibits drought rings. Willow (*Salix* sp.), the root wood of cypress (*Taxodium* sp.), and the base of a palm have also been identified (Newsom, personal communication, 1990). The willow, a hard wood, is not well preserved. A complete anatomical study of all wood specimens has not been conducted. The small amount of cypress at 6500 years ago appears to support other findings that cypress may not have been widespread in Florida 5000–6000 years ago.

Human Skeletal Remains

The large quantity of human bone (several hundred pounds) that was recovered during the salvage operation at the Bay West site suggests that the pond was a prehistoric cemetery. Because of the way the bones were removed from the pond, there was no way to determine what bones belonged to specific individuals. Each scoop of the dragline sampled one or more individuals and not very much of any one skeleton. A skeletal collection composed of the broken and commingled bones of so many partially represented individuals limits the nature and extent of analysis. Dr. William R. Maples (1981) analyzed the skeletal material and his conclusions are summarized in the following paragraphs.

The human bones were not well preserved. The thin articular cortex that protects the ends of the long bones that are so important for adequate study of a skeletal sample had not survived in most cases. Once that protective cortex is gone, the underlying spongy bone is exposed and rapidly decomposes. As a result

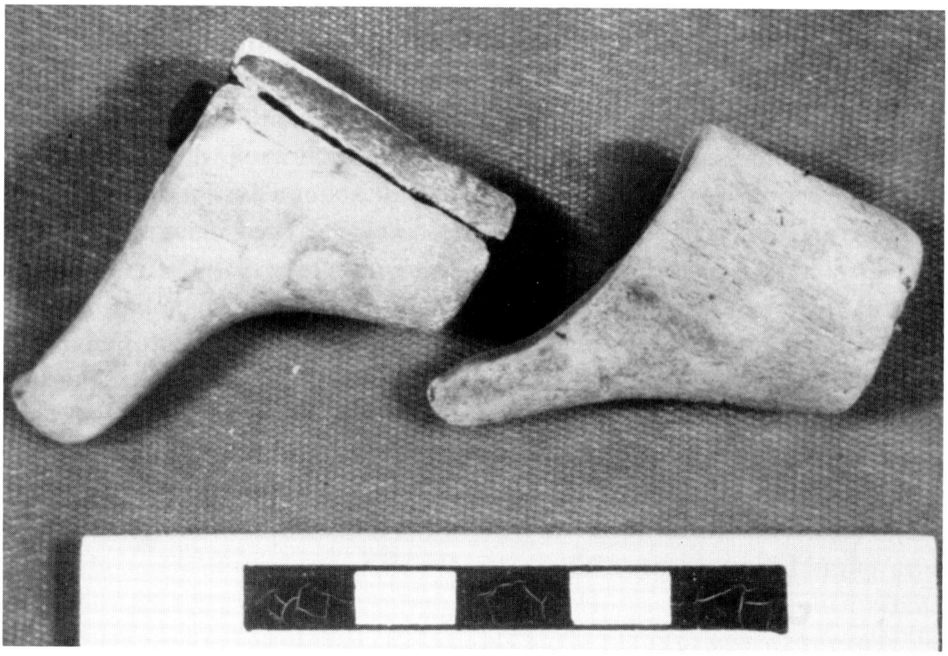

FIGURE 14 Atlatl spurs made from deer antler. Note fragment of wooden shaft in specimen on left (photo by the author).

or afterwards during a peat fire when drought conditions prevailed. All of the larger pieces of wood were fragmented, many with fresh breaks that occurred when they were removed from the pond during the dredging operation. Beriault et al. (1981:45) believe the poles were intentionally shaped and were in some sort of association with the burials. Their explanation is supported by the fact that the diameter of the pole material is remarkedly uniform, averaging 4.5 cm.

Beriault et al. (1981:46) discuss the wooden specimens that exhibit some degree of modification. The most obvious are four fire starters (Fig. 15), a broken ladle made of pine with shark teeth finishing marks (Fig. 13), and a smoothed spear shaft fragment incised with 12 paired parallel lines (Fig. 16). An unfortunate incident affecting this object occurred shortly after the Bay West materials were delivered to the Florida Museum of Natural History. Preliminary cataloging had been completed but preservation in PEG had not been initiated when this artifact was found lying out on a counter one morning. Its initial weight and measurements were 10 g, 4.38 cm long, and 1.25 cm thick. After a matter of hours out of its waterbath it weighed 1 g (90% loss) and measured 4.1 cm long (9% reduction) and .70 cm thick (>50% reduction). Efforts to rewaterlog this specimen failed (see Chapter 5 and Fig. 122).

Age of the Site

The Middle Preceramic Archaic stone spearhead styles from the Bay West site have been recovered in securely dated contexts elsewhere in Florida and

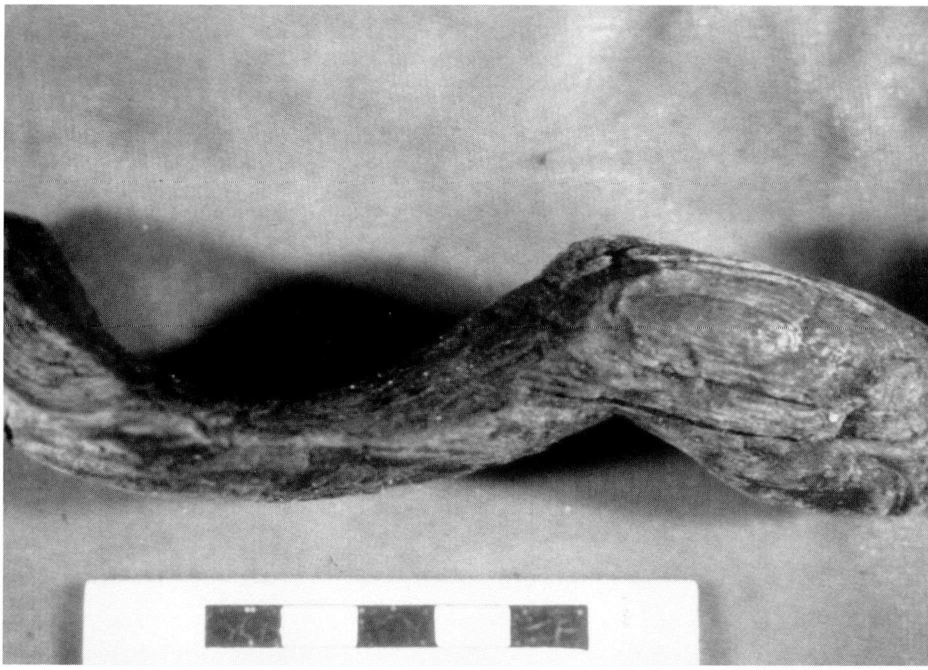

FIGURE 13 Ladle made from compression wood of pine. Specimen has shark teeth finishing marks on the handle (not visible). Photo by the author.

exhibits drought rings. Willow (*Salix* sp.), the root wood of cypress (*Taxodium* sp.), and the base of a palm have also been identified (Newsom, personal communication, 1990). The willow, a hard wood, is not well preserved. A complete anatomical study of all wood specimens has not been conducted. The small amount of cypress at 6500 years ago appears to support other findings that cypress may not have been widespread in Florida 5000–6000 years ago.

Human Skeletal Remains

The large quantity of human bone (several hundred pounds) that was recovered during the salvage operation at the Bay West site suggests that the pond was a prehistoric cemetery. Because of the way the bones were removed from the pond, there was no way to determine what bones belonged to specific individuals. Each scoop of the dragline sampled one or more individuals and not very much of any one skeleton. A skeletal collection composed of the broken and commingled bones of so many partially represented individuals limits the nature and extent of analysis. Dr. William R. Maples (1981) analyzed the skeletal material and his conclusions are summarized in the following paragraphs.

The human bones were not well preserved. The thin articular cortex that protects the ends of the long bones that are so important for adequate study of a skeletal sample had not survived in most cases. Once that protective cortex is gone, the underlying spongy bone is exposed and rapidly decomposes. As a result the ends of most long bones are the first structures to be lost and this is the case with the Bay West skeletal remains. Only 10 complete mature long bones remain of all the material recovered. Four intact long bone diaphyses are also present.

There are a few elements that are better preserved. Sectioned cortical surfaces of some of the bones still show some bluish fluorescence when examined under an ultraviolet light source. Normally, this does not occur on bones more than one hundred years old and indicates the existence of more organic remains in these bones than is normally found in bones after the passage of only a century.

It was difficult to estimate the number of individuals represented in the salvaged material. Some bones usually survive better than others because of their more durable structure and these were counted as indicators. Frontal bones, temporal bones (the petrous parts), and mandibles were selected. Based on these counts, at least 37 individuals were represented and there were probably several more.

Maples did not attempt to determine sex. He felt that too much error would be involved because of the fragmentary nature of the post-cranial bones and the inaccuracy involved using incomplete skulls. He provides some insights on paleopathology and age at death. Relatively little trauma was observed. There were healed fractures of a scapula, rib, and right humerus. Several subperiosteal hemorrhages were seen which can result from direct trauma to the bone (e.g., contused and lacerated shins) or from bleeding associated with scurvy or vitamin C deficiency. Maples considered vitamin C deficiency unlikely because of the availability of leafy vegetables such as polkweed.

Some examples of Schmorl's herniations (nodes) were seen in the lumbar vertebrae. In modern populations these are often associated with the lifting of great loads and are common among weightlifters.

The most outstanding evidence of skeletal pathology in the collection concerns arthritis and other inflammatory or degenerative joint diseases. The majority of the bones of the elbow (distal humerus and proximal ulna and radius) show some arthritic lipping. The high frequency of changes in the elbows, along with an unusual development of the muscle attachment areas on the humeri, suggests that the elbows of these people were subjected to stress such as from canoe paddling. Vertebral lipping (osteophytosis) was also very common but arthritic maladies are widely distributed in all joints.

The dental remains are well preserved. Little or no evidence of dental caries is present but attrition is relatively great. The teeth show considerable wear by the mid-20s and is pronounced by middle or old age. Extreme wear leads to apical abscesses and 75% of the mandibles with more than half of the alveolar area preserved show abscesses. Older individuals show extreme periodontal bone loss but there were no edentulous (toothless) individuals.

Eight individuals in the sample died by 9 years of age: 2 were less than 1 year old, 2 were from 3–4 years old, and 4 were 7–9 years old. Of those individuals achieving maturity, approximately one third died before the mid-20s and the rest of the adults died in the late 20s through the early 50s. It appears that there was no differentiation of burial practices based on age and, because it was not possible to determine sex, we can only assume that both males and females were included in the cemetery. The less than ideal conditions of recovery also precluded the ability to learn if grave goods were more prevalent with child burials as has been observed at similar Archaic Period sites.

There is some evidence that the skeletal material was not immediately interred upon death. Tooth marks from rodents can be seen on a number of bones and are extensive on a few specimens. Partial cremation may have occurred, or a peat fire may have caused the charring observed on some bones and other materials in the pond (Beriault et al. 1981:43). When the skeletal material was first salvaged from the dredge piles, three crania appeared to have brain tissue intact. Dr. Michael Zimmerman (1980) determined through microscopic analysis that brain tissue was not present. If the bodies had been exposed on the surface before interment, brain material would not be expected since the brain is one of the first tissues to decompose and liquify after death (Maples 1981:2).

Artifacts

Stone—Five Middle- to Late-Preceramic Archaic spearheads were found at the site, two from the mortuary pond and three from the adjacent upland area. One of the points from the pond was in close association with a bone point, an atlatl hook, and human skeletal material. These points, along with radiocarbon analysis, confirm the Archaic age of the site. Beriault et al. (1981:52) illustrate three of the stone artifacts in their Figs. 7–9. Nondiagnostic stone tools and a considerable amount of chert debitage were also recovered from the pond and the upland area. Most of this material was silicified coral which suggests trips to or trade with the Tampa Bay area since chert materials do not outcrop as far south as the Bay West site. The only ornamental object found at the site was a large stone bead manufactured of non-Florida green stone. It is pictured as Fig. 11 in Beriault et al. (1981:52).

Antler and Bone—Two antler artifacts are thought to be atlatl hooks. They consist of cylindrical sections of the main portion of deer antler with one of the branching tines shaped into a spur. Fragments of what had probably been wooden shafts still remained in a hole that had been drilled through the middle of the cross section exposed when the longitudinal portion of the antler was cut. The atlatl hooks were recovered in different locations at the site but both were in association with skeletal remains and other artifacts. They are pictured in Figs. 3 and 4 of Beriault et al. (1981:51) and in Fig. 14. Two bone points were recovered during the salvage operation. One is shown in Fig. 10 of Beriault et al. (1981:52).

Wood—In most cases it was difficult to determine if the wood from the Bay West site had been naturally deposited or had been parts of scaffolds or platforms associated with the mortuary complex. Some of the wood is charred which may have resulted from a fire-setting technique used to shape it or from a conflagration that burned the structures before they collapsed into the pond

FIGURE 14 Atlatl spurs made from deer antler. Note fragment of wooden shaft in specimen on left (photo by the author).

or afterwards during a peat fire when drought conditions prevailed. All of the larger pieces of wood were fragmented, many with fresh breaks that occurred when they were removed from the pond during the dredging operation. Beriault et al. (1981:45) believe the poles were intentionally shaped and were in some sort of association with the burials. Their explanation is supported by the fact that the diameter of the pole material is remarkedly uniform, averaging 4.5 cm.

Beriault et al. (1981:46) discuss the wooden specimens that exhibit some degree of modification. The most obvious are four fire starters (Fig. 15), a broken ladle made of pine with shark teeth finishing marks (Fig. 13), and a smoothed spear shaft fragment incised with 12 paired parallel lines (Fig. 16). An unfortunate incident affecting this object occurred shortly after the Bay West materials were delivered to the Florida Museum of Natural History. Preliminary cataloging had been completed but preservation in PEG had not been initiated when this artifact was found lying out on a counter one morning. Its initial weight and measurements were 10 g, 4.38 cm long, and 1.25 cm thick. After a matter of hours out of its waterbath it weighed 1 g (90% loss) and measured 4.1 cm long (9% reduction) and .70 cm thick (>50% reduction). Efforts to rewaterlog this specimen failed (see Chapter 5 and Fig. 122).

Age of the Site

The Middle Preceramic Archaic stone spearhead styles from the Bay West site have been recovered in securely dated contexts elsewhere in Florida and

FIGURE 15 One of four fire hearths or starters recovered from the Bay West site (photo by the author).

have yielded ages of around 6000 years old. Radiocarbon analysis of organic materials from Bay West conducted at the University of Miami (now Beta Analytic) add supporting evidence for the antiquity of these stone tool types and for the age of the site itself. Beriault et al. (1981:50) tabulate these results as follows:

UM #	FS #	Material	RC Years BP
2085	577 (Bag 18 of 25)	Wooden post	6520 ± 135
2087	578 (Bag 14 of 14)	Wooden post	6675 ± 85
2088	578 (Bag 11 of 14)	Wooden post	6630 ± 80
2169	515	Peat	6780 ± 135
2170	Sample #2	Peat	5500 ± 80
2226	Core #2 (121–131 cm)	Basal peat	5855 ± 120
2227	Core #1 (253–263 cm)	Basal peat	7550 ± 125

Discussion

Beriault et al. (1981) are effusive in their praise of the Bay West Nursery owner who allowed the Southwest Florida Archaeological Society to pick through the spoil piles dredged from the mortuary pond and who donated all of the salvaged materials to the Collier County Museum and Archives (document dated May 2, 1980). The press was not as kind. In articles published off and on for 6 months after the site's discovery, reporters told of the bulldozers and drag line

FIGURE 16 (A) Shaft fragment incised with 12 paired parallel lines (photo courtesy of John Beriault); (B) following accidental degradation (see also Chapter 5 and Fig. 122).

that were turning up skulls and long bones by the dozens. They said that volunteers left work and family to race ahead of the 'dozers and seeping water. They told how the dredging had destroyed the skeletons and other information that might have been gained by painstaking excavation. The landowner was not excited about 6000-year-old brains and thought the wooden artifacts looked like stray sticks that anyone but an archaeologist would throw away. Apparently, no one requested that he delay his project for a week or two. He had obtained his permit, hired equipment operators to create his pond, and would have lost considerable time and money had he altered his schedule to save a unique 6000-year-old piece of Florida's heritage. The destruction of the Bay West site is tantamount to desecrating King Tut's tomb. Stronger legislation and greater compensation or incentives to property owners are needed to avoid similar situations.

There seems to be little question that the Bay West site was a cemetery. It is curious that so many animal species representing about 35 vertebrates and 11 invertebrates were found in the pond along with the human skeletons. Not many artifacts were recovered but their scarcity may be more apparent than real because of the poor conditions that prevailed. The recovery of the stone bead indicates that decorative items were being manufactured by 6000+ years ago. Since the bead was made from non-Florida stone, its presence suggests at least limited out-of-state travel or trade. The bead plus the atlatl hooks, and a few

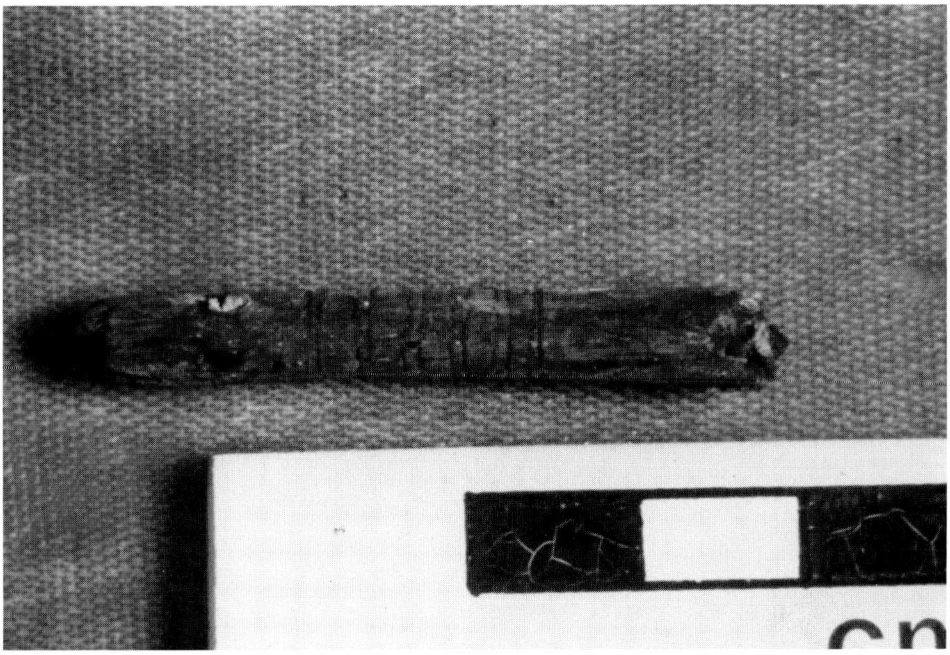

FIGURE 16B

other items, suggest that it was a cultural practice to include grave goods with burials. Perhaps some of the animals were grave goods also, like someone's guardian spirit, but this idea is strictly conjectural on my part.

The Bay West site shares similarities with the slough at Little Salt Spring and the Republic Groves site.

All of the records, skeletal remains, and artifacts are at the Collier County Museum in Naples except for the wood records and the wooden ladle fragment that are in storage at the Museum of Natural History, University of Florida along with the nonartifactual wood.

References

Beriault, John, Robert Carr, Jerry Stipp, Richard Johnson, and Jack Meeder 1981 The Archaeological Salvage of the Bay West Site, Collier County, Florida. *The Florida Anthropologist* 34(2):39–58.

Maples, William R. 1981 Human Skeletal Remains from the Bay West Nursery Site. Manuscript on file, Florida Museum of Natural History, Gainesville.

Purdy, Barbara A. and Sharon Hall 1981 Investigation and Stabilization of Wooden Artifacts in Florida. *Proceedings of the Southeastern Archaeological Conference* Bulletin 24:16–17.

Purdy, Barbara A. 1981 Cultural Resources at Wetland Sites in Florida: Investigations of Sites and Preservation of Artifacts. Final report of Survey and Planning Grant to Division of Archives, History, and Records Management, Tallahassee

1982 Survey, Recovery, and Treatment of Wooden Artifacts in Florida. *Proceedings of the ICOM Waterlogged Wood Working Group Conference*, Ottawa,1981, pp. 159–169.

Zimmerman, Michael R. 1980 Indian Skull Specimens: Analysis of Three Crania from the Bay West Site. Report on file with Collier County Museum, Naples.

Belle Glade

Introduction

The Belle Glade site is located in Palm Beach County near the southeast shore of Lake Okeechobee approximately a mile and a half west of the town of Belle Glade (Fig. 1). The site consists of a habitation midden (8-PB-40), a burial mound (8-PB-40, the Chosen Mound), and a linear earthwork (8-PB-91). As part of the Civil Works Administration under the Federal Emergency Relief program during The Great Depression, the Smithsonian Institution proposed archaeological projects that would employ approximately 1500 men as relief labor in the winter of 1933—1934. Salaries were provided also for trained archaeologists since proper supervision was essential for successful scientific results. The Belle Glade site was an ideal choice for excavation under this program because it required a large crew which usually would not have been available or practical. G.M. Sterling was in charge of the Belle Glade project. The results of the excavations are summarized briefly by M.W. Sterling but a complete study of the notes and collections had not been finished by 1935 when he wrote his report (Sterling 1935:371—372). Willey (1949) furnishes an excellent analysis of the Belle Glade site using all information available to him including that provided by personal communication with G.M. Sterling. Certain materials that are essential for an interpretation of environmental conditions were not collected during excavations however and, because radiocarbon analysis had not been developed yet, a relative chronology had to be established by comparing the ceramic styles from Belle Glade to other sites. The wooden artifacts that survived at Belle Glade closely resemble some of those recovered at Key Marco and at Fort Center, as do tools and decorative items of shell and bone.

In 1975, a house was built on the the Belle Glade burial mound. Drs. William R. Maples, H.K. Brooks, and I visited the area when the land was being prepared for construction. At that time the burials and the stratigraphy remained pretty much as described by Sterling (1935) and Willey (1949). In 1977, Florida Atlantic University, the Palm Beach County Archaeological Society, and the Broward County Archaeological Society recovered material from the site. In 1980, as part of a statewide survey to assess Florida's archaeological wet sites, Dr. George F. MacDonald and I returned to the burial mound area and discovered that the deposits and everything in them, including the burials, were completely desiccated as a result of the drainage necessary to stabilize the land for building. No evidence whatsoever remained of human bones or the fibrous peat in which they had been embedded. The deposits had turned to dust.

Excavations and Stratigraphy

The habitation midden was approximately 91 × 137 m (100 × 150 yards) and consisted principally of shell refuse. In aboriginal times, the midden was situated between two branches of the Democrat River, a stream which, by 1933, had been largely obliterated by modern drainage operations. The cultural deposits were more than 2 m (7 feet) deep in places and extended more than .60 m (2 feet) below the surface of the surrounding land. On the outer margin of the shell there was an accumulation of over .30 m of recent muck. It was clear that the muck soils had built up around the mound since its desertion. The much smaller burial mound was located on the opposite side of the stream about 91 m (100 yards) from the habitation midden (Sterling 1935:374; Willey 1949:19).

Habitation midden—The major excavation was a trench approximately 21 × 6 m (70 × 20 feet) sunk in the northwest part of the midden. The trench was excavated in seven sections arranged in four rows. Excavations were conducted in .30-m (1-foot) levels. Most of the sections extended to sterile muck but water occasionally filled the bottom of the trench before this level was reached. A smaller trench, about 12 × 3–4.5 m (40 × 10–15 feet), was also opened. If an overall grid system was employed, this trench was located approximately 24 m (80 feet) from the first and oriented parallel to it.

Visible stratification was noted in the midden but these strata and lenses do not appear to be correlated with any structural features or changes in cultural material although Willey (1949) was able to show stylistic changes in certain ceramic types that tended to correspond with depth. Sterling (1935) noted a number of houseposts during excavation that revealed little beyond indicating a rectilinear type of construction. On the south part of the midden there was a slight elevation above the general level that Sterling thought might be a platform upon which a structure was built.

Burial mound—The burial mound was circular in groundplan and dome shaped in cross section (Fig. 17). Only a small portion of the top of the mound showed above the level of the surrounding ground prior to excavation. Muck soil had almost covered the mound in the same way that it had encroached around the base of the shell habitation midden. The following description of the stratigraphy encountered in the burial mound comes from Stirling (1935:375) and Willey (1949:20–23).

At the base of the mound is an old occupation stratum similar to the refuse from the habitation midden. The Old Muck Burial Mound was constructed on top of the habitation layer and muck accumulated around its margin for an unknown period of time. The First Sand Mound was superimposed on the muck mound and a small plaza or pavement of irregular limestone slabs was placed between the fork of the Democrat River and the burial mound. The First Sand Mound was partially destroyed by flooding which scattered human bones and artifacts. Many of the wooden objects that appear to be of ceremonial use were found in the water-washed sand. Evidence of brief periods of habitation occur within and above the flooded deposit. The Final Sand Burial Mound was placed roughly centered over the previous outwashed mound (Fig. 17).

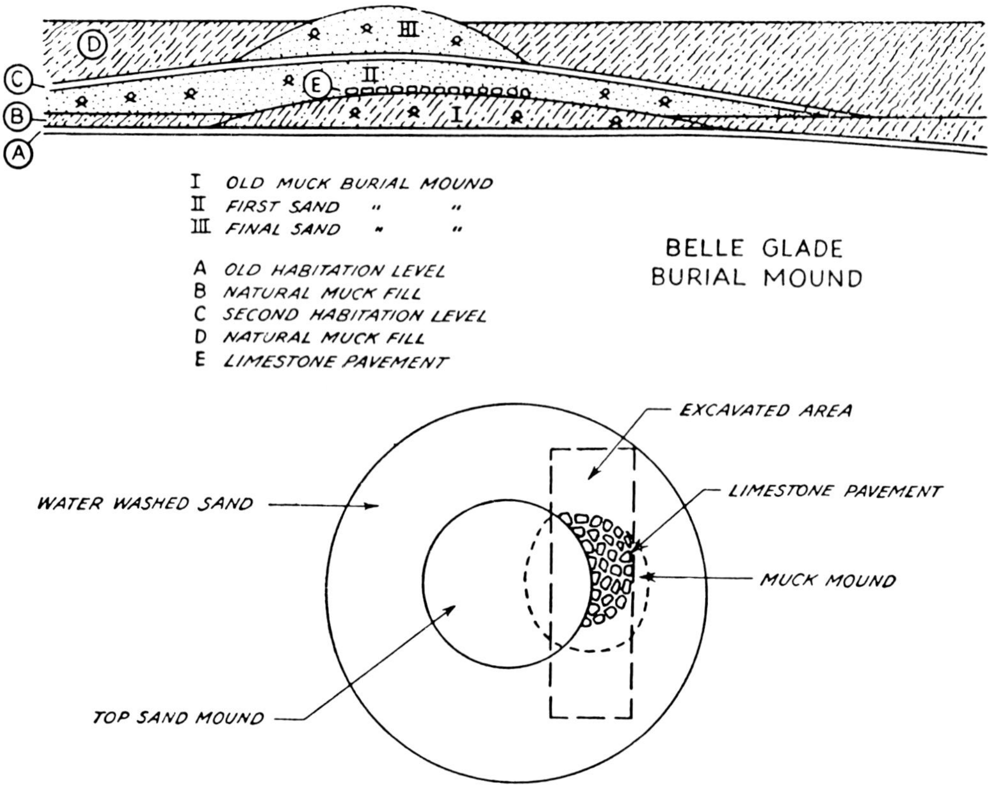

FIGURE 17. Planview and stratigraphy of the Belle Grade burial mound (from Willey 1949:21).

The burial mound excavations were confined to a fairly large rectangular trench made on the side of the mound facing the river and the habitation site. Although there is no record of the burial mound excavatons, G. M. Stirling informed Willey that the trench was excavated in "successive five-foot slices beginning at the edge and working in toward the center of the mound" (1949:22). Vernon Lamme worked at the Belle Glade site with Stirling in 1933–1934. In 1978 and 1979 the Palm Beach County Archaeological Society, Inc. published transcripts of taped interviews with Lamme in their newsletter. About the excavations in the burial mound, he said: "We cut a little face at one edge of the mound and bit by bit we kept moving it in so the face got wider and higher as we moved in toward the center. But I don't think we dug in more than 1/6 of the distance across the mound."

Fauna and Flora

Stirling made a collection of animal bones and shell. Willey (1949:61) lists the following species.

Myroca affinis (lesser scaup duck)
Coragyps atratus (black vulture)
Cathartes aura (turkey vulture)
Andea herodias (great blue heron)
Canis (dog or red wolf)
Fish, various types
Odocoileus virginianus (white-tailed deer)
Lynx rufus (bobcat)
Alligator
Reptile
Procyon lotor (raccoon)
Crab
Green turtle
Turtle
Trichechus latirostris (manatee)
Sylvilagus floridanus (rabbit)
Euarctos floridanus (black bear)
Shark
Lutra canadensis virginanus (otter)
Mustela vison lutensis (mink)
Canis (domestic dog)

Stirling (1935) does not mention the floral composition of the levels excavated and Willey states only that the principal vegetation cover in the Everglades-Okeechobee Basin is sawgrass (1949:17). Several coprolites are listed in United States National Museum Accession Memorandum 160249 (the Belle Glade site). If these are human and if they still exist in a well-preserved condition, they might provide insights into the diet of the people at Belle Glade.

Artifacts

The artifacts receive more complete treatment than the faunal and floral remains. Stirling provides a summary account of the artifacts recovered at Belle Glade and Willey (1949:24–69) describes them thoroughly.

Ceramics—Pottery at Belle Glade was very abundant. No whole vessels were found and sherds were recovered mainly from the shell midden rather than the burial mound (only 39 sherds). Most of the pottery was plain and Willey believes it was intended for cooking and other utilitarian functions rather than in a ceremonial context. Large open bowls and deep pots were the usual form. Since many plain sherds were discarded, the percentage of plain to incised, check-stamped, or red-slipped varieties was even greater than the numbers indicate. The types identified were Belle Glade Plain, Glades Decorated, and Glades Plain (now called Sand-tempered Plain), Biscayne Check-Stamped (now called St. Johns Check-Stamped), Okeechobee Plain (now called St. Johns Plain), Okeechobee Red (now called Dunns Creek Red), and a few nonlocal or unidentifiable types. I am indebted to Ann Cordell (personal communication, 1990) for informing me of the modern names for these ceramic types. Of interest is the fact that 39% of the sherds from the top level were St. Johns Check-Stamped but this type is virtually absent in lower levels. Sears (1982) and Widmer (1988) also note the recent occurrence of this type, probably postdating AD 1200.

Three fragmentary pipes and 14 potsherd pendants were the only other artifacts made of clay recovered at Belle Glade. Willey believes that the absence of a variety of artifacts made of fired clay probably reflects a lack of skill and interest in ceramic art at this site.

Stone—All of the chipped stone artifacts pictured in Willey (1949:Pl. 6) appear to me to be Preceramic Archaic types. Since the points date to a time period much earlier than the major habitation at Belle Glade, they must have been incorporated into the mound fill accidentally. No small triangular points were found.

Other stone artifacts are made of locally occurring Miami oolite, coquina, limestone, and sandstone. They consist of 2 weights or sinkers, 1 celt, 5 abrading stones, 1 smoothing stone, 7 pendants or plummets, 3 pipes, 6 tubes (beads?), 2 bars, 1 conical stone, and 1 fluted stone. The only stone artifact made of nonlocal material, besides the chert projectile points, is a piece of worked pumice that Willey believes drifted from the Lesser Antilles.

Willey is unimpressed with the stone remains and concludes that the stone industry competed poorly with the manufacture of shell artifacts which was a major medium of production at Belle Glade.

Shell—Belle Glade craftsmen were adept at fashioning all manner of implements and ornaments from a wide variety of marine shell species. "The general term 'conch shell' applies to the bulk of the shells used. . . . only a few other univalves and bivalves were employed" (Willey 1949:46). *Busycon, Strombus, Fasciolaria, Cassis, Olivella, Eontia,* and *Navicula* were used to produce picks, adzes, hammers, celts, scrapers, cups, pendants, beads, and a ring. Willey lists 43 celts; 58 picks, hammers, or adzes; 16 tools of conch shell; 11 cups or vessels; 39 pendants or plummets; 1648 beads; plus worked shell scrap and miscellaneous specimens. A freshwater mussel, native to the Mississippi River region, presumably was traded into the area and was made into an engraved gorget. The interior of this shell has a lustrous quality in contrast to the less lustrous local marine shell.

Bone—Implements and ornaments of bone are numerous, diverse, and made from a variety of species of animals including human, deer, fox, raccoon, bear, bird, alligator, fish, turtle, manatee, porpoise, and whale. Deer antler and shark teeth were used also. Bone points were probably the standard projectile for hunting and warfare, and bone pins were used commonly as ornaments. Stirling believed that the bone pins found in the burial mound were in the hair of the deceased when they were buried. Some specimens are engraved, such as a deer antler headdress, but Willey is surprised that so few artifacts were decorated considering how skilled the craftsmen were in working bone.

Bone preservation was excellent at the Belle Glade site and the quantities available made it possible for Willey to establish categories and describe manufacturing techniques. There were 197 socketed projectile points, 95 bipointed projectile points, 12 daggers or dagger fragments, 110 bone pins of many different varieties, 9 awls, 3 antler adzes, 7 beads, 2 ear ornaments, and other types of bone artifacts represented by only 1 or 2 specimens. An interesting artifact is a circular plate made from the vertebral disk of whale. It was trimmed around the edges and ground on both surfaces but not polished.

Four fossil shark teeth, 3 porpoise teeth, and 2 bear(?) teeth all had holes drilled in them. These were probably used as ornaments although teeth from various species of shark were used in south Florida as engraving tools also.

In addition to several weapons or ornaments made of human bone, a human skull cap of an adult male was made into a bowl or gorget. The skull was found in the Old Muck Burial Mound. A dagger made of a human femur had been placed with one individual near the surface of the mound (the Final Sand Burial Mound) along with European artifacts (Stirling 1935:376).

The bone industry at Belle Glade did not overlap greatly with the shell industry. Willey considers this an important point. Beads and pendants are abundant in shell but rare in bone. Most of the other items manufactured of shell were not duplicated at all in bone, and the reverse is true also.

> This functional division along the lines of materials further indicates the importance of bone and shell in the material culture and the relative lack of importance of stone. The artifact types of shell and bone complement each other in a well-adjusted pattern of use; stone, employed much less frequently, competes unsuccessfully with both (Willey 1949:47).

Wood—Woodcarving was one of the most developed of native crafts in Florida. In a region such as south Florida with an absence of stone suitable for carving, wood was even more important than it was elsewhere in the southeast. Both utilitarian and ceremonial objects were manufactured using a variety of woodworking techniques. No painted specimens were found at Belle Glade but Willey believes that, as at Key Marco, some of the objects were painted. Wood species were not identified but the pestles were obviously made of a hard dense wood and the flat pieces were probably made of softwoods such as cypress or pine. In view of the fact that nearly all of the large carvings found in Florida were made from pine, one could speculate that the Belle Glade specimens were made of pine also. The flat pieces from Key Marco were made from cypress and perhaps those from Belle Glade are made of cypress since the two sites share a number of other similarities. Since many of the wooden objects are still in the United State National Museum, it is possible that they will be identified someday. Carving tools were probably shell celts and adzes and hafted shark teeth. The fine striations made by shark teeth were still evident on many of the specimens when Willey conducted his analysis (1949:53).

A canoe paddle (Fig. 18A), two stirring paddles (Fig. 18B), and a stool (Fig. 19) were found in the muck near the river, unassociated with the midden or burial mound. Three intricately decorated bone pins were recovered there also (Stirling 1935:374). Willey (1949:54) says the stool (Fig. 19) was in a disintegrating but preservable condition. It is larger but otherwise nearly identical to one from Key Marco (Purdy 1974). The stirrers were small (27.5 cm and 29 cm), paddle-shaped artifacts made of a softwood (Fig. 18).

The houseposts, two dumbbell-shaped pestles (Fig. 20), and two broken pestles came from the shell midden. No location is indicated for a wooden fire hearth or 11 other wooden objects. The rest of the wooden artifacts came from the water-deposited sand of the first sand burial mound. These included 1 animal effigy (dog, fox, or deer), 5 bird effigies, 2 human carvings, and 3 plaques.

The animal effigy (Fig. 21) displays considerable realism very similar to

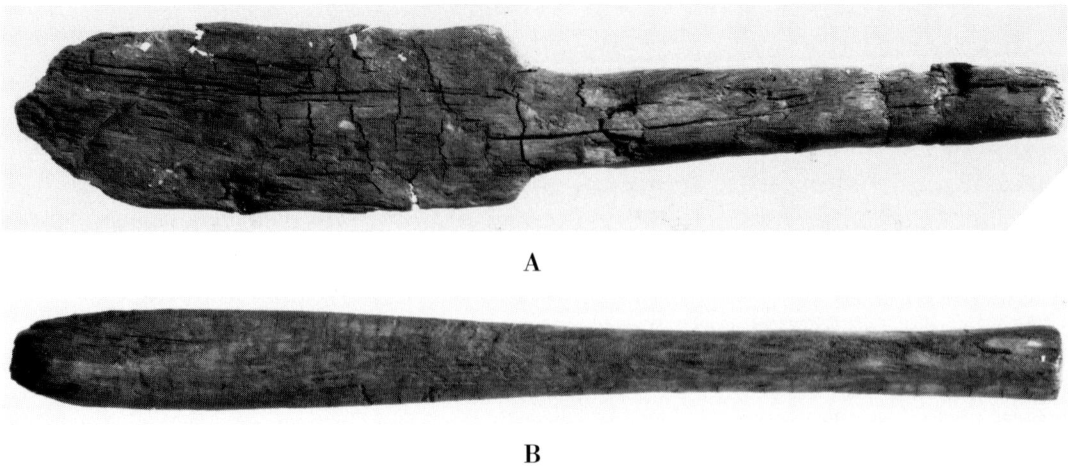

FIGURE 18 (A) Canoe paddle (Smithsonian Institution Cat. No. 383873, Neg. No. 31647-C); (B) Paddle-shaped wooden artifact (Smithsonian Institution Cat. No. 383872, Neg. No. 31647-C).

specimens from the Fort Center site. According to Willey (1949:56), four branches growing from a log were shaped to serve as legs of the animal. Its rump and tail also followed the natural shape of the log and the same is true of the neck and head. The animal was sculptured as though facing to the side, with head and neck twisted sideways. The body has a slightly sway-backed, full-bellied aspect.

The 5 bird effigies are described by Willey (1949:56). Fig. 22 shows a head fragment of a long, hook-beaked bird of prey. Sculpturing is expert. The eye and delineations of the beak are done with deep grooves. Another bird of prey, possibly an eagle, consists only of the top of the bird's head. On one side there are deep grooves depicting a beak and eye.

A beautifully conceived and carved duckhead is seen in Fig. 23. Willey believes this is the finest piece of art in the Belle Glade collection. The artist capitalized on the natural curve of a tree fork. A raised median crest extends up the back of the neck to the top of the head. The eye is a raised circle within an intaglio pointed ellipse. The bill is flat and convincingly ducklike. This specimen is reminiscent of one of the Key Marco wooden artifacts.

A full-round head of a woodpecker or similar bird is seen in Fig. 24. The eye and bill are outlined in deep grooves. On the back of the head are two protuberances which have been hollowed out.

The extended wing of a bird was made from a thin, flat board. It was carved on only one side with deep grooves representing feather markings. Presumably, the wing was broken from a complete effigy.

The 2 human figures were crudely made or incomplete. The more completely finished is the figure of a standing human with head, nose, folded arms over stomach, legs, knees, feet, buttocks, calves, and a large hair-knot on the back of the head all distinguishable. Warping and erosion have destroyed the face except for the nose (Fig. 25). A standing human was intended for the other specimen but only the head, shoulders, arms, and feet have been blocked out.

A

B

FIGURE 19 Two views of a stool similar to one recovered at Key Marco (Smithsonian Institution Cat. No. 383860, Neg. No. 31647).

The 3 plaques or tablets are similar to others found in south Florida (Allerton et al. 1984; Fewkes 1928). The purpose and function of these plaques or tablets has been speculated upon. They undoubtedly were of ceremonial or religious significance. Some of them were probably set upright on buildings or posts (Willey 1949:57). These plaques are usually made in two sections which are connected by one or three small center pieces. One plaque from the Belle Glade site consists of the upper half of a large, decorated tablet that is carved with deep, broad grooves. The design is composed of two reversed concentric capital D-shaped figures placed in the lower corners. On the top edge of the tablet is remnants of two birds' claws. Apparently a bird figure was perched on top of the plaque. This specimen was carved from a single piece of wood. The peculiar bulge at the bottom of the upper section of the plaque is characteristic of similar examples from south Florida.

Another example from Belle Glade is the lower half of a double-sectioned plaque made from a single piece of wood with only one central connecting piece. It is smooth, flat, and plain on both surfaces.

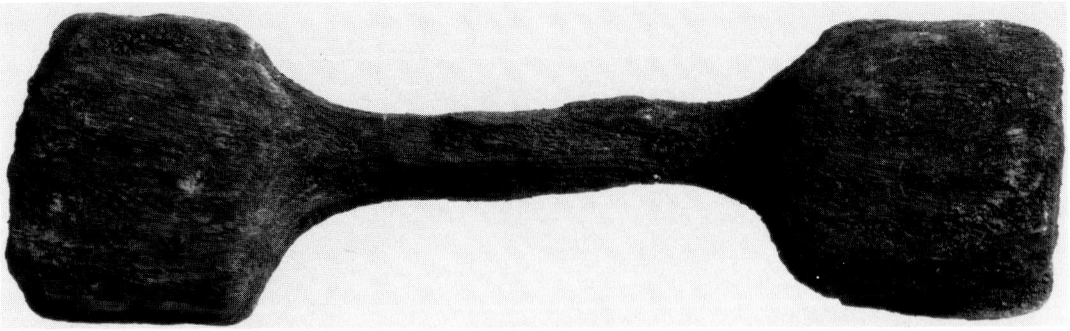

A

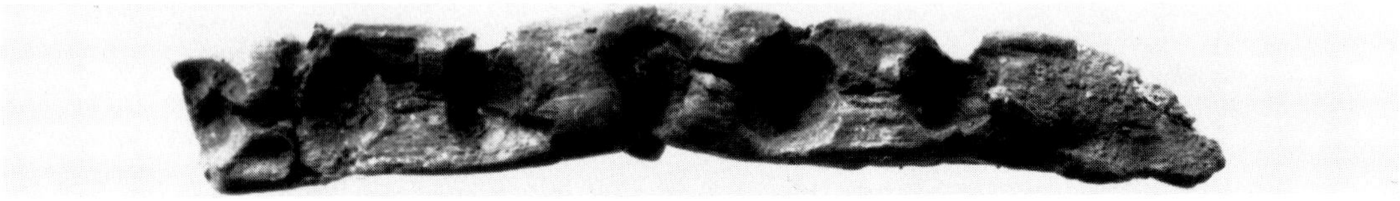

B

FIGURE 20 (A) Dumbell-shaped pestle similar to those recovered at the Fort Center site and the Windover site (Smithsonian Institution Cat. No. 383864, Neg. No. 3164; 7-C; (B) Fire hearth (Smithsonian Institution Cat. No. 383870, Neg. No. 31647-H).

FIGURE 21 Animal effigy (53 cm long) probably represents a dog, deer, or fox (Smithsonian Institution Cat. No. 383863, Neg. No. 31647).

The third specimen is a large plaque carved from a single piece of wood with a post base (Fig. 26 from Stirling 1935, Pl. 2). It is 1.16 m long. The cylindrical post part of the plaque is 54 cm long and tapers at the bottom. The face of the plaque is decorated in low relief. A solid circular element at the base of the plaque is connected to a similar element at the top of the plaque by a raised ridge. This is thought to be a thigh bone design.

Eleven additional wooden artifacts are described by Willey. Two of these are trays(?) that were fragmentary and badly charred; the others are miscellaneous or broken objects of indeterminate function.

Stirling (1935:376) mentions three logs, 7 feet long and 4 inches thick, that were located deep in the burial mound near the northeast side.

Lamme (1979) tells how the wood was preserved:

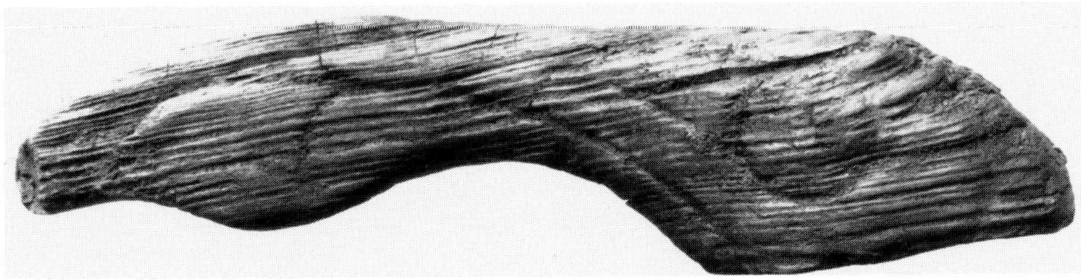

FIGURE 22 Head fragment of a long, hook-beaked bird of prey (Smithsonian Institution Cat. No. 383869, Neg. No. 31647-K).

> We got a lot of wood artifacts. I'll tell you how we saved them. You see, when we first find it, it's so dried up it looks like a branch or a piece of root or something. Well at Belle Glade we got all the paraffin we could find and we had a big tub of water at the dig where we were working and when we would get these things out of the ground, we'd put it right in the water and we'd carry it down to the hotel at night and melt the paraffin down and cover it all over. Then it was sent to the laboratories in Washington and they put it in containers with half water and half alcohol and wait until the water evaporates down to one part water to two parts alcohol and wait again to cut it down a little more, then they take it out and leave it till all the water evaporates.

European artifacts—As at a number of sites in Florida, objects of European materials appear at the Belle Glade site during the final period of utilization of the site. Stirling says the second sand burial mound extends into the historic period as a few of the burials near the surface were accompanied by glass beads. "Only those burials that were near the surface were accompanied by cultural remains. As these were mostly articles of a post-European nature, a change in burial custom may be indicated (Stirling 1935:375). Willey lists the European artifacts as 2 gold beads, 1 silver bead, 1 copper bead, 1 iron spike, 2 iron fragments, 1 lead pendant or plummet, 3 sherds of Spanish olive jars, and hundreds of glass beads. Additional European objects were found by the landowners. Willey believes all of the objects probably date to the sixteenth century and came from ships wrecked off the Florida coasts while on their way from South America to Spain (Willey 1949:59-60). Stirling should be congratulated on the small mesh screen that he must have used during excavations in order to recover hundreds of tiny seeds beads averaging 2 mm in diameter.

Human Remains

Undisturbed burials from the Old Muck Burial Mound were extended on the back but most of the burials were disturbed by subsequent intrusive burials which resulted in a solid, tangled mass of human bones. Undisturbed burials from the bottom of the First Sand Burial Mound were also extended on the back but this mound had been destroyed by a flood (hurricane) and many complete and broken human bones were found in the water-washed sand that surrounded it. Burials were scattered through the Second or Final Sand Burial

FIGURE 23 Carved duck head (Smithsonian Institution Cat. No. 383868, Neg. No. 77-11222).

FIGURE 24 Woodpecker or similar bird (Smithsonian Institution Cat. No. 383867, Neg. No. 31647-K).

Mound and those near the surface contained European artifacts (Stirling 1935:375; Willey 1949:22).

There is no record of the number of skeletons removed at Belle Glade. Willey mentions that Hrdlicka measured and described skulls of 17 males and 26 females but these must represent only a small part of the individuals who were buried in the combined levels of the mound. Lamme (1978) gives some indication of bone density: ". . . stick your hand down there and there was nothing but bones. You couldn't take a pencil and stick it anywhere in the wall but what you'd find bones." Lamme's description fits precisely the impression that Maples, Brooks, and I had when we visited the site in 1975.

Deformation was virtually absent. The cephalic index of Belle Glade males averages 80.72 and 81.45 for females, a mesocephalic to brachycephalic, high-vaulted Gulf type of American Indian. Willey was informed by T.D. Stewart that a few bones indicating dwarfism were present in the Belle Glade skeletal collections (Willey 1949:23).

Discussion

Stirling commented that only the burials in the top of the mound contained intentionally included cultural remains. Nevertheless, Willey was able to compare the artifacts from the habitation midden to those from the burial mound and make tentative suggestions about relative time or functional differences between the two structures. Willey noted differences in ceramic, bone, shell, and wood artifacts. As mentioned earlier, St. Johns Check-Stamped ceramics occur only in the top levels of the habitation mound. This type does not occur at all in the burial mound. This suggested to Willey that the Final Sand Burial Mound may have been built prior to the late midden period. The small number of sherds from the burial mound, however, made it impossible to draw conclusions about ceramic trends or contemporaneity between the midden and the mound.

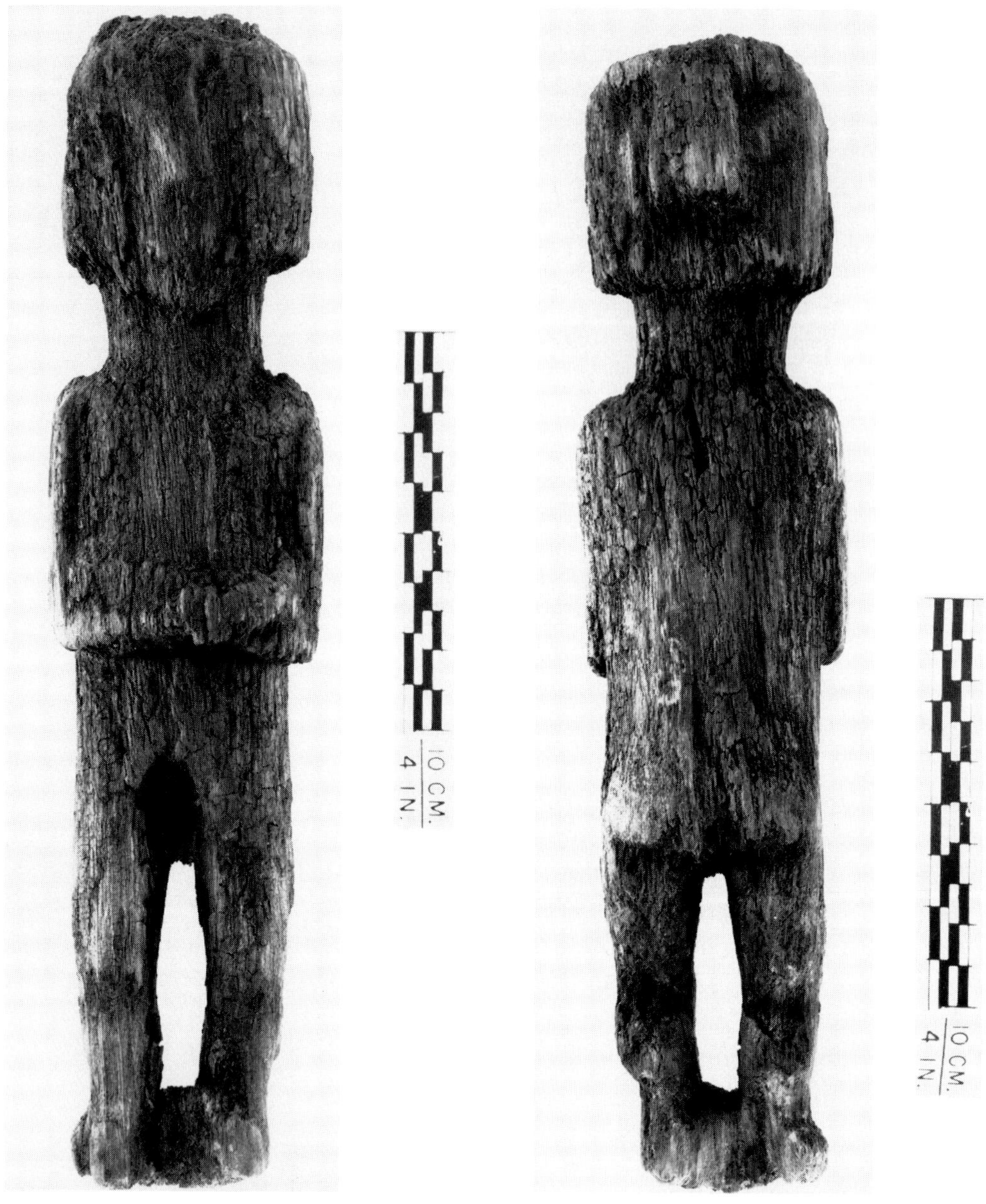

FIGURE 25 Human figure (Smithsonian Institution Cat. No. 386067).

Both socketed and bipointed bone points occur in all levels of the midden. In the burial mound, more than 20 socketed bone points and only 1 bipointed point came from the Old Muck Burial Mound. No bone points were found in other levels of the burial mound. This suggests either that the socketed points were earlier than the bipointed bone projectiles or that the socketed points had a special function in burial practices.

Concave surface shell celts appear in all levels of the habitation midden but flat surface shell celts occur only in the late midden levels. No concave celts were recovered in the burial mound but nine flat surface celts came from the

FIGURE 25C

Old Muck Burial Mound and below suggesting a special burial function for this type. All of the conch shell drinking cups are associated with the burial mound again implying a special ceremonial use.

The association of the carved wooden objects with the outwash of the First Sand Mound led Willey to conclude that the First Sand Mound marked the peak of religious or political elaboration at the Belle Glade site. The prepared platform or pavement of limestone slabs further substantiates this speculation. Since many of the wooden objects are similar to the Ft. Center carvings, a tentative date of AD 200–600 can be suggested for the First Sand Burial Mound.

European objects were found near the top of the habitation midden and

FIGURE 26 Large plaque carved from a single piece of wood (Smithsonian Cat. No. 383876A, Neg. No. 90-10264; permission no. 90-1022).

in the Final Sand Burial mound. Willey says it is possible the burials with European trade goods were intrusive into the Final Sand Burial Mound and may represent a reoccupation or reuse of the site rather than a terminal date for the mound occupation proper (Willey 1949:70). Sears (1982) drew the same conclusion for Ft. Center. Widmer (1988), however, believes that Belle Glade was the location of the village called Guacata visited by Fontaneda. If so, it must have been occupied during the early Historic Period.

Willey's overall conclusion is that "all that can be safely said is that the various levels of the burial mound were constructed sometime during the occupation of the habitation site" which he felt had been more or less continuously occupied (Willey 1949:70). Willey was more successful in demonstrating that some of the artifacts served special burial functions than he was in comparing the chronology of the habitation midden to the burial mound. His thoroughness in describing and interpreting the position of the artifacts in the habitation midden and various strata of the burial mound, however, has made it possible to place this site chronologically in relation to sites that have been investigated and dated since that time.

Stirling recognized that the importance of the Belle Glade site "lies in the wide range of the material collected. Here, for the first time in Florida, there is a representative collection of habitation shell mound artifacts, burial furniture, and skeletal material. . . Another important feature here is the correlation that is possible with the wooden material found by Cushing at Key Marco, which tends to link the two sites together culturally" (Stirling 1935:376).

As at many other wet sites, the wood carvings have attracted a great deal of attention. Wood was the only medium available to the Indians of south Florida in which they could create large artistic sculptures. It is lamentable that other botanical information was not recovered which would have furnished a more complete picture of the environment and the diet of the people at Belle Glade.

References

Allerton, David, George M. Luer, and Robert S. Carr 1984 Ceremonial Tablets and Related Objects from Florida. *The Florida Anthropologist* 37(1):5–54.

Fewkes, J. Walter 1928 Aboriginal Wooden Objects from Southern Florida. *Smithsonian Miscellaneous Collections* 80(9):5 pages.

Lamme, Vernon 1978–1979 Taped Interview about the Chosen Burial Mound. *Palm Beach County Archaeological Society Newsletter*, Palm Beach, Florida.

Sears, William H. 1982 Fort Center: *An Archaeological Site in the Lake Okeechobee Basin.* University Presses of Florida, Gainesville.

Stirling, M.W. 1935 Smithsonian Archeological Projects Conducted Under the Federal Emergency Relief Administration, 1933–34. *Annual Report of the Board of Regents of The Smithsonian Institution* for the year ending June 30, 1934. Washington, D.C.

Widmer, Randolph J. 1988 *The Evolution of the Calusa.* The University of Alabama Press, Tuscaloosa.

Willey, Gordon R. 1949 Excavations in Southeast Florida. *Yale University Publications in Anthropology* No. 42. Yale University Press, New Haven.

Fort Center

Introduction

The Fort Center site, located in Glades County west of Lake Okeechobee, extends for approximately 1 mile along Fisheating Creek (Fig. 1). The site gets its name from a small blockhouse built in the area during the Seminole Wars. No trace of the fort remains and Sears (1982:ix) believes it was washed away by Fisheating Creek. Because of the military activity there, the site was the focus of considerable attention in the 19th century (Benson 1856; Ives 1856; Kenworthy 1875, 1883; LeBaron 1884).

In 1926, Mr. Nathan F. Norton dredged a wood carving of an eagle (Fig. 27) from under 3 feet of muck near a then unexplored Indian mound in the vicinity of Lakeport (Anonymous 1933). The carving was used as a yard decoration for some time and then was displayed in a small grocery store. It was loaned to the Florida Museum of Natural History in 1932 and became a permanent acquisition in 1957. It has been the symbol of the museum for many years. The original head of the eagle figure was missing when the carving was found in 1926 and one was fashioned for it. An eagle head recovered at Ft. Center by William H. Sears in the 1960s was thought to be the missing head but an effort to place it on the museum specimen was not successful. Although the size was compatible, the fit was not good. A new head that conforms more closely to the one excavated by Sears has been created for the eagle (Schwehm 1983:9–12,31). The carved portion of the specimen is 26.5 inches and the total length, including the post, is 62 inches.

Goggin (1951) conducted tests at Fort Center, and in 1961 an amateur organization secured permission to "survey" the site. Instead of surveying, these individuals dug extensively into the top of one of the large mounds (Mound B) and looted items of silver and gold, as well as other artifacts, that had been placed there as grave goods in the early historic period. The group's access to the site was terminated when their actions were discovered. Steps were taken to recover the artifacts and those that were recovered are described by Sears (1982:59–67) and Leader (1985).

Sears surveyed and tested the Fort Center site in 1964 and 1965 with funds provided by the Florida Museum of Natural History and by Lykes Brothers, Inc. who own the land where Fort Center is located. Sears and crews from Florida Atlantic University, Colgate University, and the University of Florida excavated various portions of the site from 1966 to 1971 with grants provided primarily by the National Science Foundation. The site includes middens, house mounds, ceremonial mounds, earthworks, and a charnel pond. As a result of Sears' careful excavations and analyses, he was able to identify four periods of aboriginal occupation at Fort Center spanning more than 2000 years: Period I, 450 BC to AD 200; Period II, AD 200 to AD 600–800; Period III, AD 600 to AD 1200–1400; Period IV, AD 1200–1400 to AD 1700. We will

FIGURE 27 Eagle carving recovered at Fort Center in 1926 (picture courtesy of Roy C. Craven, Jr.).

be concerned particularly with Period II and with Mounds A, B, and the charnel pond (see Fig. 28 taken from Sears 1982:4, 146; and Fig. 29 from Morgan 1980:34–35).

About AD 200 the inhabitants at Ft. Center constructed a special area to prepare and bury their dead (Fig. 31). This consisted of a low mound upon which was probably built a charnel house (Mound B), a special activity area that may have served also as residences for specialists and their families (Mound A), and an artificial pond with a platform where bundled burials were placed (the charnel pond). Mounds A, B, and the pond functioned as an integrated ceremonial area and the entire complex was separated from the rest of the village by a low sand embankment. Sometime around AD 500, the platform burned and collapsed into the pond carrying with it approximately 300 bundled burials and between 100 and 150 carvings or fragments of carvings thought to represent a total of 69 complete specimens. Many of these carvings supported or decorated the platform; others may have accompanied the dead as grave goods. Rotted wood was found in the pond as well (Sears 1982:164). After the platform collapsed, 150 of the burials were retrieved from the pond, interred on the top of Mound B, and then covered with white sand. The other 150 remained in the pond.

The charnel pond and associated mounds are situated in a hammock with an elevation of 21 feet above sea level. Sears excavations at the site were constantly adapted to expose stratigraphy that would reveal the physical and chronological relationships of these features. The pond had been dug to 5 feet below the present-day land surface making its lowest point 16 feet above sea level. The pond was 60–70 feet long and 40 feet wide and appeared as an obvious feature or depression between Mounds A and B.

Excavations

The "green, scummy, and evil smelling" pond was not physically inviting (Sears 1982:147; Fig. 30). Sears first season of excavation at Fort Center was a "gingerly excursion into this part of the site" (Sears 1982:163). A corner of one square was taken down quickly and kept low so that the hose intake for a pump could be placed there and water from all other excavations in the pond directed into it. Sears learned at this time that the lowest parts of the wood carvings and the human bones were at a relatively constant elevation of 18 feet above sea level (i.e., 2 feet above the pond bottom) and were resting on 6 inches or more of midden that had washed in or was thrown in from the adjacent mounds before the platform collapsed. The stratum containing the carvings and bones was 18–24 inches thick and was about 2 feet below the surface of the pond.

Sears next tackled the pond and Mounds A and B simultaneously, a strategy he concluded was a necessity for working out the prehistoric erosion-building-deposition-excavation sequence. A large area of Mound B adjacent to the pond was planed down carefully with a tractor so that elevations that

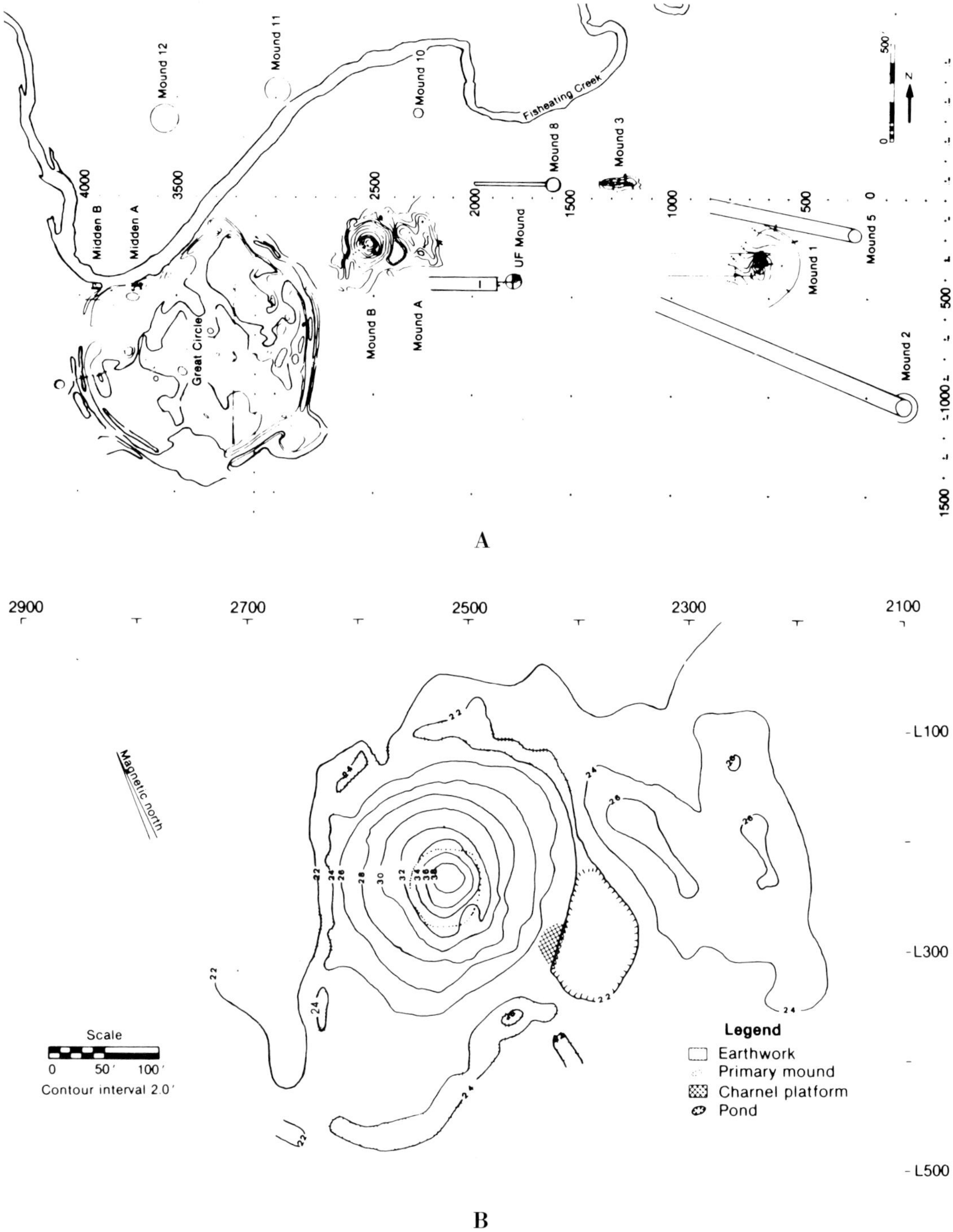

FIGURE 28 (A) Map of the Fort Center site (taken from Sears 1982:4). (B) Mounds A and B and the pond complex (taken from Sears 1982:146).

threatened to slump were consistently reduced in height as he pumped and dug. He excavated a major part of the pond using basically the techniques developed during the first season. Twenty-five to 30-foot sections were opened and muck was cleared from materials in a way that allowed drainage into a sump.

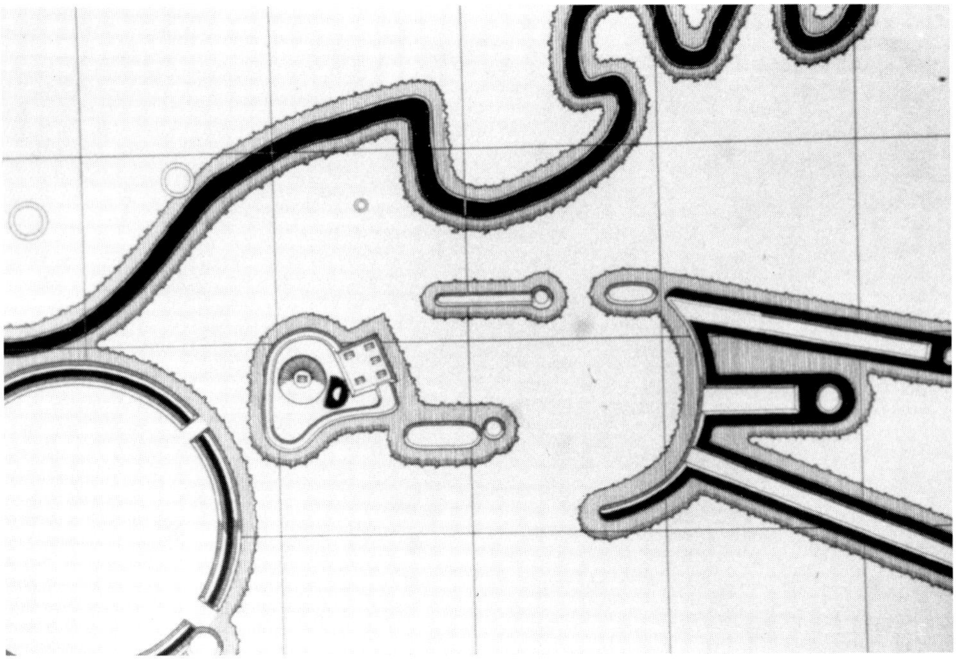

A

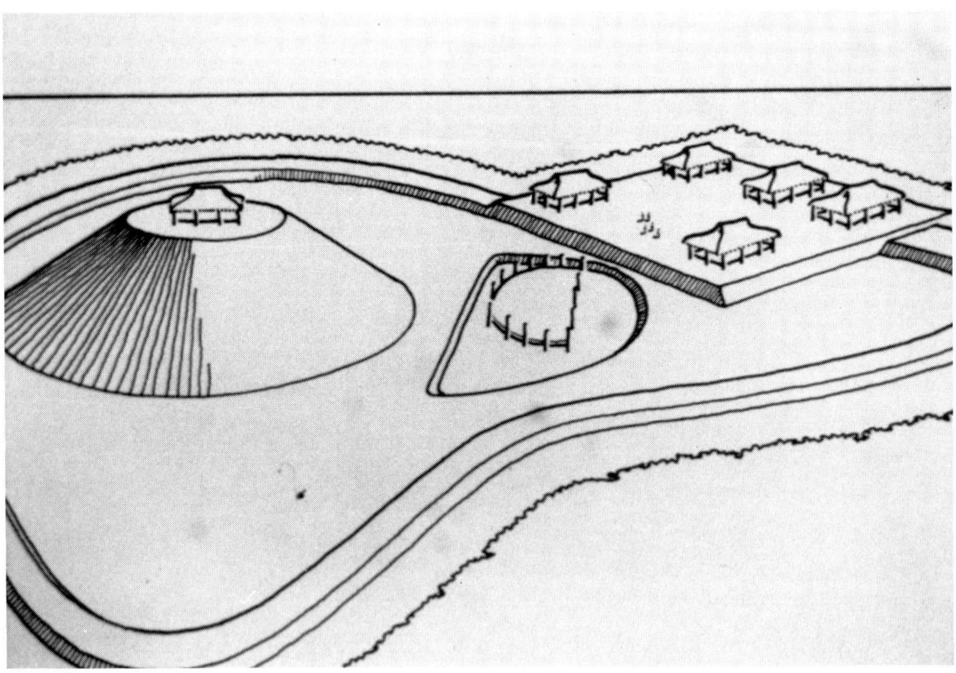

B

FIGURE 29 (A) Stylized illustration of the Fort Center site. (B) Stylized illustration of mounds A and B and the charnel pond (taken from Morgan 1980:34–35).

Each morning 1 to 2 feet of water had to be pumped from the area (2 to 4 feet after the weekends). Sears ran pumps throughout each day but it was not possible or desirable to dry the deposits out completely for fear of ruining the specimens. One of the pumps had a valve system at the output end that allowed water to be sprayed on strata or artifacts that were becoming too dry and was used also for the excavation of the fragile wood and bone (Sears 1982:163).

Wood

When the wood carvings collapsed into the pond, they fell in all directions and most of them probably floated until they became waterlogged. Many had rotted completely leaving their surfaces and even deeper layers permeated by palmetto and cabbage palm roots. Some specimens were burned as a result of the fire that started on the north end of the structure and ended its function as a charnel platform (Sears 1982:38).

Two double-ended pestles recovered from the pond were manufactured of lignum vitae (Fig. 32). All other wooden objects were made of pine. Sears believes that pine was used instead of the equally available cypress because it has characteristics that make it more suitable for shaping into the desirable styles produced by the artisans. Knots and twisted areas in the grain of the wood were used for detail as if the craftsman, having a figure in mind, found a piece of wood to which the idea could be adapted (Sears 1982:39). Schwehm, after consulting with a wood sculptor, adds that the long interlocking fibers of pine would be more desirable for carving than cypress which is a short-fibered wood with a tendency to break and splinter. In addition, the heartwood of pine is easily carved when green and hardens with age to provide a durable insect-resistant sculpture. Only the heartwood was used because the sapwood deteriorates with time. The resin cells of pine also may have played a role in the production of these figures if the fire-setting method was employed as part of the manufacturing process. An alternate explanation for the use of pine might be that it had ritual significance (Schwehm 1983:49–50) but this hypothesis is difficult to confirm at present.

Shell, stone, and shark teeth tools found at the site were probably used to carve and create surface detail. Unfortunately, surface detail has been obliterated from all but a few specimens. A large unattached wing was incised with a series of parallel lines curved across the grain of the wood to represent feathers (Sears 1982:46), and markings on the wings of a smaller bird were accentuated by the addition of a rubbery white pigment which contained large quantities of pollen (Sears 1982:48,51). The lack of surface decoration can be best explained as weathering or erosion by wind and rain over the 300-year period that the effigies guarded the charnel pond before it burned. Sears mentions that wings were more numerous and better preserved than bodies or heads. He reasons that the thin vertical wings seasoned and stabilized quickly thus escaping degradation whereas heads and bodies deteriorated from the top

A

B

FIGURE 30 (A) The charnel pond before excavation, (B) excavations in progress (pictures courtesy of Roy C. Craven, Jr.).

A

B

FIGURE 31 (A and B) Artists's depiction of mortuary pond and activities associated with it (pictures courtesy of Roy C. Craven, Jr).

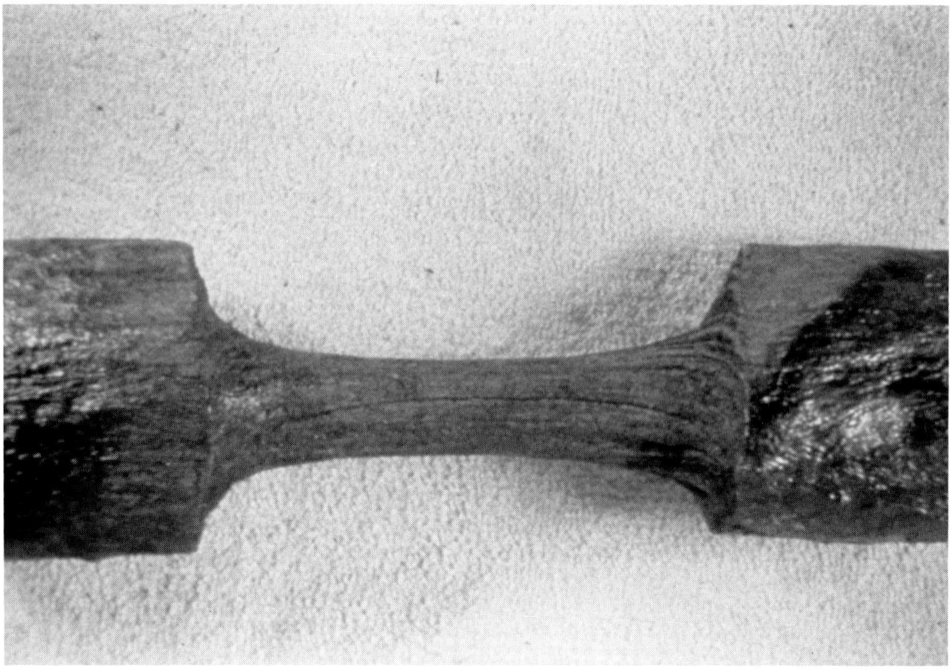

FIGURE 32 Double-ended pestle made from lignum vitae (picture courtesy of the Florida Museum of Natural History).

down and from the inside out, like fence posts, because growth rings and grain were facing up and trapped water.

Sears (1982:38–58) and Schwehm (1983) describe the wooden artifacts from the Fort Center site. Sears separates the wooden objects into two categories that he calls ceremonial and utilitarian but the only utilitarian items found were the two doubled-ended pestles manufactured of lignum vitae, a bowl carved with an owl head, and what may be a fragment of a cylindrical mortar fragment. The ceremonial category (Sears 1982:42-58) has three subdivisions:

1. *Large birds and beasts* were structurally a part of the eastern straight front edge of the platform. These kinds of carvings are about 10 feet long including a 6-foot post section and a 4-foot carved portion. There were at least seven large birds (Fig. 33a), a bear (or dog), and a cat.
2. *The two-legged style* of carving consists of nine animals including three birds. The creatures seem to be paired. There are two cats, two dogs or bears, two foxes, two birds, and one eagle. Other specimens too fragmentary or decayed were observed in the field and may have fit into this two-legged class. The common feature of this type is that the front legs in the quadrupeds or the legs in the birds have been cut free of the body from shoulder to foot. Rear legs and birds' tails remain attached and the rear legs are proportionately undersized. The specimens making up this class (Fig. 33b), excluding bases, are 18 inches for owls, 24 inches for dogs or bears, and 30 inches for foxes. All of these carvings were located at the rear or western arc of the platform.

A

FIGURE 33 (A and B) Examples of two styles of carving from the Fort Center site (pictures courtesy of the Florida Museum of Natural History, Gainesville.).

3. ***Birds with a distinct tenon*** that fitted into sockets that had been worked into the tops of long poles make up the third subcategory of ceremonial carvings. Their portability suggests part time use elsewhere. These objects are small, measuring 7–9 inches with tenons 13–19 inches long. The largest, an eagle or hawk, is 22 inches and is a minature of the large birds with erect wings. It was a wing of this bird that had white pigment mixed with pollen adhering to it (Sears 1982:54). Other carvings in this class—ducks, long-legged water birds, a hawk or eagle—had tenoned bases as

FIGURE 33 B

long as 4 feet and their diameters were too wide to fit into the post sockets. Sears postulates that they may have been forced into the center of inverted root masses of oak that were used in platform construction (Sears 1982:52). Ten socketed poles and the portable tenoned figures were clustered at the north end of the platform. Almost all remains of children recovered from the pond were located in the northern end of the platform and the positioning of the tenoned figures there may have had some connection with the child burials. Also located at the north end of the platform were several carved bird heads and five unrelated small sculptures representing two otters, a running panther (Fig. 34), a woodpecker head, and a pair of carved wood deer antlers. Schwehm suggests that these objects may have been playthings or toys. "The woodpecker head and the better preserved of the two otter figures both exhibit an appealing liveliness and personality and are attached to short handles that would adapt them especially well to the role of a child's hand-held toy" (Schwehm 1983:39–40).

Preservation

Sears knew that the eagle carving found in 1926 probably came from Fort Center and he was prepared to excavate and preserve organic materials. His first tests of the pond were designed to "develop excavation techniques for a permanently wet medium of mud, sand, and cow manure and to develop techniques for the possible collection and preservation of wooden and other perishable materials" (Sears 1982:15). A 55-gallon steel drum was cut in half lengthwise to make two vats in which wood specimens were treated. These were replaced in subsequent years by wooden vats the size of which was dictated by the largest specimens recovered. The vats were made from 3/8-inch plywood, 8 × 4 × 4 feet. All seams were sealed inside with fiberglass tape and resin, and the entire interior was resin coated to waterproof it. As many as four vats were needed at one time depending on the rate of recovery. One or more of the vats was kept full of water and changed frequently. Carvings that had been removed from the pond and rinsed off were placed in the water-filled vat to prevent them from drying out while they awaited treatment. Another vat was filled with a 50% solution of commercial grade white glue (Weldwood). Sears originally had PEG on hand but preferred the glue although it posed a problem because it tended to settle out becoming thicker at the bottom of the vats. "At first we simply removed all of the specimens quickly, stirred up the glue, and put the specimens back, a technique that produced tangled tags and glue-coated personnel. . . The solution was a manual marine diaphragm pump with 1-1/2 inch intake and outlet. The intake hose was placed at the bottom of the tank and the outlet hose just under the surface at the top" (Sears 1982:16). Two or three 15- to 30-minute pumpings daily maintained a uniform solution. Specimens were kept in the treatment tank from several days to several months depending on estimates of thickness, porosity, and other factors affecting penetration. After removal from the vats and careful brushing to remove surface buildup of glue, specimens were dried in the shade.

Sears chose white glue, a plastic emulsion, over water-soluble waxes for two reasons. After treatment the specimens when dried, "regained strength and abrasion resistance to a level approximating that of fresh wood; plasticizing the wood by replacing water in pore spaces with plastic not only prevents checking but also strengthens the object. Second, after drying, resoaking the

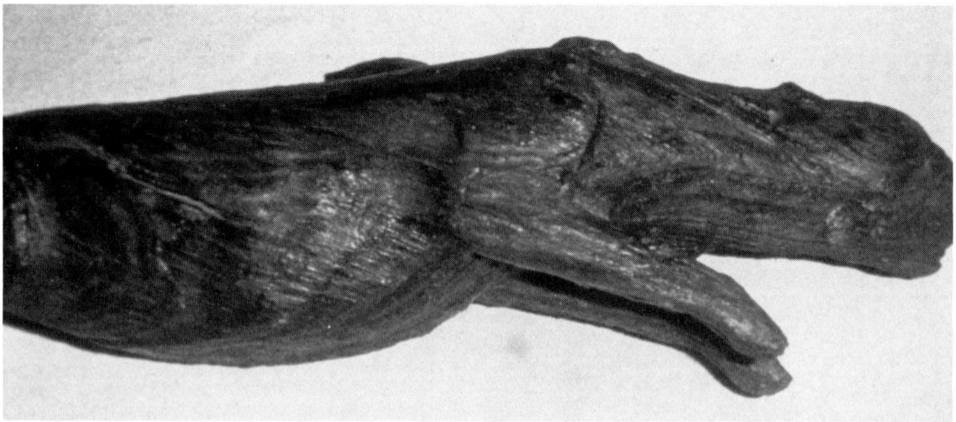

A

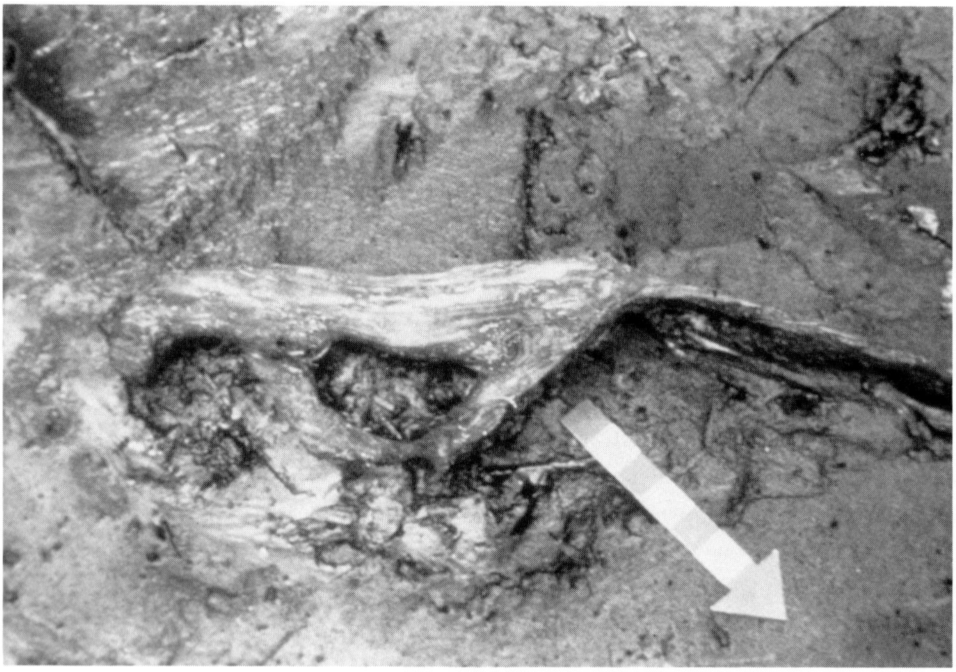

B

FIGURE 34 (A) Wood carving of an otter, (B) wood carving of a panther. Both objects may have been children's toys (pictures courtesy of the Florida Museum of Natural History, Gainesville).

material in a white glue solution softens the plastic already present and permits further absorption . . ." (Sears 1982:17). As mentioned in Chapter 5, white glue is not a recommended product for the preservation of degraded waterlogged wood. As of 1990, however, the Fort Center carvings, 20 years after treatment, appear to be in good condition. It would be informative to core one of the specimens to determine by microscopic examination whether or not the white glue has replaced the water in the lumina and cell walls.

Sears mentions that generally the carvings recovered from the pond were either intact or that connecting pieces were not found so that physical restoration of fragmentary specimens was not possible. A few of the Fort Center carvings were mended or reconstructed by applying "mixtures of white glue, sawdust, and sanding dust of various grades and pigment troweled on and finished to the desired surface after drying" (Sears 1982:41). Sears does not describe specific objects that were repaired.

Human Remains

Audrey J. Sublett, a physical anthropologist at Florida Atlantic University during the years that excavations were conducted at Fort Center, carried out an in situ analysis of the charnel pond's human skeletal remains. She supervised a separate crew in recording, measuring, removing, and cleaning the burials. After Dr. Sublett's death, the human bones and all data pertaining to them remained in the Florida Atlantic University anthropology laboratory (Sears 1982:52,149,164–65) and a selection of them was eventually studied by Shaivitz (1986) under the direction of Dr. M.Y. Iscan.

Through the human remains, Shaivitz set about to test Sears' hypothesis that Mounds A, B, and the charnel pond formed a ceremonial complex and that the burials placed on the platform were those of high status individuals: "Our evidence suggests that, during Period II, the site was used as a ceremonial center with a resident population [on Mound A] of ceremonial specialists of several families" (Sears 1982:175). Shaivitz's sample included 43 males, 50 females, 11 adults of unknown sex, and 17 subadults of unknown sex. The crania resemble most closely the Iswanid type described by Neumann (1952) but the Fort Center inhabitants may not have been a totally inbreeding population because it shared traits with all four of the groups classified by Neumann. Shaivitz used postcranial data from humeri, femora, and tibiae to obtain sex, age, and height. The average height for males was 168.1 cm (about 5 ft, 5 in) ranging from 167.2–169.8 cm which is considered typical of other Florida Indians. Shaivitz looked for paleopathologies that would affect growth, and for evidence of traumatic lesions and dental problems. Chipping, caries, abcesses, peridontal disease, and attrition were present but tooth wear was slight to moderate compared to that usually seen in Florida Indian populations. One female had probable syphilitic lesions and five individual crania had evidence of trauma. Two adult crania were perforated. Skull and long bone fractures were also noted. Crania showed a high degree of porotic lesions in both sexes with parietal thickening indicating people may have suffered from anemia resulting from bacterial and worm infestations as well as iron-deficient diets. Osteoarthritis was present also. Sex and age of death were determined for a sample of 64 adult individuals (33 males and 31 females). Of these, 3 males and 7 females died between ages 18–20, 25 males and 19 females died between ages 21–35, and 5 males and 5 females died between ages 36–55. They did not live to a ripe old age by present-day standards. Shaivitz found no evidence of malnutrition that would have been manifest by Harris or transverse lines and concludes that the individuals buried on the platform at Ft. Center did not suffer greatly from nutritional or pathological stress and were generally pretty healthy. Because of the absence of major bone diseases, Shaivitz believes her findings support Sears' archaeological interpretation that this location at the Ft. Center site was a ceremonial complex where certain people residing there and buried there were of high social status.

Another major hypothesis set forward by Sears was that horticulture had been practiced at the site and would account in one way or another for the several kinds of earthworks (Sears 1982:18). Teeth from the Fort Center human remains were examined to determine if dental patterns agreed with what would be expected from an agricultural population. The conclusion was that, at Fort Center, horticulture played only a minor role in the subsistence base and that the corn (pollen only) identified at the site was utilized in a nonnutrient way (Isler 1985).

Pollen Analysis

Sears felt that pollen analysis provided the only available and most useful test to determine if cultigens were grown at Fort Center and would also produce data on other plants of economic value as well as information about environmental changes. Elsie Sears (1982:118–129) conducted palynological studies of soil samples, coprolites, and the white pigment found on one of the charnel pond wood carvings. The primary objective of pollen analysis was to determine if the occupants practiced agriculture. Several hundred slides of pollen grains were examined for pollen of maize, manioc, squash, and sweet potato. Of these cultigens, only maize pollen was identified; it was found in the white pigment, in three coprolites out of 100, and in soil samples from the Great Circle, the single linear mound between Mounds 8 and 9, and the Mound 1 linear mound. Cheno-am pollen (Chenopodiaceae and Amaranthaceae) comprised 11.5 to 51.0% of the pollen from 50 coprolites studied contrasting sharply with percentages of 2.0 to 11.0 from soil samples. It seems possible, therefore, that pigweed-goosefoot plants formed part of the diet; however, "they tend to concentrate on highly nitrogenous wastes around villages, thus will be present in pollen and charred seed counts" (Sears 1982:198–199; personal communication, 1989). One sample contained 14.5% of elderberry pollen. The results of pollen analysis to determine environmental change indicated that "the river meander belt is essentially unchanged from prehistoric times, that the savannah was the same during Period IV as it is today, and that the dense oak hammock growth present today is of fairly recent origin, probably moving in on an area that was savannah during the occupation of the site" (Sears 1982:129).

Hogan (1978) conducted pollen studies of coprolites that had been recovered from the bottom of the charnel pond and from the quicklime deposits in

Mound A. She selected at random 40 coprolites that had not been analyzed previously. Nineteen of these were unuseable. Pollen was extracted from the remaining 21 samples. She found no corn pollen.

Sears (1982:189) suggests that the ditch surrounding the Great Circle would have had to be cleaned out occasionally in order for it to function as a drained field for maize cultivation. A recent study of the soils from this circular earthwork at Fort Center concludes that it was not intended for this purpose because particle size distribution and organic carbon content indicate that the ditch was dug once and no periodic cleaning took place. In addition, the naturally occurring soil is too acidic and its aluminum content too high to allow annual harvests of maize (Johnson and Collins 1989; Johnson 1989). Sears responds:

> The circle clearly was redug, in new or partly new locations, several times. This is clear on aerial photos and in a number of long and deep profiles from trenches dug for the purpose . . . . My statements relevant to digging and growing were with respect to cleaning out, not re-digging. Corn is a heavy nitrogen feeder, and this could have provided nitrogen rich muck to gardens and/or corn hills. . . . corn was grown in gardens on the site, including one in the circle, in the 1960–70 period and earlier. Most of the Seminole gardens are in the same soil type and environmental situation. Perhaps the point is *gardens*, not fields (Sears, personal communication 1989).

There is little question that corn pollen was found at Fort Center but because plant-processing artifacts, macroplant remains of corn, and ethnographic accounts of corn cultivation were not found or do not exist, it is unlikely that this cultigen was the food staple that supported high status individuals and an elaborate ceremonial complex. About the nature of corn and its pollen, Sears reponds:

> The heavy pollen will only be found where corn grows or possibly where it was prepared or stored, especially if the husks are kept on it during storage, not the best or commonest way. . . . (Sears, personal communication 1989).

About the lack of plant processing artifacts, Sears says

> All "plant processing artifacts" . . . such as log mortars and wooden pestles are perishable. We hoped for them, of course. But, this kind of evidence . . . hasn't even been found on Mississippian sites full of charred corn cobs.
>
> No other tenable hypothesis has been presented . . . for the presence of large amounts of lime, burned from shells to quicklime, in the midden. It was burned there, had to contact fecal material there as quicklime, and was produced and used all the way through the development of the Mound A midden. Pigment use would only account for a small percentage, if that. The large quantities do require a testable hypothesis for explanation of its presence. You have mine, which also serves to magnify the evidence for corn growing in the soil samples and ingestions, fortuitously, in the few fecal specimens (Sears, personal communication 1989).

At the present time, however, there is no evidence for maize cultivation anywhere in the nearby Caloosahatchee or Everglades archaeological areas (Griffin 1988:298) but its absence may be more apparent than real.

Evidence for the utilization of plants by the human inhabitants at the Fort Center site is inconclusive in that it has come solely from an examination of pollen from 121 human fecal samples recovered from the charnel pond and Mound A. A variety of wild plants in all samples suggests a random selection from the local habitats with minimal dietary significance for any particular wild plant (Hogan 1978:v). Arrowhead and waterlily, for example, are edible and available in the area. The story might be different had macrobotanical remains been preserved.

Fauna

The faunal picture is more revealing than that provided by plants. Sears systematically recovered and transferred a large assemblage of faunal remains to the Florida Museum of Natural History. This collection was analyzed by Hale (1984, 1989) under the supervision of Dr. Elizabeth S. Wing. Numerous species of shellfish, fish, reptiles, amphibians, birds, and mammals were identified and quantified. Aquatic species made a major contribution to the diet of the inhabitants at Fort Center but, interestingly, deer and some species of birds played an important part in the ceremonial life of the people. The mound-pond complex contained a disproportionately large number of deer bones which may support the contention that ritual consumption of food took place on Mound A. There was an unusually large concentration of unworked deer antler in the pond and Mound A sample. Despite the great numbers of real deer antlers, a pair of carved wooden deer antlers was found in the pond also (Sears 1982:55). The bones of several kinds of raptorial birds were recovered from the ceremonial complex area. These include a hawk, barred owl, three black vultures, a turkey vulture, Audubon's caracara, and a bald eagle. Only one raptorial bird was found in the rest of the site. Raptorial birds such as eagles, osprey, owls, and hawks were common subjects for the wood carvings on the charnel platform (Hale 1984:177). A small number of bear and large cat remains were identified among the fauna and these animals also are represented by carvings.

The presence of a large variety of edible fauna at Fort Center, known to have been utilized by Florida Indians for thousands of years, leads to the conclusion that a major portion of subsistence was supplied by animals. The Indians were particularly successful in harvesting the products of the aquatic environment. As on the Northwest Coast of North America, the food supply from the waterways was secure once the technology had been developed to exploit it. Elaborate cultures evolved in that area, as at Fort Center, with a subsistence base centered on aquatic species and the gathering of a wide diversity of wild plants. Although corn pollen was identified at Fort Center, there is no concrete evidence that cultivated crops were a major part of the diet.

In addition to food, fauna functioned for ritual and practical purposes at Fort Center. Bone, teeth, shell, skins, and sinew were used for tools, decoration, warmth, costumes, and all manner of other needs satisfied in modern times by metals and synthetic materials. In a study of these artifacts and the stone remains from Fort Center, Steinen demonstrated that Mound A and the mortuary pond contained an overwhelming majority of implements that probably were used for woodworking and other special activities. He thus supported Sears hypothesis that the mound-pond complex was a ceremonial center. These classes of objects recovered from the pond were found under the platform and had been deposited as a result of slump or housecleaning activities from Mound A before the platform collapsed. Mound A continued to be used into Period III after the mortuary pond no longer functioned as a burial area (Sears 1982:98–101).

Discussion

The Fort Center site was composed of a myriad of mounds, middens, and earthworks. Some of these structures were utilized at the same time as Mounds A, B, and the charnel pond; however, none of them contained the quality and quantity of artifacts as those from the ceremonial complex and nowhere other than the pond were wood carvings or other organic materials recovered. Since the purpose of this book is to call attention to the survival of usually perishable materials, the discussion of Fort Center has concentrated particularly on the ceremonial complex and the objects from the pond.

Originally, Mound B was a low earthen brown mound constructed from dirt obtained when the pond was dug; a charnel house used for the preparation of bodies was built on top. As mentioned earlier, when the platform burned and collapsed around AD 500, about 150 individuals were retrieved from the pond, interred on Mound B with conch shell drinking cups, and covered with white sand. Streaks of black muck found in Mound B at the level of the burials were identified by Sears as coming from the feet of those who recovered their dead kin from the pond (Sears 1982:155). The skeletal material from Mound B was very degraded compared to that remaining in the pond. Mound B was not used again until 1000 years later when early historic period burials were placed there along with aboriginal and European grave goods. The two episodes of burial on Mound B elevated it tremendously and brought it to the attention of 19th and 20th century visitors and collectors.

Mound A contained the largest number of nonperishable prehistoric artifacts found at Fort Center. The nature of the artifacts and the presence of quantities of lime made in pits by burning shell supports Sears' conclusion that Mound A was an area where special activities were carried out that did not take place elsewhere at the site. Many of the artifacts, such as 441 shell tools and 197 shark teeth representing 7 species, were probably used to carve the wooden effigies. Sears believed the lime was used to process corn. Also, corn pollen was mixed with a lime-based pigment and painted onto at least one of the wood carvings found in the pond. It is quite possible they were all painted. According to Fontaneda ". . . they are great painters, and whatever they see, they paint" (True 1944:31). Sears also notes that human feces were preserved in the lime deposits and suggests that the chemical action of quicklime kills odor and was responsible for preserving the fecal material. Mound A was subjected to a "housecleaning" sometime before the pond platform collapsed. The trash was thrown into the pond beneath the platform and was subsequently excavated by Sears and his crew. Many human coprolites and animal bone scrap were recovered but no botanical remains were preserved.

The survival of the wooden objects at Fort Center, as at other archaeological wet sites in Florida, has attracted the most attention. Schwehm (1983:6) makes an excellent point when she says that "the artistic tradition of most Florida Indians was manifested in their beautifully carved wood sculptures, not in their utilitarian ceramic bowls." The Indians of Florida did not have stone suitable for sculpting large objects. Sears (1982:168) illustrates his interpretation of the arrangement of the effigies on the pond platform. Schwehm concurs except she believes the posts were not buried in the ground but attached to the platform. Her scheme elevates the carvings considerably over that depicted by Sears. Schwehm notes that the majority of figures on the mortuary platform were of birds and says that bird images, especially those with upraised wings, lend themselves to a position of elevation. Elevated sculptures would be more visible, more effective and dramatic, and there would be more space on the platform floor for burials. Early historic accounts state that shelters were placed over Indian burial platforms and Schwehm speculates that a thatched roof covered the Fort Center bundles and many of the carved effigies were visible above the roof of the shelter. The platform so decorated, especially if the carvings were painted, would have been impressive indeed.

According to Schwehm, the placement on the mortuary platform of some of the large sculptures, such as the eagle, bear, and panther, which represent many of the stronger, more aggressive creatures might be explained tentatively by their obvious protective capabilities. All of the animals represented on the platform are native to south Florida and were naturally portrayed. The Indian artists knew their subjects' anatomical forms and mannerism intimately. The foregoing facts led Schwehm to make deductions about the ritual meaning of the platform's animal guardians although she readily admits that it will never be possible to gain a full understanding of the symbolism involved. Early historic accounts mention that the Calusa believed that when a man dies his soul was thought to enter an animal. The concept of a transference of the human soul to an animal might account for the realistic appearance of the carved figures associated with the Fort Center burial bundles. Schwehm also discusses the stylized eye that appears on otherwise naturalistic carvings and incised bone and stone objects found in south Florida such as those from Key Marco and Belle Glade. She infers that this stylized eye probably was present on the charnel pond wood carvings as well but has been obliterated by weathering processes. Her interpretation is based again on historic narratives. Father Juan Rogel wrote that the Calusa claimed "that each man has three souls . . . and when he dies two of the souls leave the body and the third one, which is the pupil of the eye, always remains. . ." (Schwehm 1983:112). Sears says that the stylized eye discussed by Schwehm is "present on one eagle(?) head if light is directed correctly; I neglected to mention this in the report" (personal communication, 1989).

The ritual nature of the pond cannot be denied. The sole use of pine for the carvings suggests that the wood of pine had special significance. It is important to note that carving debitage was not preserved, only the end products of their manufacture using a single species. The only exceptions, two double-ended pestles that accidentally or intentionally ended up in the bottom of the pond, indicate that other species of wood were used for utilitarian implements but, like all other botanical remains, did not survive. By contrast, at Hontoon Island, thousands of pieces of wood debitage, representing more than 30 species, were recovered but very few finished products. The finished products probably were used on the terrestrial portion of the midden and, when discarded, perished quickly.

The field notes, maps, artifacts, and fauna from Fort Center are in storage at the Florida Museum of Natural History, Gainesville. The human skeletal material is at Florida Atlantic University, Boca Raton.

References

Anonymous 1933 Totem Pole from Florida. *Scientific American* 148:292–3.

Benson, Henry 1855 Lieut. Benson's Route on Okeechobee and Examination of the Kissimmee River. Records of the War Department. The National Archives, Washington, D.C.

Goggin, John M. 1951 Archeological Notes on Lower Fisheating Creek. *Florida Anthropologist* 4(3–4):50–66.

Griffin, John W. 1988 *The Archeology of the Everglades National Park: A Synthesis*. National Park Service, Southeast Archaeological Center, Tallahassee.

Hale, Stephen H. 1984 Prehistoric Environmental Exploitation Around Lake Okeechobee. *Southeastern Archaeology* 3(2):187.

1989 Prehistoric Subsistence Strategies and Settlement Patterns in the Lake Okeechobee Basin of the South Florida Peninsula. Ph.D. dissertation, Department of Anthropology, University of Florida, Gainesville.

Hogan, Jacqueline 1978 Palynology of Fort Center: Environmental Interpretations and Cultural Implications for a Central Florida Hopewellian Ceremonial Center. M.A. thesis, Department of Anthropology, Florida Atlantic University, Boca Raton.

Isler, Robert 1985 Dental Pathology: Fort Center Indian Population. *American Journal of Physical Anthropology* 66:184 (Abstract).

Ives, Lieut. J.C. 1856 *Military Map of the Peninsula of Florida, South of Tampa Bay*. M.B. Wynkoop, Book & Job Printer, New York.

Johnson, William G. 1989 The Role of Maize in South Florida Aboriginal Societies: An Overview. Paper presented at the 46th Annual Meeting of the Southeastern Archaeological Conference, November 9–11, Tampa.

Johnson, William G. and Mary E. Collins 1989 A Study of Soils from the Fort Center Archaeological Site in South Florida. Paper Presented at the 41st Annual Meeting of the Florida Anthropological Society, April 28–30, Jacksonville.

Kenworthy, Charles J. 1875 Indian Mounds and Canals. *Camp Life in Florida*, pp. 313–321. Edited by Charles Hallock. New York.

1883 Ancient Canals in Florida. *Annual Report of the Smithsonian Institution for 1881*, pp. 631–635. Washington, D.C.

Leader, Jonathan M. 1985 Metal Artifacts from Fort Center: Aboriginal Metal Working in the Southeastern United States. M.A. thesis, Department of Anthropology, University of Florida, Gainesville.

LeBaron, J. Francis 1884 Prehistoric Remains in Florida. *Annual Report of the Smithsonian Institution for 1882*, pp. 771–790. Washington, D.C.

Morgan, William N. 1980 *Prehistoric Architecture in the Eastern United States*. MIT Press, Cambridge, Mass.

Neumann, Georg K. 1952 Archeology and Race in the American Indian. In *Archaeology of Eastern United States*. Edited by James B. Griffin. The University of Chicago Press.

Sears, W.H. 1971 Food Production and Village Life in Prehistoric Southeastern United States. *Archaeology* 24(4):322–329.

1974 Archaeological Perspectives on Prehistoric Environment in the Okeechobee Basin Savannah. In *Environments of South Florida*, pp. 347–351. Edited by Patrick J. Gleason. Miami Geological Society, Coral Gables.

1982 *Fort Center: An Archaeological Site in the Lake Okeechobee Basin*. University Presses of Florida, Gainesville.

1989 Personal communication to Barbara A. Purdy, December 1989. Notes on file, Department of Anthropology, University of Florida, Gainesville.

Schwehm, Alice Gates 1983 The Carved Wood Effigies of Fort Center: A Glimpse of South Florida's Prehistoric Art. M.A. thesis, Department of Fine Arts, University of Florida, Gainesville.

Shaivitz, Patricia Miller 1986 Physical and Health Characteristics of the Prehistoric Indians from the Fort Center Site. M.A. thesis, Department of Anthropology, Florida Atlantic University, Boca Raton.

True, David O., Ed. 1944[1575] Memoir of D° d'Escalente Fontaneda Respecting Florida. Buckingham Smith, transl. (1854). University of Miami and Historical Association of Southern Florida, Coral Gables.

Hontoon Island

Introduction

The Hontoon Island site (8-VO-202), Volusia County is midway along the St. Johns River (Figs. 1 and 35). René Laudonnière in 1564 and Pedro Mendendez de Aviles in 1565 may have been the first Europeans to visit Indian villages near this location. Laudonnière ascended (south) the river to a village of the Mayacas at Volusia Landing, approximately 16 miles north of Hontoon Island. When Menendez arrived at this point a year later, a hostile cacique (chief) refused the Spaniards further passage upriver (Gold 1927:13). Menendez succeeded in establishing friendly contact with a cacique near Lake George and left six men there to teach christianity to the Indians. After these initial journeys, the Europeans very quickly dominated the Indians along the river. By 1570, a Spanish colony was established at DeLeon Springs and by the mid-1600s two missions were erected in Volusia County. Fifty years later, none of the original Indians remained in the area.

William Bartram may have put in at Hontoon Island when he traveled in east Florida during the Revolutionary War:

> Soon after ascending this branch of the river, on the right hand presents itself to view a delightful little bluff, consisting chiefly of shells, and covered with a dark grove of red cedar, Zanthoxylon, and myrtle. I could not resist the temptation to stop here . . . . From this grove appears to view an expansive and pleasing prospect. The beauteous long lake in front [Lake Beresford], about North East from me . . . (Bartram 1928:128–132).

Europeans who first described the Indians of Florida and their cultural practices paid particular attention to what the Indians' ate because the intruders often had to survive on the same foodstuffs. It is interesting to note, therefore, that neither Laudonnière nor Menendez mention the Indians' use of freshwater shellfish resources despite the fact that dozens of huge refuse middens of shell exist along the St. Johns River in the vicinity that they traveled. Wyman says:

> Unfortunately, history gives us but little or no assistance in tracing these antiquities to their origin. For two centuries following the arrival of the French and Spaniards they are only incidentally noticed, and were considered as of natural and not artificial origin. A century ago the Bartrams, father and son, ascended the St. John's River, and mentioned them somewhat particularly in the published accounts of their travels, but neither of them appears to have understood their true nature, and no one since their time has studied them (Wyman 1875:3).

In the 1860s, Wyman visited many of the mounds on the St. Johns River and dug into some of them. He provides excellent descriptions of these structures and speculates about the human activities that created them. He says of the area of Hontoon Island where 8-VO-202 is located:

> A few hundred feet above its union with Huntoon Creek the St. John's passes between two mounds. That on the left bank, on Huntoon Island, is of gigantic size, covers several acres, and consists of two ridges both parallel to the river. That nearest the river is over six hundred feet in length, and rises from twelve to fourteen feet in the highest parts; much has been washed away, leaving an abrupt slope towards the water, mostly exposed to examination, but here and there covered with recently fallen trees. The lower or western end of this ridge ends in an abrupt bluff facing a bay or lagoon lying between it

> and the mouth of Huntoon Creek.The upper or easterly end slopes gradually into a very large lagoon now overgrown with water plants and willows.
>
> About one hundred feet in the rear of this ridge, and separated from it by a deep valley is a second, of much larger size [8-VO-202]. This at its eastern or upper end rises to the height of at least twenty-five feet, by measurement, and were it not for the trees would command a very extensive view, including Lake Beresford, and a long stretch of the river as well as of recently formed swamps.
>
> In the rear of the eastern end of the inner ridge are two conical mounds [8-VO-182 and 8-VO-183] the larger, twenty-five feet high and both made of a mixture of sand and shells . . . . Both were probably burial mounds, but from inability to procure the necessary labor we did not fully explore them. An excavation made on the top of the smaller, to the depth of five feet, brought up only small fragments of human bones and pieces of earthern vessels. More extensive excavations were made in the larger mound. A trench from four to six feet wide was dug from one of the sides to the apex, gradually ascending, and to a depth varying from two to four feet. All the objects discovered were buried within two feet of the surface. Lower than this there were only Paludinas [*Viviparus georgiana*, freshwater gastropod] and Ampullarias [*Pomacea paludosa*, apple snails]. . . .
>
> Of all the mounds we have examined none have yielded more abundantly fragments of earthern vessels and bones of animals than the one on the left bank . . . . The front washed by the river gives a large surface for inspection, much larger than could be had from mere excavation (Wyman 1975:28–30; Fig. 35).

C.B. Moore also traveled the St. Johns River and visited many of the same areas as Wyman. He "demolished" (his wording) a burial mound (8-VO-36) associated with the Thursby midden (8-VO-35) located on the right bank of the river. The Thursby midden had been described by Wyman as being 6 feet high and 600 feet long. When Moore finished his work at Thursby he looked longingly at Hontoon Island, on the opposite side of the river, hoping to examine the nature of the middens there (Fig. 36). Mrs. Thursby had recently sold the property to a Mr. G.A. Dreka of Deland who refused permission to excavate.

Moore was a prolific writer and he published clear accounts, all accompanied by excellent illustrations, of his investigations at Thursby (Moore 1893, 1894) and other archaeological sites in Florida. We may wish that C.B. Moore had been more precise in his excavation techniques 100 years ago; nevertheless, his methods were far superior to those of the operators of heavy machinery who scooped the heart out of many of the middens in Florida and used the shell for road beds and septic tanks. Such was the fate of the large mounds at Hontoon Island that were "excavated" in this manner in the 1930s leaving only an apron of undisturbed midden on the edge of the lagoon.

The story of Hontoon Island and vicinity may have gotten lost in the literature except in 1955 a 12-foot wood carving of an owl was found by a dragline operator in the river adjacent to the old Thursby midden (Bullen 1955; Fig. 37) and, during drought conditions in 1977, two smaller totems, an otter and a pelican, were recovered from the same location (Purdy 1987b; Fig. 38 a,b).

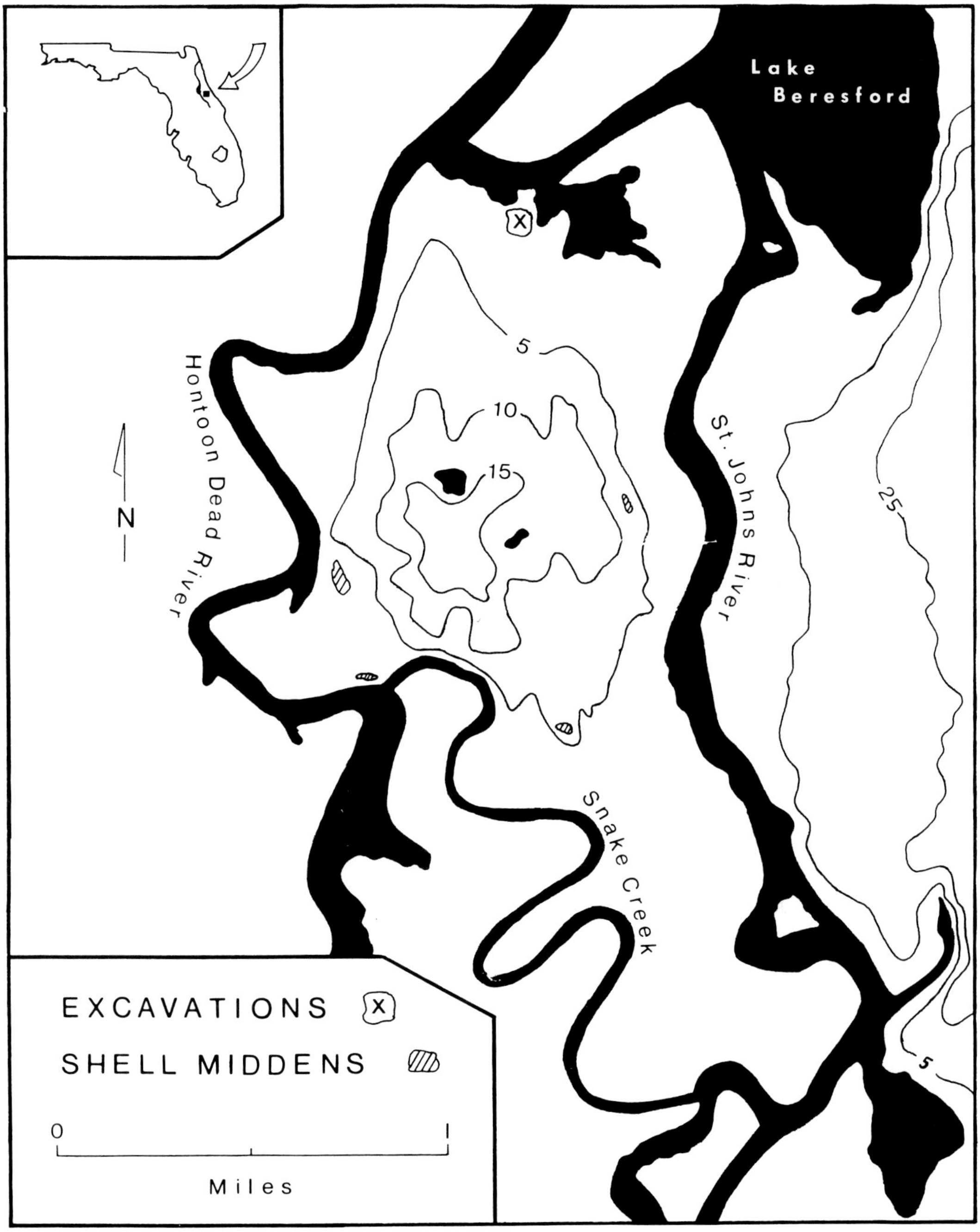

A

FIGURE 35 (A) Map of Hontoon Island (drawn by L.A. Newsom). (B) Enlargement of area marked with an X in Fig. 35a; illustration is a reconstruction of two large middens and two burial mounds based on descriptions by Wyman (1875) and on surveys conducted by B.K. Nodine and R.M. McGee (drawn by R.M. McGee).

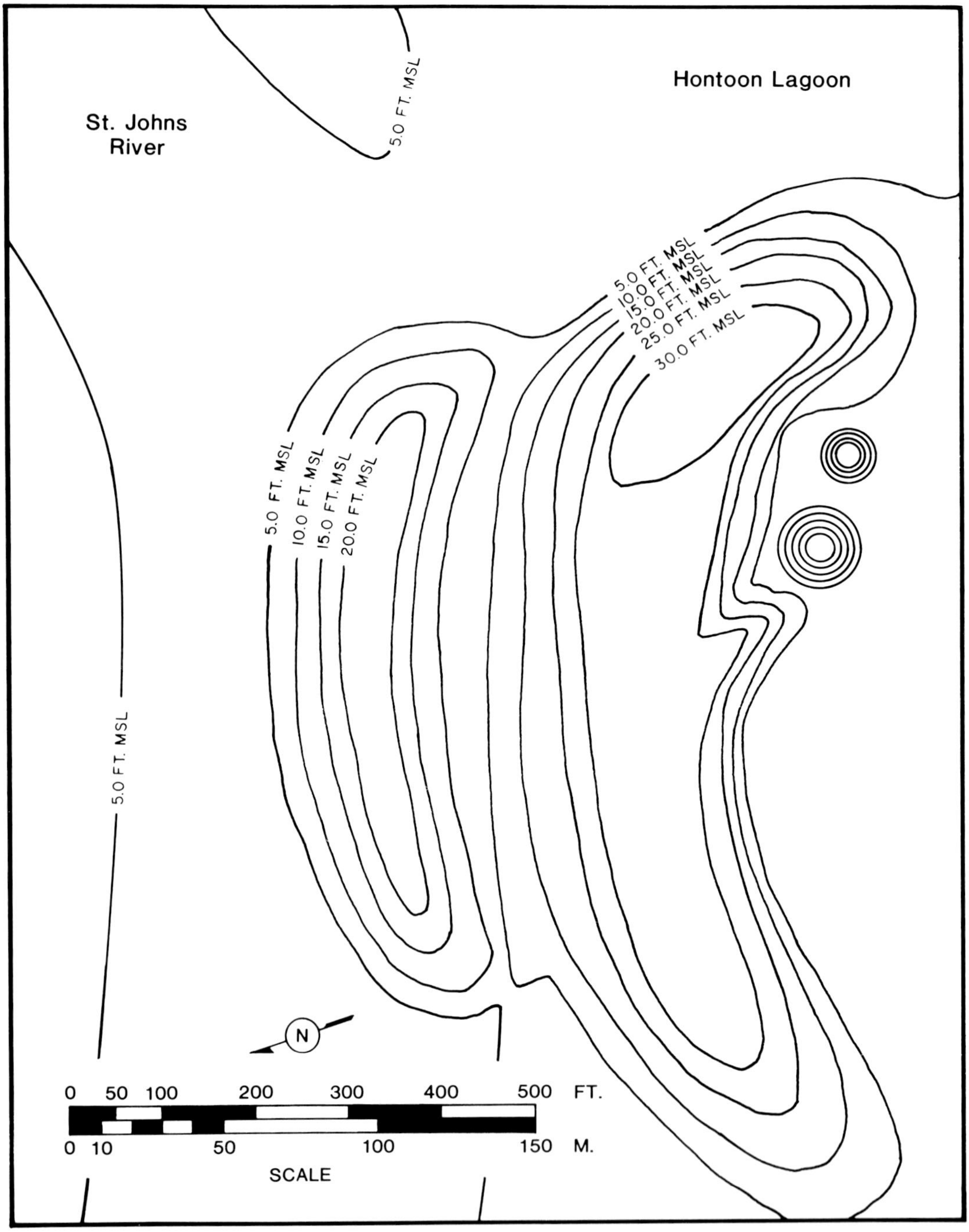

FIGURE 35B

When a statewide survey was conducted in 1980 to determine the extent of Florida's archaeological wetsites and assess their potential for fullscale investigations, Hontoon Island was one of the first areas visited by George F. MacDonald and me. Using a soil auger, we examined the shell deposits that extended into the lagoon beyond the eastern end of the largest midden described by Wyman. The samples contained bone and plant remains preserved

FIGURE 36 The great shell deposits on Hontoon Island shown in an 1894 photo (Moore 1894).

in such excellent condition that leaves were still green and berries still red. This kind of survival was what we were seeking in order to recover usually perishable materials that would expand our knowledge of past cultures and environments in Florida.

Excavations and Stratigraphy

Excavations were conducted at Hontoon Island during several field seasons from 1980–1988. The 3-m^2 test pit opened in 1980 established the stratigraphy which was observed in all subsequent excavations in water-saturated squares: Zone I, 50 cm of culturally sterile organic overburden; Zone II, 10 cm of shellfree midden containing large quantities of plants, animals, and cultural material including early historic period artifacts of European manufacture; Zone III, 30–60 cm of freshwater mussel shell (*Elliptio buckleyi*) with abundant remains of plants, animals, and cultural material; Zone IV, 60+ cm (depending on the location of the unit being excavated) of freshwater snail shell (*Viviparus georgianus*) with abundant remains of plants, animals, and cultural material; and Zone V, an unexcavated sand deposit that appeared to be culturally sterile based on several core samples (Fig. 39a,b). When excavations were placed in dry, upland portions of the midden, Zones I and II were missing and Zone III was either missing or severely truncated. A 2-m^2 test pit was excavated in 1982,

FIGURE 37 Owl totem found in the St. Johns River adjacent to Hontoon Island in 1955 (see Bullen 1955). See also Frontispiece.

a 26-m^2 long trench with a 6-m^2 lateral extension during a 1984 University of Florida field school, a 4-m^2 extension of the trench in 1985, additional exploratory tests in 1986 and 1987, and a 2-month EARTHWATCH project in 1988 when a further extension of the trench was excavated. Beginning in 1982, 30 × 30-cm or 30 × 15-cm column samples were removed in 10-cm segments from the profiles of all units after excavations were completed. Since the column samples were unbiased by sorting techniques or by the loss of small animal

FIGURE 37 (continued)

bones and plants through large mesh screens, their analysis made it possible to obtain a more complete picture of the relationships between animals and plants throughout the occupation of the site. The contents of the column samples were run through a series of nested geologic screens and analyzed. Continuous cores were taken in various locations adjacent to the trench and they provided a portable, archival record of the strata which extended to depths not accessible through excavation (Purdy 1988a).

A

B

FIGURE 38 (A) Otter carving (original and reproduction), (B) pelican carving (original and reproduction). Both found in the St. Johns River adjacent to Hontoon Island in 1978 (Photos: Museum of Florida History, Tallahassee).

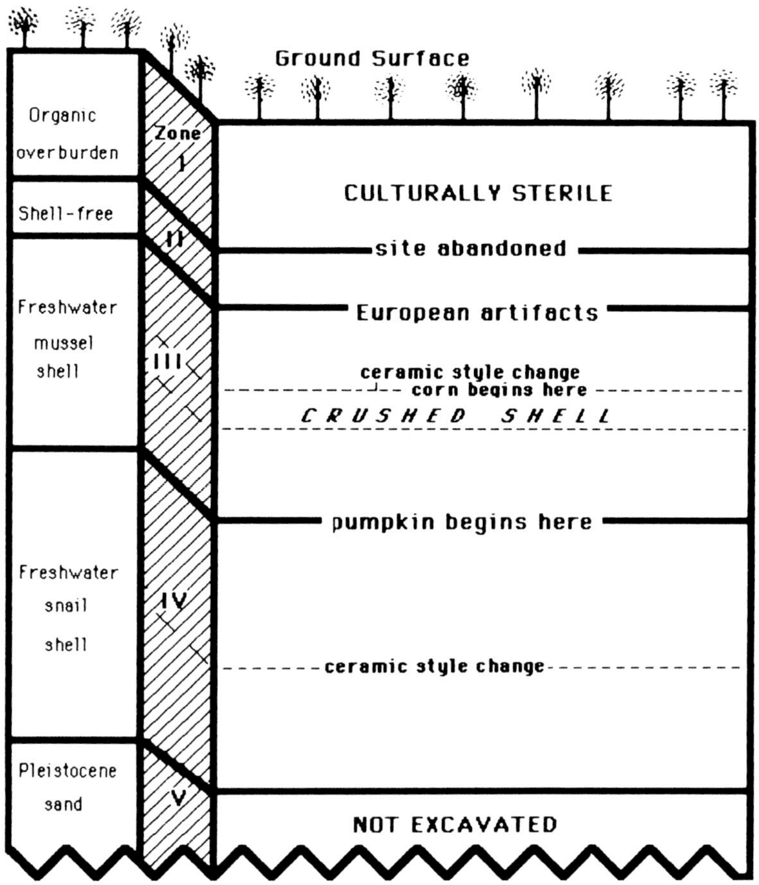

A

FIGURE 39 (A) Diagram of stratigraphy at Hontoon Island, 1988 field season. (B) Photo of strata described (drawing and photo by the author).

Since most of the excavations at Hontoon Island were conducted below the watertable, a combination of hydraulic excavation and evacuation techniques, using several types of pumps, was the primary recovery method.

The procedures implemented in 1988 were designed by my field supervisor, Ray M. McGee, and could be used as a model for this type of wet site excavations. The primary objective of the investigations was to recover materials from deposits that had always been totally submerged. The two squares selected for excavation were an extension of the trench begun in 1984 and expanded in 1985 (Fig. 40). The entire area was covered with about 15 cm of water when the project began. The squares to be excavated plus the 1985 units were surrounded with sandbags. The backfill from the 1985 units was partially removed to create a sump area for the pump intake. A balk was exposed between one wall of the 1985 project and one of the new squares to be excavated. The purpose of this procedure was to pinpoint the exact location of the stratigraphic units and ensure precise recovery of the material from the vari-

FIGURE 39B

ous zones previously identified. Then, each 2-m² was divided into 1-m quadrants and each quadrant was totally excavated by zone and by 10-cm levels within each zone before a new quadrant was begun. Using this strategy, we were able to excavate without stepping on any of the units to be removed. This method eliminated the problem of displacing the fragile waterlogged strata downward and mixing the materials temporally. It had the disadvantage of not being able to view an extensive area at one time. Through notes, measurements, illustrations, photographs, and materials processing we were able to reconstruct the total surfaces of each level within the two squares and the balk. The logistics problem of beginning systematic excavation of a new 1-m quadrant, while at a depth of 2+ m in the bottom of one previously completed, was solved by standing on a 1-foot-wide board placed over the top of two large trash cans. The cans were removed when the new unit was taken down to a point where the deposits could be reached easily (Fig. 41).

All material was water-screened through 1/2-inch, 1/4-inch, and 1/8-inch mesh. The materials remaining in the 1/2-inch screen after washing were sorted in the field and placed in bags labeled pottery, bone, wood, or miscellaneous (usually seeds, polypores, corn cobs, coprolites, and anything else that could not be assigned to one of the major categories). Artifacts were bagged separately. The material remaining in the 1/2-inch screen was then discarded except from one quadrant where we bagged all of the shell separately from each 10-cm level in order to quantify the total number of shells in relation to all other specimens. After water screening, the 1/4-inch and 1/8-inch screens were

FIGURE 40 Sandbagged area at Hontoon Island in 1988 prior to excavations. Individual is an EARTHWATCH volunteer (photo by the author).

scanned for artifacts, corn cobs, and gourd and pumpkin seeds and rind, and bagged according to screen size. All of the bagged material was subsequently rewashed at the field laboratory. Each major category was treated as follows: pottery was washed, dried, and rebagged; bone was soaked in a weak solution of water and Elmers Glue, dried, and rebagged; wood was washed gently and placed in bags with water to prevent degradation until it could be preserved with PEG. The 1/4- and 1/8-inch screens contained thousands of small fish vertebra and other small bones and plant materials that provided clues to prehistoric diet and environment.

Column samples were cut from the east wall of one of the 1988 squares and from the west wall of the other square.

Fauna

Identification of faunal remains from Hontoon Island was made under the direction and guidance of Dr. Elizabeth S. Wing, Curator in Zooarchaeology, Florida Museum of Natural History, Gainesville. The hundreds of thousands of bones recovered at the site from column samples and larger excavation units have furnished a clear picture of species available and utilized by the Indians over a 1500-year+ period. Viewing the faunal assemblage as a whole, 11 species predominate. These include the two freshwater molluscs (*Elliptio buckleyi*, mussel and *Viviparus georgianus*, snail), five freshwater vertebrates (*Lepisos-*

FIGURE 41 Solution to the problem of beginning excavation of a 1-m quadrant while at a depth of 2+ m below ground surface in a previously excavated quadrant (photo by the author).

teus sp., gar; *Ictalurus* sp., catfish; *Micropterus salmoides*, bass; *Mugil* sp., mullet; and *Pseudemys* spp., slider turtles), and four primarily terrestrial vertebrates (*Gopherus polyphermus*, gopher tortoise; *Anatidae*, ducks; *Sylvilagus* sp., rabbit; and *Odocoileus virginianus*, deer). Twenty-nine other vertebrate species were identified but are considered relatively less important numerically (Wing and McKean 1987:41). Most of these animals exist around the island today except gopher tortoise and wild turkey that prefer drier terrain.

There is little or no evidence from the faunal remains that the site was inhabited only seasonally. Most of the fishes are year-round residents of the St. Johns River but some species are easier to procure during the dry season when they are more concentrated in deeper pools and lakes. Mullet become increasingly abundant in freshwater drainages in late summer, fall, and early winter. Gopher tortoises and alligators are not frequently encountered during cold weather (Wing and McKean 1987:41). Wing and McKean (1987) provide insights into the kinds of cultural paraphernalia needed to exploit some of these species successfully, e.g., nets, hooks, and a knowledge of animal behavior.

Fishes and turtles were probably the most important species of animal to the people at Hontoon Island throughout all time periods. An obvious change in the faunal assemblage was a switch in shellfish utilization from freshwater snail in the early horizon (Zone IV) to freshwater mussel around A.D. 1500 (Zone III). About this time also a greater emphasis was placed on terrestrial vertebrates, particularly gopher tortoise and deer. A complete disuse of shellfish occurred just prior to site abandonment (Wing and McKean 1987:45). Deer may have had a greater representation than is apparent from the tabulations because the bones used for artifacts were placed in the artifact category and were not counted with the faunal remains.

The problem with sampling is knowing if the universe has been sampled. Wing and McKean (1987) discuss this situation noting that the biological remains from the columns provide an idea of how important small organisms are that may have been lost in screening. All species, however, may not be represented because of the small size of the columns. On the other hand, it would take the lifetime of an investigator to examine all of the remains from a single 1 m^2 of a wet site as prolific as Hontoon Island. Even then, we could not be certain that it was representative of the whole. It is important to understand that archaeologists as anthropologists must always consider the "typical" from a site, not the "exotic" as interesting as it might be. I believe Dr. Wing and her assistants have identified the major fauna used at Hontoon Island and their relative importance to the inhabitants.

Flora

Newsom's (1987) report on the botanical remains from Hontoon Island is comprehensive and informative. The diverse species identified represent a lush river basin and hammock environment similar to that which exists in the area today. Most of the plants used by the inhabitants grew in five principal ecological communities occurring within 10 km of the site.

Thirty species of wood and 82 species of seeds or other plant parts were identified. The native flora from the snail and mussel shell zones were similar although their frequency and distribution differed through time. Differences between the two zones were observed in cultivated plants. Corn was found only in Zone III (mussel) and above as were new varieties of *Cucurbita pepo*.

Several plants listed by early explorers as being in the environment and/or used for food by the Indians were found in the deposits at Hontoon Island. Since many of the plants occur naturally in the marsh/river swamp ecotone near Hontoon Island, it is difficult to know if they were used as food. Most of the seeds show no evidence of human manipulation or alteration, such as charring, nor were they found in caches (Newsom 1987:59–60). The recovery and analysis of human coprolites would shed light on plants used by the Indians. The few coprolites found at the site have not been analyzed and they may not be human. To determine which plants from Hontoon Island were of economic value to the Indians, Newsom used ethnohistoric and ethnobotanical sources, and logical reasoning in her analysis and interpretation. Her detective work is very convincing.

Wild plant resources that were used as food are hickory, oak, may pop, grape, saw palmetto, and cabbage palm. Acorn, hickory, and cabbage palm were the most abundant wild plant resources recovered. Hickory was more plentiful in Zone IV (snail zone). At least 13 other plants, and possibly 9 more, may have contributed to subsistence.

Cabbage palm and 11 other plants increased substantially in Zone III (mussel) and above at the site. All of these are potential food plants. A sudden marked reduction in these same species occurred toward the termination of occupation at Hontoon and they were succeeded by plant species, such as buttonbush and willow, indicating a return to natural conditions. The fact that 12 plants which may have been used as food underwent a substantial reduction at the same time that animals and artifacts disappear is a good indication that these plants were being used by the Indians residing at Hontoon Island (Newsom 1987:63–66).

While some species identified by Newsom are plants that normally inhabit the river swamp/hammock environment, there are others, such as amaranth, that characteristically inhabit disturbed sites. They may have occurred nearby but the presence of humans permitted the enhancement and expansion of their habitat. Their niche was reduced and their numbers declined when the site was abandoned and the natural habitat returned.

Seeds and rind of bottle gourd (*Lagenaria siceraria*) and an ornamental gourd (*Cucurbita pepo*) were recovered in Zone IV (snail zone) and above at Hontoon Island. The bottle gourd may have been used for a container rather than for the edible seeds. Beginning in the bottom level of Zone III (mussel) a large seeded variety of *Cucurbita pepo*, similar to Mexican pumpkin, appears in the deposit. Other varieties may be present also. Newsom (1987:69–74) and Decker and Newsom (1988) provide an indepth discussion of these species. The presence of pumpkin in the lower levels of Zone III is significant because its introduction to the midden takes place along with other major changes and modifications.

Corn (*Zea mays*) occurs at the site, also in Zone III, but only above a crushed shell lense that was identified precisely during 1988 excavations. This lense lies 30 cm above the bottom of Zone III and thus corn was introduced more recently than pumpkin. All of Zone III, however, was deposited during a brief period, perhaps 150 years or less. A corn cob recovered in situ from the lowest stratum in which corn was found was dated by AMS (atomic mass spectrometry or accelerator dating). The corn has been identified tentatively as northern flint.

Along with other botanical remains from Hontoon Island, 30 species of wood were identified. These specimens supply information about environment and cultural activities that seldom survive.

Human Remains

Approximately 200 m^3 of midden was excavated at Hontoon Island from which hundreds of thousands of bones were removed and identified. Only 23 of these were human. These included several teeth, cranial and pelvic fragments, and finger or toe bones. Wyman (1875) encountered human remains in some of the shell middens he dug and concluded that the people must have been cannibalistic because the bones were mixed in with those of butchered animals. His explanation seems unlikely for Hontoon Island. Such a scarcity of human bones indicates that *Homo sapiens* was not a large part of the diet. The bones probably became incorporated into the midden accidentally during subsistence activities, possibly while digging clams from the river sands. The two burial mounds on Hontoon Island, described by Wyman (1875:29), were obliterated by shell-mining activities in the 1930s.

Artifacts

Ceramic, wood, stone, shell, bone, and historic period artifacts were recovered at Hontoon Island. These are discussed in Purdy (1987a:27–39). The 1988 excavations yielded additional examples of the same materials bringing the totals to 177 bone tools or decorative items plus hundreds of bone fragments that could have been utilized, 147 stone artifacts, 80 wood artifacts plus thousands of adzed wood chips and naturally deposited wood debris, 112 shell tools or beads, 82,000+ ceramics, and 53 historic period objects. Many of the artifacts, including those of the historic period, are similar to implements and decorative items recovered elsewhere in Florida.

Ceramics—Most of the pottery is St. Johns paste. Undecorated ceramics predominate in the lower levels of Zone IV and check stamping and plain ware coexist above Level 3 of Zone IV. Another change in ceramics takes place in the same level of Zone III where corn first makes its appearance (see Fig. 39). At this level, checks become smaller and there is a change in the rim thickness. The change is abrupt with no transition. An apparent innovation in ceramic technology also occurs at this time. After the vessels were formed and scraped to thin and smooth them, they were covered with a very heavy slip before they were check stamped. No other level at the site exhibits this technology (McGee 1990, personal communication). There were a few sherds recovered that differed from the typical in paste, decoration, or shape. Some of these are similar to ceramics from other areas in Florida and suggest at least limited contact with people who did not live along the St. Johns River. Unique specimens were also recovered that may represent nothing more than individual expression (Fig.

FIGURE 42 Bird head incised on sherd. This specimen was recovered from the deepest cultural level at Hontoon Island (photo by the author).

42). Clay objects other than sherds are scarce and include two pendants or plummets, a spindle whorl (?), a squash blossom or pipe, a sculpted head (Fig. 43), and several unidentified shapes. Willey's (1949) conclusion about the ceramics at the Belle Glade site might be appropriately applied to Hontoon Island also: ". . . the absence of a variety of artifacts made of fired clay probably reflects a lack of skill and interest in ceramic art . . . ."

Wood—No large wood carvings like the owl (Bullen 1955; Purdy 1987a), pelican, and otter (Schwehm 1983; Purdy 1987a; Figs. 37 and 38) were recovered during excavations from 1980 to 1988. These sculptures are made of the heart wood of southern pine as are the large carvings from the Fort Center site which also portray animals. The art styles are very different, however. The Fort Center carvings are naturalistic while those found near Hontoon Island are more stylized (Schwehm 1983).

Bullen's (1955) description (paraphrased) of the owl follows. The overall length of the owl totem was 12 feet, 2 inches long before a rotted portion of the bottom was sawed off. The cross section is 17 inches wide and 11 inches deep at the middle of the carving. The lower portion of the owl is rectangular in cross section and 5 to 5.5 inches deep. It tapers slightly from a width of 17.5 inches at the feet of the owl to a width of 15 inches just above the rotted portion. Evidence of several kinds of tools is present: shark teeth knives were used to make innumerable medium-fine grooves on the back and sides, stone knives were used to make 1/8- to 1/4-inch-wide decorative grooves that appear to have

FIGURE 43 Human head sculpted from clay. This is the only specimen of its kind found at Hontoon Island (photo by the author).

been lined with a dark substance, shell gouges were used to make concave marks on the lower portion of the totem, and sand or sandstone was used to smooth all surfaces of the lower part of the pole. The totem pole represents an owl standing at rest on top of a prepared "perch". The craftsman clearly had in mind one of the owls with tufted ears, the screech owl or the great horned owl. Bullen believes it was intended to depict the great horned owl. From the top of the horns to the bottom of the talons, the owl measures 6 feet, 3 inches. The horns are nearly round measuring 7 × 8 inches. The end of the tail is at the same elevation as the talons. The middle portion of the owl is more or less rectangular in cross section with rounded corners. Each leg is also a rounded rectangle in cross section. The beak and horns are exaggerated. The eyeballs are set back into the head. The iris is a 3.5-inch diameter circle incised within a slightly curved diamond-shaped area representing, presumedly, half-closed eyelids. Nostrils and feathers are also represented. The incisions depicting feathers have been carefully and symmetrically planned. The artist has carefully indicated the difference between wing and tail feathers and has even differentiated the under feathers of the tail. The bird has five instead of four talons suggesting anthropomorphism.

The owl, otter, and pelican totems were all recovered from essentially the same area (Dunbar et al. 1978) and all of them could have been carved by the same artist. The otter and pelican are much smaller specimens and their features depict those of the animals they represent, but they are "perched" exactly as the owl described above. The owl totem was on display in the Florida Museum when the museum was still in the Seagle Building in Gainesville. It was in storage for several years after the museum was built on the University of Florida campus. It is now at Fort Caroline. The owl has not been preserved but it may have been coated with varnish, shellac, or a similar substance. The otter and pelican are at the Museum of Florida History, Tallahassee. They were stabilized by the conservators at that museum. The three specimens have been replicated. Like wood carvings found elsewhere in Florida, these objects provide insights about native cultures not represented in other media. Because their surfaces have not eroded away, like those from Fort Center, they divulge information about aboriginal design elements that is extremely rare.

A small carving of a bird was recovered during the 1988 excavations at Hontoon Island, (Fig. 44). It was made of red cedar (*Juniperus* sp.). It is the only wooden "art" object recovered and may have been a hair or clothing ornament that was lost while its owner worked along the edge of the lagoon. The bird is similar in size to one from Key Marco pictured by Gilliland (1975) that was made of cypress (Newsom 1990, personal communication). Another wood artifact may have been a ceremonial or decorative item (Fig. 45). It is similar in shape to specimens described by Cushing (1897) and Gilliland (1975) for the Key Marco site and by Stirling (1935) and Willey (1949) for the Belle Glade site. In addition to wood, these objects were made of other materials, including metal (Allerton et al. 1984). The one from Hontoon Island was from Zone II (the shell-free historic period) as was the bird carving. The other wood artifacts from Hontoon Island are probably utilitarian items although the function of most of them is not known. These include posts, firestarters, shafts, notched sticks, plugs, a wedge, a broken canoe paddle, tool handles, small points, a bowl, and miscellaneous objects. The notched sticks (Fig. 46) could have been parts of atlatls, traps, or "tent" pegs. The canoe paddle (Purdy 1987b; 1988b) has all manner of adze and shark teeth finishing marks on its surface. The wood species used to produce the artifacts were identified by Lee A. Newsom as cypress (*Taxodium* sp.), pine (*Pinus* sp.), willow (*Salix caroliniana)*, ash (*Fraxinus* sp.), bay (*Persea* sp.), staggerbush (*Lyonia* sp.), oak (*Quercus* sp.), elm (*Ulmus americana)*, buttonbush (*Cephalanthus occidentalis)*, cedar (*Juniperus* sp.), persimmon (*Diospyros virginiana)*, and several unidentified species including some tropical woods not native to Florida (Purdy 1987a). Certain woods were probably valued over others to perform certain tasks or they may have had ritual significance. In some cases or for some woodworking projects, availability of a species might have been just as important as wood properties. One example from Hontoon Island pertains to the artifacts classified as posts (Fig. 47). These were made of cypress, pine, elm, ash, and bay. There is very little with which to compare the wood artifacts from Hontoon Island so it is impossible to know if the objects are typical. Even those from Hontoon Island are not plentiful enough to create classes of wood artifacts, except possibly the posts. Key Marco is the only other site where a variety of

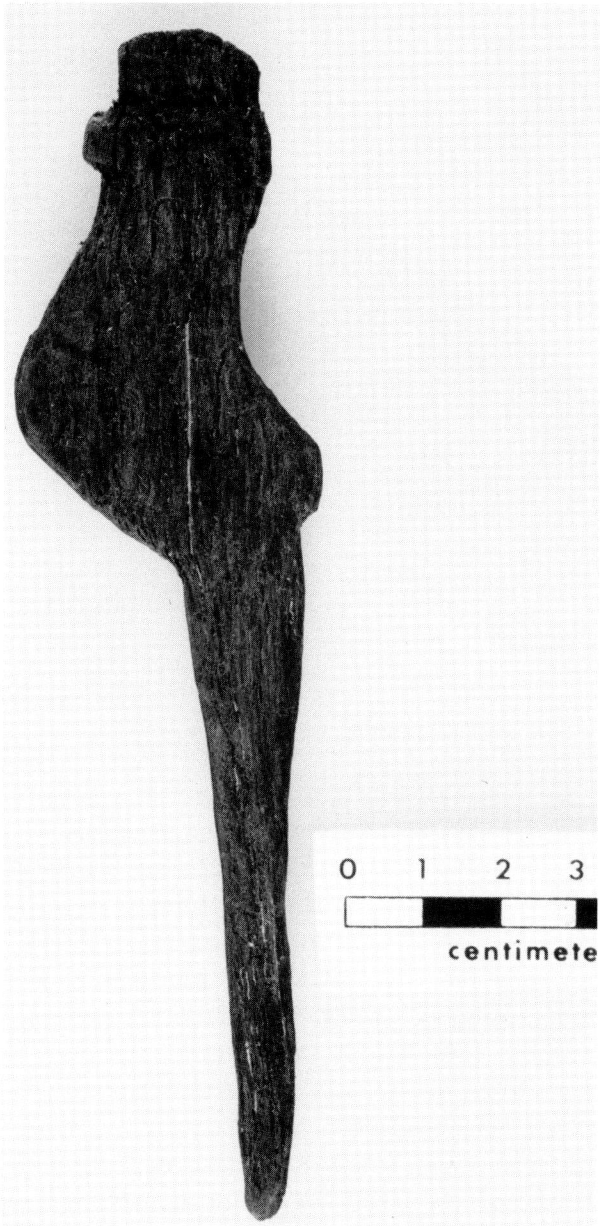

FIGURE 44 Bird effigy made of red cedar (photo by the author).

wooden artifacts has been described. They are more numerous and elaborate than Hontoon Island.

While 80 wooden objects have been labeled artifacts, the majority of the wood found at Hontoon Island is either a byproduct of artifact manufacture or was naturally deposited. Thousands of wood chips, mostly cypress, were recovered and many exhibit extensive adze marks, charring, or other evidence of manufacture (Fig. 48). They probably represent the debitage discarded into

FIGURE 45 Small wooden artifact similar to those recovered at Key Marco and described by Cushing (1897) and Gilliland (1975) (photo by the author).

the lagoon after village structures, long since decomposed, were completed. The chips survived because they became entombed into the water-saturated peat and shell matrix that borders the northeastern end of the island. This wood debitage from Hontoon Island offers an opportunity, not found at most wet sites in Florida, to study tool marks and manufacturing techniques and to learn something about accumulated knowledge of wood properties, availability of species, environment, and much more.

Shell—Tools and ornaments were produced primarily from various species of conchs and a lesser number were made from marine clams (Purdy 1987a). Marine shellfish do not live along the St. Johns River but fossil shell beds that are millions of years old do occur in certain areas along the river. Radiocarbon dates on two marine shells excavated at Hontoon Island, however, place their ages in the same time frame as the occupation of the site. Besides, the fossil shell that we examined was friable and could not have been used for implements. The shell was probably traded in or procured when visits were made to the coast.

No celts similar to those from Belle Glade and Key Marco were found. All of the unbroken marine shell tools (Fig. 49a)—gouges, hammers, scrapers, adzes—were recovered from Zone IV (snail). For reasons not completely understood, the changes that occurred beginning with Zone III (mussel) included dropping the marine shell tool category from the cultural inventory.

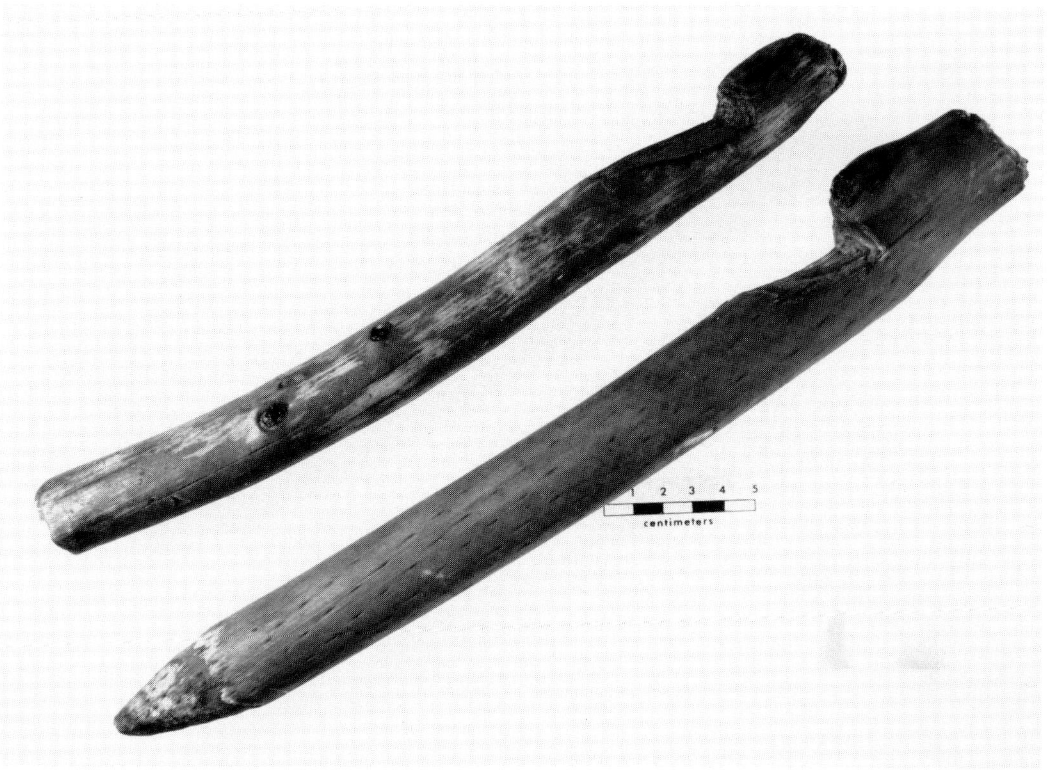

A

FIGURE 46 Four views of notched sticks recovered at Hontoon Island in 1982 and 1988, function unknown. Note the excellent preservation of the wood and the shark teeth (?) finishing marks in *d* (photos by the author).

Cups or dippers made of conch were recovered only in Zone III and above (Fig. 49b). These look exactly like those recovered at other sites, such as Key Marco.

Four ornaments, three beads, and one pendant were made of marine shell. Three fish hooks (Fig. 50) manufactured of freshwater mussel and a mussel shell that was painted with a bright red pigment were the only additional shell artifacts recovered.

Stone—Forty-seven points, most of which were found in Zone III and above, and 34 small flakes were made of Florida chert. Nearly all of the flakes had been utilized. Only six other chert specimens were recovered and most of these appeared to be "relics" incorporated into the midden accidentally, as were several of the points, because their styles are typical of time periods predating the age of the midden by thousands of years. If these points and tools are subtracted from the total of chert artifacts, it is apparent that this kind of stone was scarce at the site. Flint rock does not outcrop along the St. Johns River but is present within 50 miles or less of Hontoon Island. The small size and evidence of use on most of the chipped stone implements suggest that chert may not have been easily obtainable (Purdy 1987a). The majority of the points

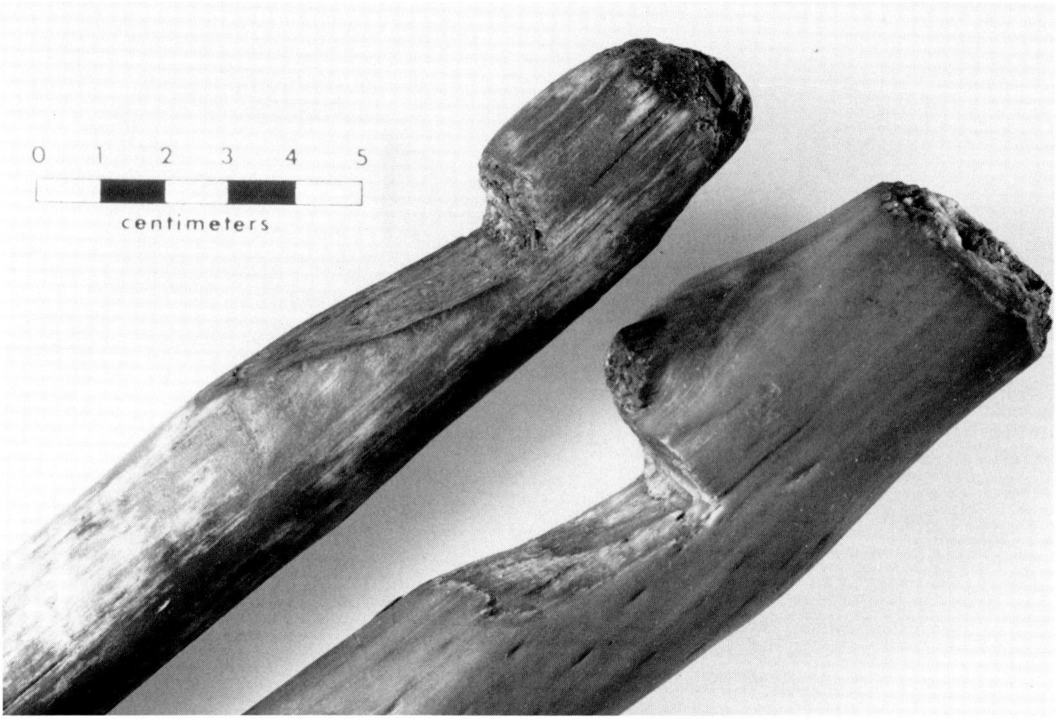

FIGURE 46B

FIGURE 46C

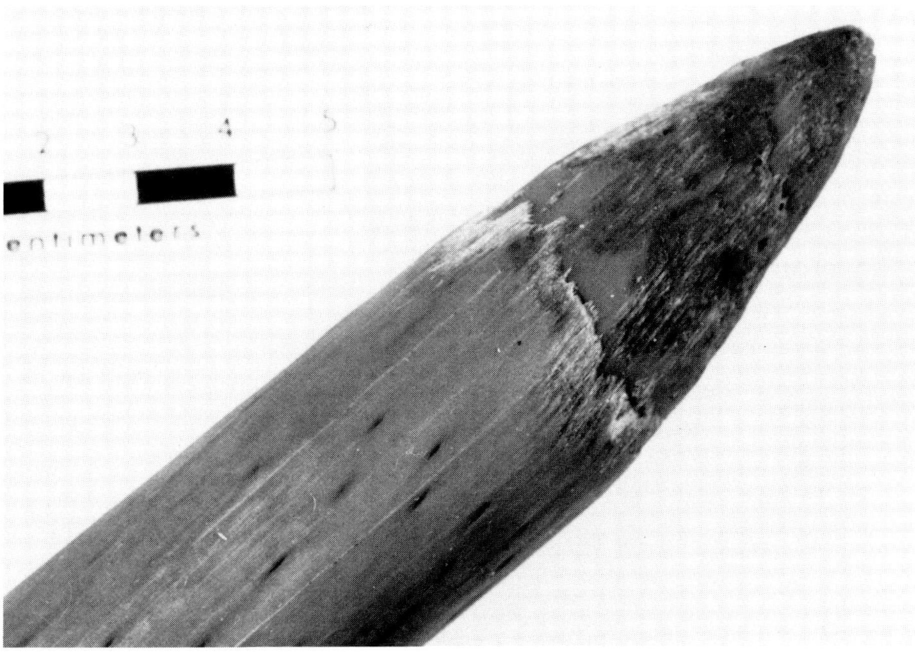

FIGURE 46D

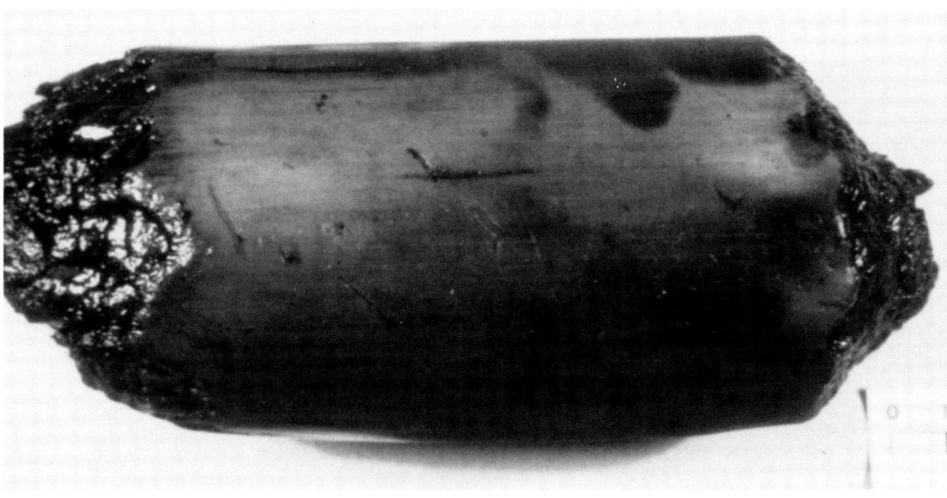

FIGURE 47 Fragment of a burned "post" recovered at Hontoon Island. (Note the excellent preservation of the wood; photo by the author.)

were Pinellas (Fig. 51), a type that is common for the time period of the Hontoon Island shell midden occupation. It is interesting that most of the points were recovered from Zones II and III although the total duration of both zones was brief compared to Zone IV.

Twenty-nine crude sandstone abraders form the second largest class of stone implements. These were probably used for a variety of tasks such as smoothing and sharpening bone implements. Two of the sandstone specimens

A

B

FIGURE 48 Wood showing (A) adze marks, (B) close up of *A*, (C) sharpened branch, and (D) close up of tool marks on *C* (photos by the author).

are smooth on all surfaces and one appears to have been "sculpted" into the shape of an animal. Its shape, however, could be fortuitous.

Fragments of red and yellow ochre were recovered. There is no indication of how the ochre was used at Hontoon Island. It was typically used as pigments and/or in ceremonies. The red pigment observed on one freshwater mussel shell may have been produced from ochre.

One metate of coquina was found toward the top of the midden.

Figure 48C

Figure 48D

Eleven greenstone celt fragments and one pendant or plummet made of a hard limestone constituent were the only non-Florida stones recovered at Hontoon Island. Most of the greenstone celt fragments came from the deeper levels as did the limestone pendant.

The stone industry at Hontoon Island is not very impressive and it probably was not an economic essential. The absence of stone ornaments suggests that stone was also not important for personal adornment or ritual activities.

Bone—It is difficult sometimes to distinguish butchered bone from slightly modified bone splinters that may have served briefly as tools leaving little or no evidence of utilization. The bone artifact count would be increased meas-

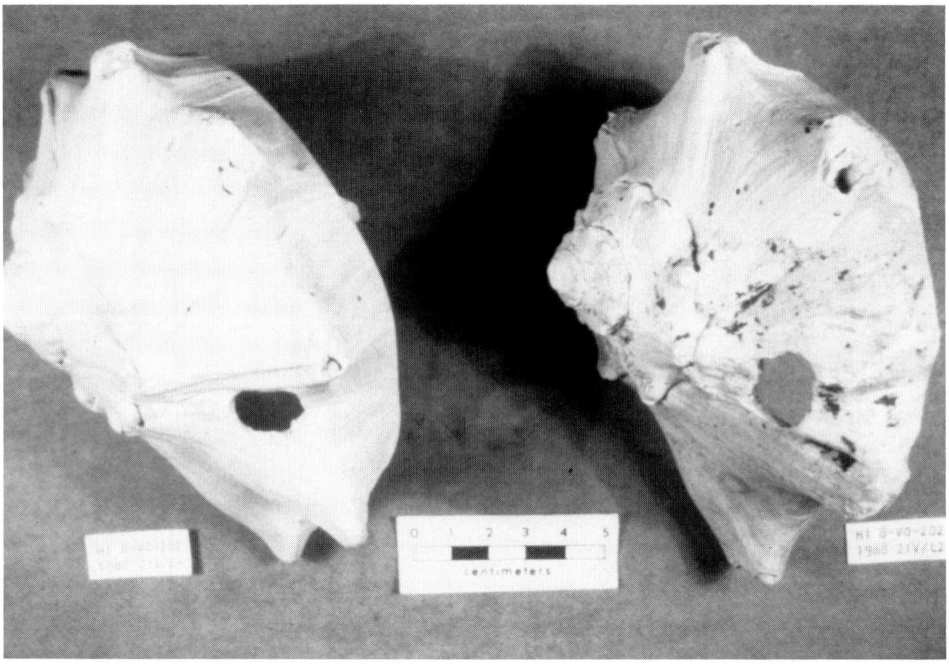

A

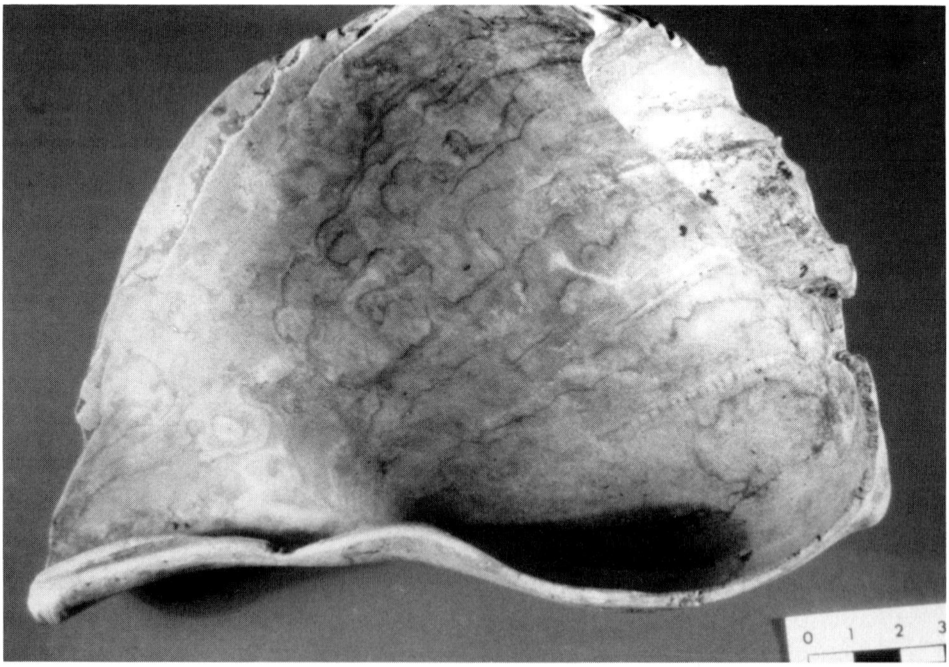

B

FIGURE 49 (A) Examples of marine shell tools recovered from Zone IV at Hontoon Island. These were probably used as hammers (photo by the author). (B) Example of a marine shell "drinking cup" from Hontoon Island (photo by the author).

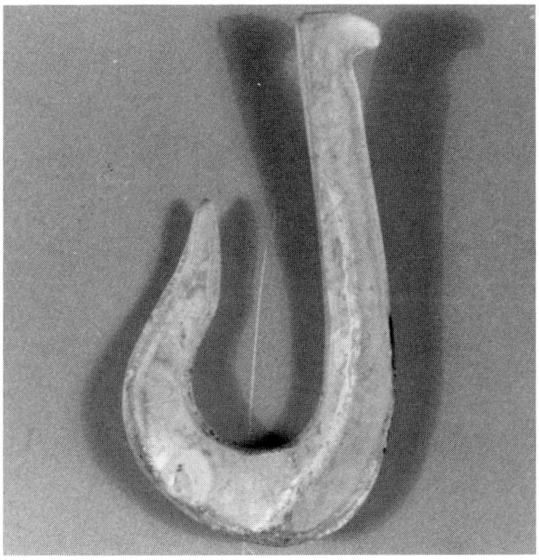

FIGURE 50 "Fish hook" made from freshwater mussel (photo by the author).

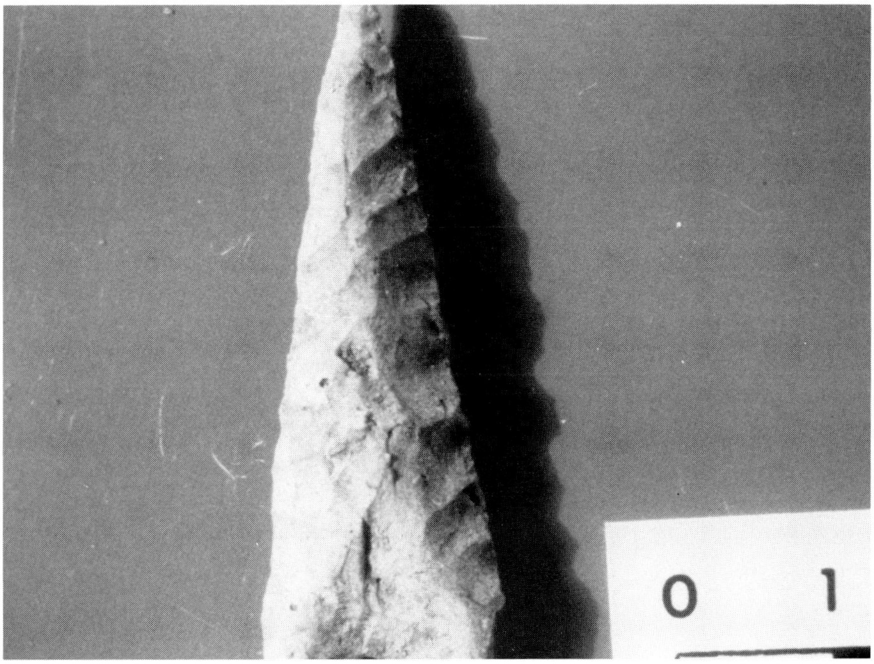

FIGURE 51 Example of a Pinellas point from Hontoon Island (photo by the author).

urably if it included these fragments lacking formal shapes and sustained use. Possibly because these questionable specimens were not considered as artifacts, the decorative bone items were as numerous as those that were placed in the utilitarian category (Purdy 1987a).

Seventy bone tools, 15 ulna awls, 71 pins, 3 pendants, 2 beads, a dagger, a fishhook, and 24 shark teeth (some drilled) were recovered from the excavated units at Hontoon Island. Deer antler was fairly abundant in the deposits but it was not worked into tools or elaborate tool handles as it was at Key Marco and other sites. Most of the bone artifacts were made from deer although other species were identified, such as bird and alligator.

The bone industry was important to the people at Hontoon Island. The absence of stone in the immediate area may have made it necessary to use bone for tasks normally performed with stone. Marine shell technology may have developed for the same reason. It can be concluded, in fact, that the bone and shell industries at Hontoon Island constituted a rather strict division of labor based on raw materials. Both materials were essential to the maintenance of the culture and there was very little overlap in the kinds of objects manufactured of bone with those manufactured of shell.

Bone pins, probably used for the hair or clothing, were found in all excavation units at Hontoon Island. In Zones III and IV, these artifacts are usually undecorated or contain only a small amount of incising (Fig. 52a). There were exceptions (Fig. 52b). In Zone II, along with European objects, a dramatic change occurs in decoration on several bone pins and a pendant made from deer scapula. One of the designs, rectilinear guilloche, is so distinct it must have been introduced by Europeans or by non-Florida Indians fleeing European atrocities (Purdy 1987a; 1988; Fig. 53).

Historic Period Artifacts—Objects of European materials and manufacture were recovered at Hontoon Island predominately in Zone II (shell free). Three small glass beads were recovered in the top of Level 1 in Zone III (mussel). Thirty-two pieces of metal (gold, silver, copper, and iron), 19 glass beads (Fig. 54), 6 fragments of majolica, 3 pieces of olive jar, and 1 gunflint made of non-Florida stone constitute the total assemblage. Most of these items are typical of those that date to the early historic period, i.e., within 50 years after the first voyage of Columbus. Since these artifacts arrived at the site during its terminal years of occupation, it is logical to conclude that their presence signals the beginning of events that led to the demise of the Indian culture. This trend may have been in motion considerably earlier, however, and the reasoning behind this statement is elaborated upon in the chronology section.

Chronology

Eliminating several misleading age calculations made on freshwater snail, 51 dates were received for the 8-VO-202 site at Hontoon Island, 36 by radiocarbon analysis and 15 by thermoluminescent analysis. The dates range from BC 740 to AD 1850. Human occupation at the site undoubtedly began much earlier than BC 740. The area mined out during the 1930s lies west of the 1980–1988 excavation activities. Limited testing in that location revealed that pottery was not present in the lower strata of shell undisturbed by the mining operation. We suspect that during the Preceramic Archaic Period 5000–6000 years ago, the original habitation began here on a sand bank that was slightly

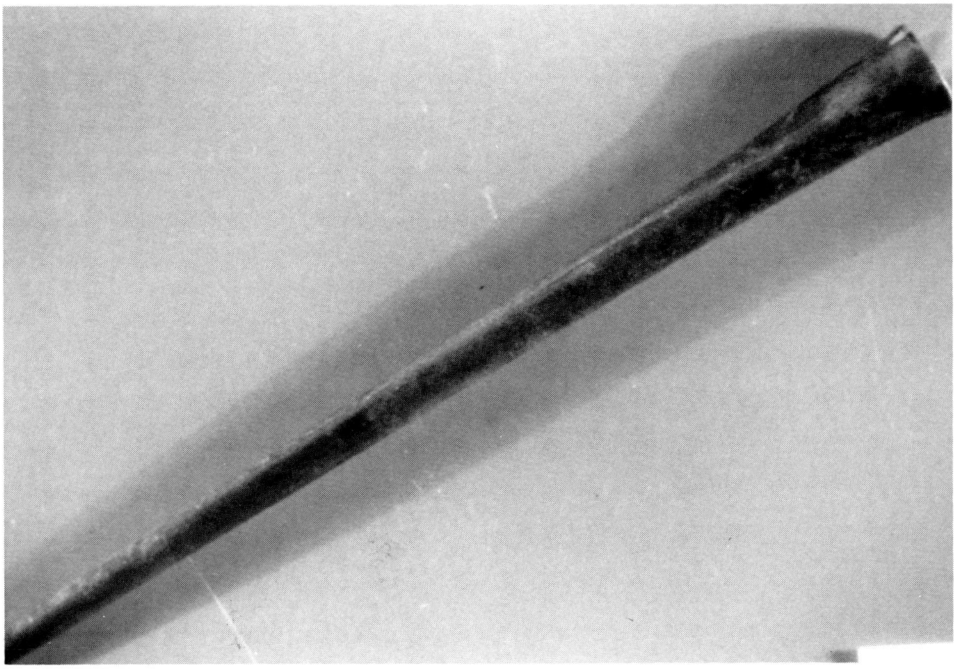

A

FIGURE 52 (A) Example of a plain bone pin from the lower zones at Hontoon Island, (B) an elaborately incised bone pin from the lowest zone at Hontoon Island (an exception to the usually undecorated types found in the lower zones), (C) close up of *B* (photos by the author).

elevated above the water level. As the midden grew, it expanded toward the lagoon. Thus, our excavations that mostly fringed the lagoon recovered a record of only the more recent residents.

The water-saturated deposits in which excavations were concentrated had not been trampled nor had the organic material degraded compressing the strata as occurs at terrestrial sites. The stratigraphically discrete zones that survived contained hundreds of thousands of well-preserved environmental and cultural materials. These zones can be seen easily in the profile (Fig. 39a,b).

Zone IV (snail midden) of the water-saturated excavations is approximately 60 cm thick. Average dates of AD 900 were received for the lowest level (Level 6) and average dates of AD 1350 for the upper level (Level 1). All of the dates received for Zone IV have been consistently prehistoric even when calculating the range of variation. Minimum and maximum extremes extend from AD 645 to AD 1470. Thus, the 60 cm of cultural and environmental deposit of Zone IV accumulated in a minimum of 400–500 years and possibly as long as 700–800 years.

Zone III (mussel midden) of the water-saturated excavations is also 60 cm thick. Average dates of AD 1429 were received for the bottom level (Level 6) and the average of only one date received for the top level (Level 1) was AD

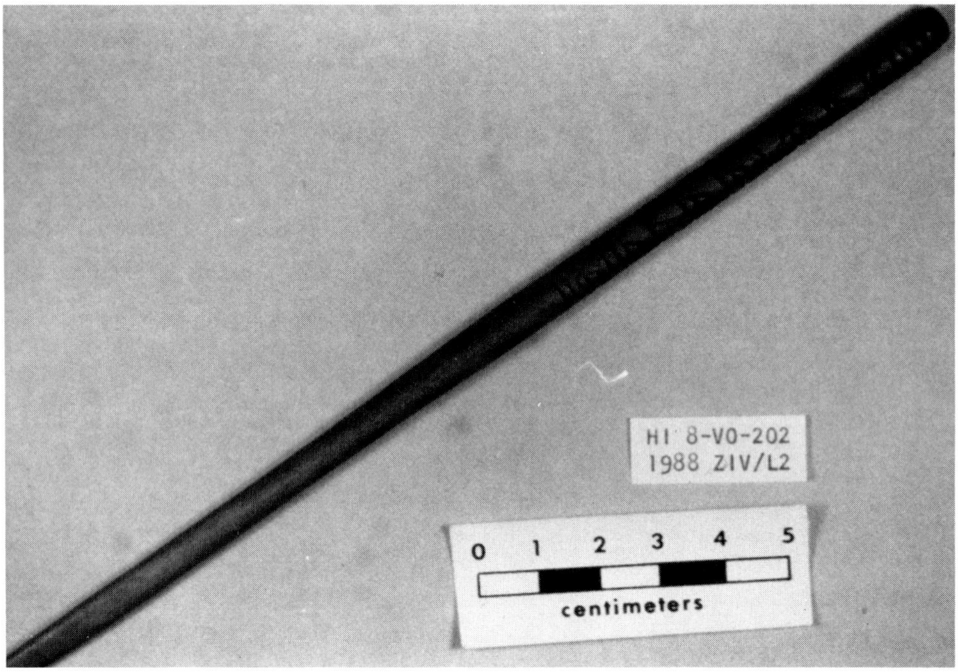

Figure 52B

Figure 52C

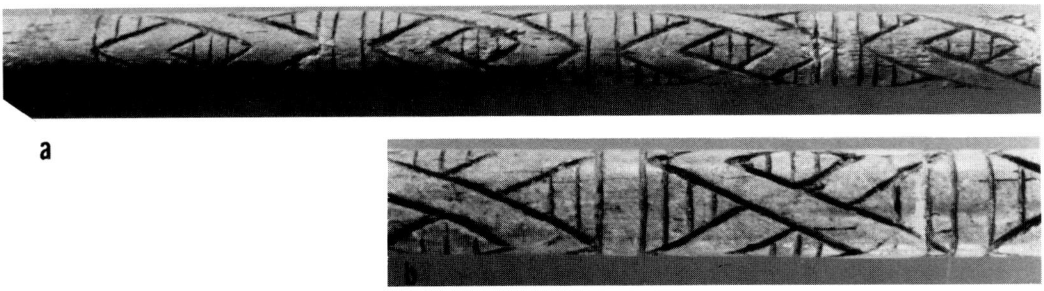

FIGURE 53 (A, B) Bone pin and close up of the rectilinear guilloche design that is introduced in Zone II, the Historic zone, at Hontoon Island (photo by the author).

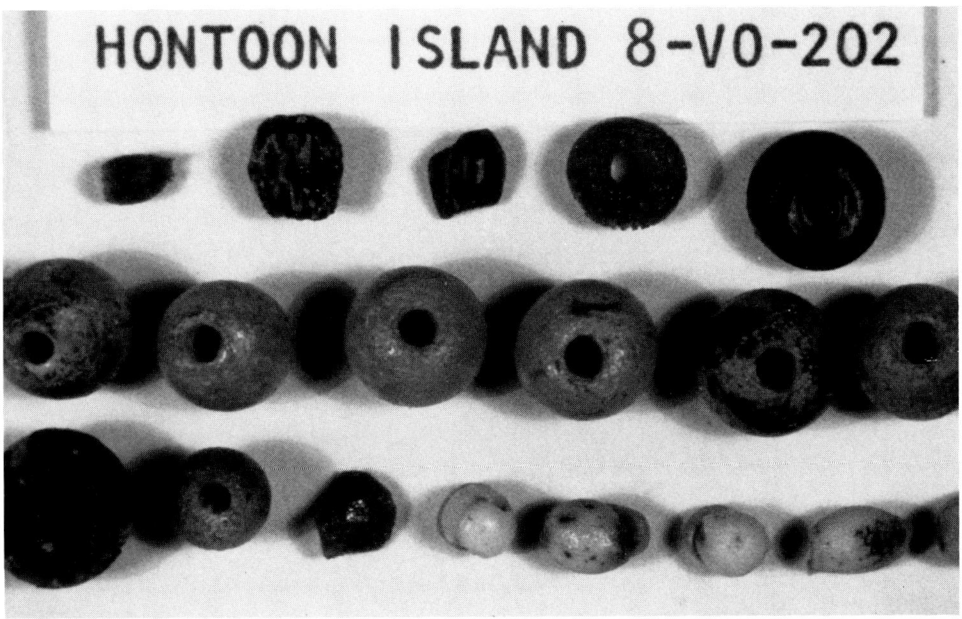

FIGURE 54 The glass beads recovered from Zone II, the Historic zone, at Hontoon Island (photo by the author).

1572. When calculating the range of variation, all of the dates received for Zone III fall within the historic period, i.e., after AD 1492. Minimum and maximum extremes extend from AD 1260 to AD 1678. Thus, the 60 cm of cultural and environmental deposit of Zone III accumulated in a minimum of 150 years and a maximum of about 400 years. The fact that 60 cm of deposit accumulated in Zone III in a much shorter period of time than it took for the 60 cm of deposit to accumulate in Zone IV suggests that a change in discard patterns took place in Zone III or a larger population is represented.

Zone II (historic midden) of the water-saturated deposits at Hontoon Island is 10–20 cm thick. The average of eight dates received for Zone II is AD 1517. Minimum and maxium extremes extend from AD 1369 to AD 1703.

It is clear there are problems with the chronology at Hontoon Island and these are discussed thoroughly in Purdy (1990). The chronology has a crucial bearing on critical changes that occurred during the late prehistoric/early historic time periods. Of considerable interest is the presence of corn cobs and the seeds and rind of pumpkin. The position of these species in the profile beginning in Zone III and above (Fig. 39) and the time frame for their introduction are significant because these cultivated species have not been reported in an archaeological context in peninsular Florida predating the historic period. If cultigens occurred after historic contact, it might be easy to explain their presence as the result of movements of displaced people. If they occur prior to AD 1492, other explanations are required. Changes coinciding with the introduction of pumpkin in Zone III, Level 6 include the absence of marine shell tools and the almost total abandonment of the use of freshwater snail (Zone IV) in favor of freshwater mussel. As noted above, the dates received for the top of Zone IV have been consistently prehistoric and there is some evidence to suggest that the site may have been deserted for a short time (maybe 50 years) before the mussel-eating, pumpkin-raising people arrived. Corn occurs only above a 10-cm thick crushed shell layer in Zone III (Fig. 39) and is accompanied by a change in ceramic technology and typology.

When determinate errors of variation in original carbon-14 production are combined with indeterminate errors of nature, animals, and humans, the range of variation is magnified and the *real* dates may be difficult to determine. Obviously, when the dates fall within the last 500 years, interpretation of whether or not a site is prehistoric, protohistoric, or historic is somewhat dubious (Purdy 1990). Based on an evaluation of all data available at this time, I conclude that the extreme changes that took place at Hontoon Island beginning with Zone III were somehow stimulated directly or indirectly by European intrusion into the Western Hemisphere. Corn was introduced somewhat later than pumpkin and was also accompanied by major changes. Historic Period artifacts appear in the final days of occupation at Hontoon Island along with the interesting decorated bone objects. Shellfish were no longer utilized but great quantities of vertebrate remains attest to the fact that subsistence practices continued for a while. I do not believe that habitation at Hontoon Island extended to the Mission Period in the mid-17th century. There are no ceramics that are usually typical of sites that date to this period and bones of no European animals were recovered from Zone II. Most of the Spanish artifacts are typical of the 16th century. The chronology at Hontoon Island is discussed in greater detail in McGee and Nodine (1987), Newsom (1987), Nodine (1987), Purdy and Newsom (1985), Purdy (1987a, 1987b, 1988c, 1989, 1990), and Wing and McKean (1987).

Preservation

Preservation experiments using PEG were conducted with chips and branches of several wood species from Hontoon Island. Some of these experiments are described in Chapter V. All of the wooden artifacts were preserved using PEG without freeze drying because freeze drying equipment was not available. The preservation of many of the wooden artifacts and wood debitage was completed in the early 1980s and the specimens remain in good condition as of the spring of 1990. Following essentially a cookbook recipe, preservation of still waterlogged woods began with a concentration of 10% PEG and water. The concentrations were increased periodically in 10% increments up to 80%. This procedure worked well for the Hontoon Island woods but experiments should be conducted using expendable pieces before preservation is attempted on woods from other sites. Without treatment, the woods from Hontoon Island, regardless of species or age, undergo a weight loss of about 90% with accompanying shrinkage, distortion, and cracking to such an extent that the objects do not resemble their original size and shape at all (Fig. 55).

Botanical remains such as seeds, nuts, and leaves survived well when placed in formic acidic acid. These specimens now form part of a reference collection at the Florida Museum of Natural History. Bone from Hontoon Island was soaked in a weak solution of Elmers Glue.

Discussion

The people of Hontoon Island depended upon the aquatic and other natural resources available to them from the St. Johns River and adjacent uplands. The still water of the lagoon, neutral pH, and anaerobic conditions prevented the degradation of organic materials entombed in the deposits bordering the northeast end of the island.

Many of the animals and plants, including ornamental and bottle gourds, and several kinds of implements and weapons are similar to those found at archaeological sites throughout Florida. Some of the shell, bone, and stone tools are identical to those manufactured elsewhere as are pottery designs and ceramic technology. Woodworking technology at Hontoon, and where known from other locations, utilized firesetting, shark teeth, shell adzing, and pine for large carvings. Utilitarian objects produced in most raw materials are similar throughout Florida as are some decorative items in shell and bone. Wood art styles in the vicinity of Hontoon Island, however, differ from those in south Florida. The owl, pelican, and otter found near Hontoon Island are as different from the carvings found in south Florida as Easter Island statues are from those of Greece. With the survival of these unique objects, it is possible to recognize cultural divisions within the state that are not readily perceived through subsistence practices or utilitarian items (Purdy 1988b).

Quantities of wood debitage attest to the fact that wooden structures or canoes were manufactured at Hontoon Island. Bone implements, pottery, and smaller objects of wood must have been made at or near the site also. Marine shell tools and chert implements were probably brought into the site already completed.

Hontoon offers views of the past that differ from mortuary sites such as Fort Center, Republic Groves, Little Salt Spring, and Windover where the emphasis is placed on human skeletons and accompanying grave goods, and from Key Marco where the art objects have received the primary attention. Hontoon is important for the potential it provides to study, both synchronically and diachronically, a nearly complete and unlimited assemblage of environmental and cultural "trash." It is a multicomponent site with a depositional sequence that is not trampled, degraded, or compressed. It has furnished an unparalleled opportunity to document precisely when changes in subsistence patterns occurred and when the culture was first influenced by European presence. Corn cultivation, for example, is assumed to be the catalyst for stimulating sedentary living and social differentiation. Yet, in peninsular Florida, no concrete documentation for corn cultivation predating the Mission Period had been found in an archaeological context prior to the excavations at Hontoon Island. The evidence from Hontoon Island shows that, at this site, corn was not present until sometime after historic contact. There is no indication whatsoever that the Indians at Hontoon Island were dependent upon cultivated crops for survival. Most of their subsistence and maintenance items came from the natural environment very close to home. It should be noted that the terrestrial deposits at Hontoon Island did not contain the strata in which these observations were made. Thus, without the water-saturated components of the site we would not have been able to record these changes.

Archaeologists do not know what is missing from sites that they excavate but the superb preservation in the water-saturated deposits at Hontoon Island takes part of the guesswork out of interpretations. The extra pieces of the archaeological puzzle present at Hontoon Island provide a clearer picture of the way humans integrated cultural and natural resources to make life more comfortable and meaningful.

Millions of people throughout the world know about the stone sculptures from Easter Island and Greece. It is a sad commentary that the owl totem from Hontoon Island is not known by most people even in Florida. It is the largest wooden carving ever recovered from an archaeological context in the Western Hemisphere. Florida lacks suitable raw materials, like marble, to produce large stone statuary. The manufacture of the owl totem and other large carvings of wood required skills that equaled or surpassed those needed to sculpt stone. These effigies portray the values and ideals of the people who produced them.

All of the materials recovered at Hontoon Island from 1980 through 1988 are in the Florida Museum of Natural History or the Department of Anthropology at the University of Florida, Gainesville.

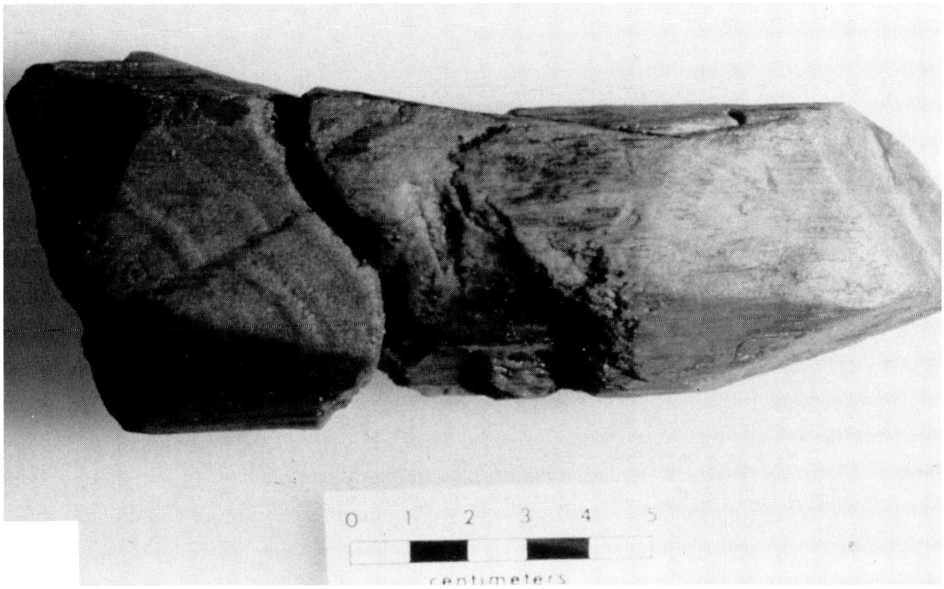

A

B

FIGURE 55 (A) "Wedge" made of persimmon photographed in 1982 when excavated, and (B) photographed in 1990. Although usually successful, in this case treatment with plyethylene glycol did not prevent cracks from developing (photos by the author).

References

Allerton, David, George M. Luer, and Robert S. Carr 1984 Ceremonial Tablets and Related Objects from Florida. *The Florida Anthropologist* 37(1):5–54.

Bartram, William 1928 *Travels of William Bartram.* Edited by Mark Van Doren. Dover Publications, Inc. New York. (Originally published in Philadelphia in 1791.)

Bullen, Ripley P. 1955 Carved Owl Totem, De Land, Florida. *The Florida Anthropologist* 8(3):61–73.

Cushing, Frank Hamilton 1897 Exploration of Ancient Key Dwellers' Remains on the Gulf Coast of Florida. *Proceedings of the American Philosophical Society* 25(153):329–448. Philadelphia.

Decker, Deena S. and Lee A. Newsom 1988 Numerical Analysis of Archaeological *Cucurbita pepo* Seeds from Hontoon Island, Florida. *Journal of Ethnobiology* 8(1):35–44.

Dunbar, Jim, Mark Russo, Tom Neilson, and Star Lockwood 1978 Huntoon Island State Park Cultural Resource Survey. Manuscript on file at the Division of Historical Resources, Tallahassee.

Gilliland, Marion Spjut 1975 *The Material Culture of Key Marco, Florida.* The University Presses of Florida. Gainesville.

Gold, Pleasant Daniel 1927 *History of Volusia County, Florida.* E.O. Painter Printing Co., Deland.

McGee, Raymond M. 1990 Personal communication. Mr. McGee is a graduate student in the Department of Anthropology, University of Florida, Gainesville.

McGee, Raymond M. and Bruce K. Nodine 1987 Sampling and Excavation Strategies at Hontoon Island (8-VO-202). *The Florida Anthropologist* 40(1):13–18.

Moore, Clarence B. 1893 Certain Sand Mounds of the St. John's River, Florida, Part I. *Journal of the Academy of Natural Sciences Philadelphia,* 2nd Series, Volume X.

1894 Certain Sand Mounds of the St. John's River, Florida, Part II. *Journal of the Academy of Natural Sciences Philadelphia,* 2nd Series, Volume X.

Newsom, Lee A. 1987 Analysis of Botanical Remains from Hontoon Island (8-VO-202), Florida: 1980–1985 Excavations. *The Florida Anthropologist* 40(1):47–84.

Nodine, Bruce K. 1987 Refuse Disposal and Midden Formation at Hontoon Island (8-VO-202), Florida. *The Florida Anthropologist* 40(1):19–26.

Purdy, Barbara A. 1987a Hontoon Island, Florida (8-VO-202): Artifacts. *The Florida Anthropologist* 40(1):27–39.

1987b Investigations at Hontoon Island (8-VO-202), an Archaeological Wetsite in Volusia County, Florida: An Overview and Chronology. *The Florida Anthropologist* 40(1):4–12.

1988a Piston Corers: Equipment, Technique, and Applications to Archaeology. *The Florida Anthropologist* 41(3):381–392.

1988b Archaeological Wet Sites: Untapped Archives of Prehistoric Documents. In *Wet Site Archaeology*, pp. 325–335. Edited by Barbara A. Purdy. Telford Press, Caldwell, New Jersey.

1988c American Indians after AD 1492: A Case Study in Forced Culture Change. *American Anthropologist* 90(3):640–655.

1990 Chronology of Cultivation in Peninsular Florida: Prehistoric or Historic. *Southeastern Archaeology* 9(Summer).

Purdy, Barbara A. and Lee A. Newsom 1985 Significance of Archaeological Wet Sites: A Florida Example. *National Geographic Research* I(Autumn):564–568.

Schwehm, Alice Gates 1983 The Carved Wood Effigies of Fort Center: A Glimpse of South Florida's Prehistoric Art. M.A. thesis, Department of Fine Arts, University of Florida, Gainesville.

Stirling, M.W. 1935 Smithsonian Archeological Projects Conducted Under the Federal Emergency Relief Administration, 1933–34. *Annual Report of the Board of Regents of the Smithsonian Institution* for the year ending June 30, 1934. Washington, D.C.

Willey, Gordon R. 1949 Excavations in Southeast Florida. *Yale University Publications in Anthropology* No. 42. Yale University Press, New Haven.

Wing, Elizabeth S. and Laurie McKean 1987 Preliminary Study of the Animal Remains Excavated from the Hontoon Island Site. *The Florida Anthropologist* 40(1):40–46.

Wyman, Jeffries 1875 Fresh Water Shell Mounds of the St. John's River, Florida. *Peabody Academy of Science,* Salem, Mass.

Little Salt Spring

Introduction

Underwater explorations using SCUBA (self-contained underwater breathing apparatus) are so common today, it is difficult to remember that SCUBA has been available for only a few decades before which the "briny deep" was essentially off limits to sport divers, treasure hunters, the armed forces, and scientists interested in marine biology, geology, and archaeology. The discovery of human remains in association with artifacts and now extinct fauna in several Florida springs has excited the press and the public, and has expanded tremendously our knowledge of late Pleistocene and early Holocene cultures and environments. Two of these springs, Little Salt Spring and Warm Mineral Springs, were discovered in the late 1950s by William R. Royal, a retired air force colonel and sport diver. Because Royal was not an archaeologist and because his findings were sensationalized by the press, his observations and their signifiance were minimized, ignored, ridiculed, or discredited by certain well-known archaeologists (Goggin 1960, 1962). Both of these springs were subsequently investigated by interdisciplinary teams composed of geologists, ecologists, biologists, and archaeologists. Time has demonstrated that many of Royal's first impressions were correct although his methods of recovery were not acceptable by archaeologists.

Little Salt Spring (8-50-18), located in Sarasota County in southwest Florida (Fig. 1), is a large flooded sinkhole 60 m deep. It is a basinlike depression 78 m across with a water surface 5 m above present mean sea level. The floor of the depression slopes at 25° from the land surface to 12 m in depth. The center of the depression is a circular opening 25 to 30 m across (Clausen et al. 1979:609; Fig. 56).

Semiarid conditions prevailed in Florida during periods of cooler world climate and were associated with lower than present sea and groundwater levels. Twelve thousand to 13,500 years ago a freshwater cenote existed at Little Salt Spring. It provided a source of water for humans and animals in an oasis-like setting that contrasted with its xerophytic surroundings that lacked abundant flowing surface water. Because of lowered sea level, the site was much farther from the Gulf Coast than it is today.

The earliest documented evidence for human habitation in Florida comes from a ledge 26 m below the surface at Little Salt Spring. This reentrant encircles the wall of the spring. On this ledge, the overturned, collapsed shell of an extinct species of giant land tortoise was found with a sharply pointed wooden stake between the carapace and plastron. Wood from the stake was dated by carbon-14 at 12,030 years ago and the carbonate fraction of a tortoise bone was dated at 13,450 years ago. Several of the long bones and portions of the carapace appear carbonized and numerous fragments of fire-hardened clay were found under and around the animal's remains. The tortoise was

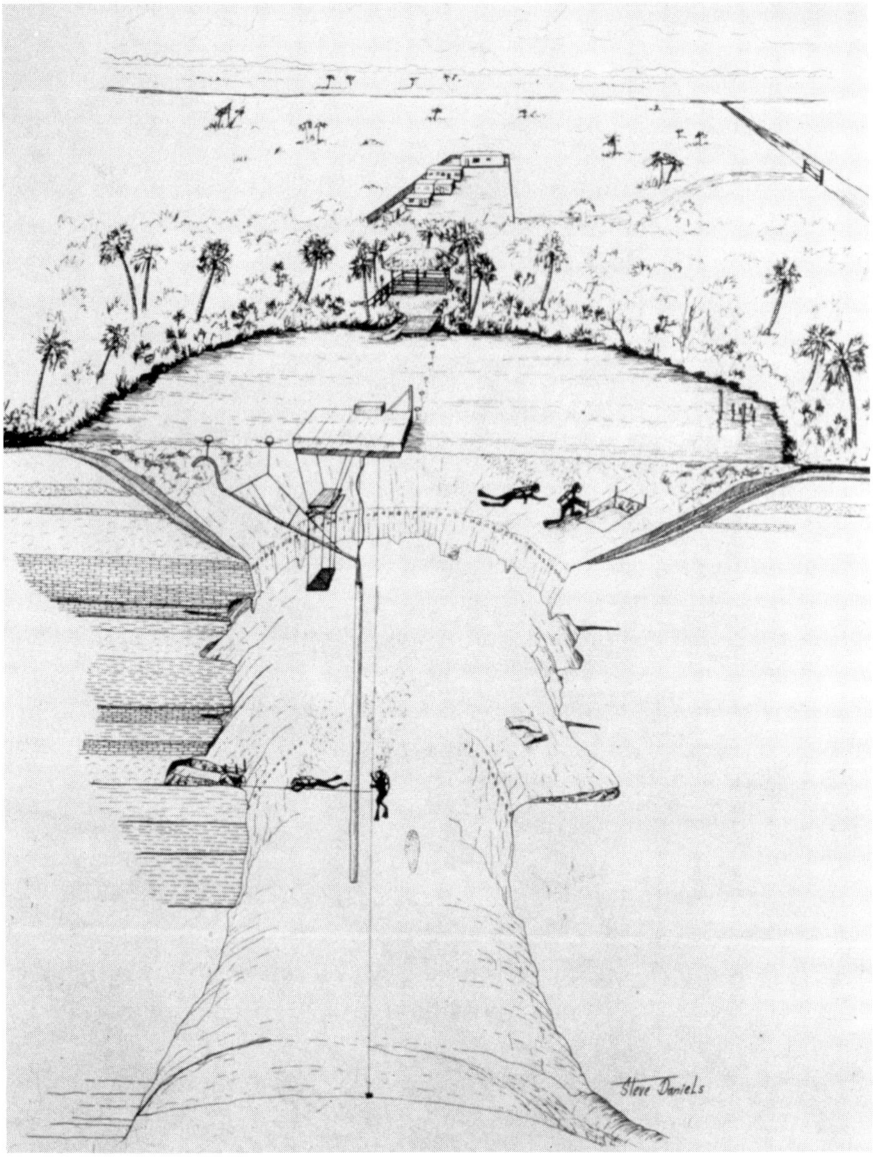

FIGURE 56 Little Salt Spring: note the –26-m ledge, the slope of the basin depression, and the sediment deposit at the bottom of the spring, all discussed in the text (drawing by Steve Daniels, courtesy of W.R. Royal).

apparently killed and cooked in an upside down position (Clausen et al. 1979:609–610). The water level in the sink was presumably at or below the –26 m (>85 feet) ledge when the tortoise was eaten there 12,000 years ago. The presence of both loose and articulated valves of the freshwater mussel *Uniomerus obesus*, which existed in isolated colonies on the reentrant when the water level was only slightly higher, indicates that the water in the cavern during the cenote phase of the feature's history was fresh (Clausen et al. 1979:610).

By 10,000 years ago the water had risen about 14 m and was only 11 to 12

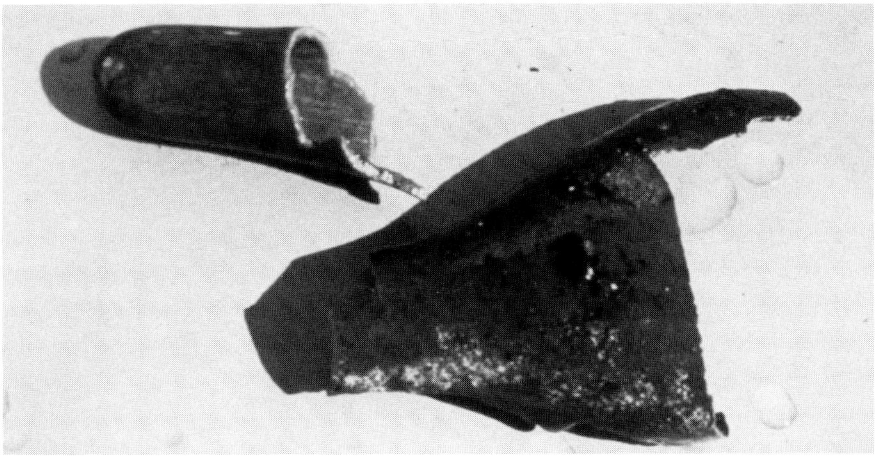

FIGURE 57 Bottle gourd recovered at Little Salt Spring (picture credit: General Development Corporation).

m below the present surface. People lived around the sinkhole, built fires on the slope of the basin, and consumed game of all modern fauna, such as white-tailed deer. Food refuse and artifacts (Fig. 57) have been found in association with drowned hearths below more recent sediments now lining the basin (Clausen et al. 1979:610). Numerous pointed wooden stakes fashioned from pine had been driven into the sediment near the bottom of the basin around the opening to the lower cavern. These stakes yielded an average radiocarbon date of 9572 years, and hickory nuts from the same sediments were dated at 9920 years. The function of the stakes is not known. They may be the basal portions of longer poles that were erected to mark the opening of the cavern. Artifacts recovered from this time period include a socketed antler projectile point with the dart shaft still preserved in its base, a carved oak mortar, and a nonreturning boomerang made of oak (Fig. 58). By 8500 years ago the water had risen rapidly and had invaded the drainage way leading into the basin from the northeast. More than 3000 years of early human utilization of the site came to an end at this time probably because of increased availability of fresh surface water in other locations (Clausen et al. 1979:611–612).

Approximately 2000 years later, the area around Little Salt Spring was reoccupied. Water levels began to drop 6800 years ago, reaching 8 m below present at 5500 years, and the spring became a focus of activity for Middle Archaic Period people. Their habitation area covered 10,000 to 20,000 square meters along the higher elevations paralleling the slough or drainage way. Their dead were interred in the moist soft peat of the slough and as the water level continued to drop the burial sites followed, progressing into the exposed pond basin. The bodies were wrapped with grass and buried in extended fashion on biers of wax myrtle or with leafy limbs placed between the arms and torso. Some of the burials are in a remarkable state of preservation because of the anaerobic environment resulting from hard water and resettling of fine

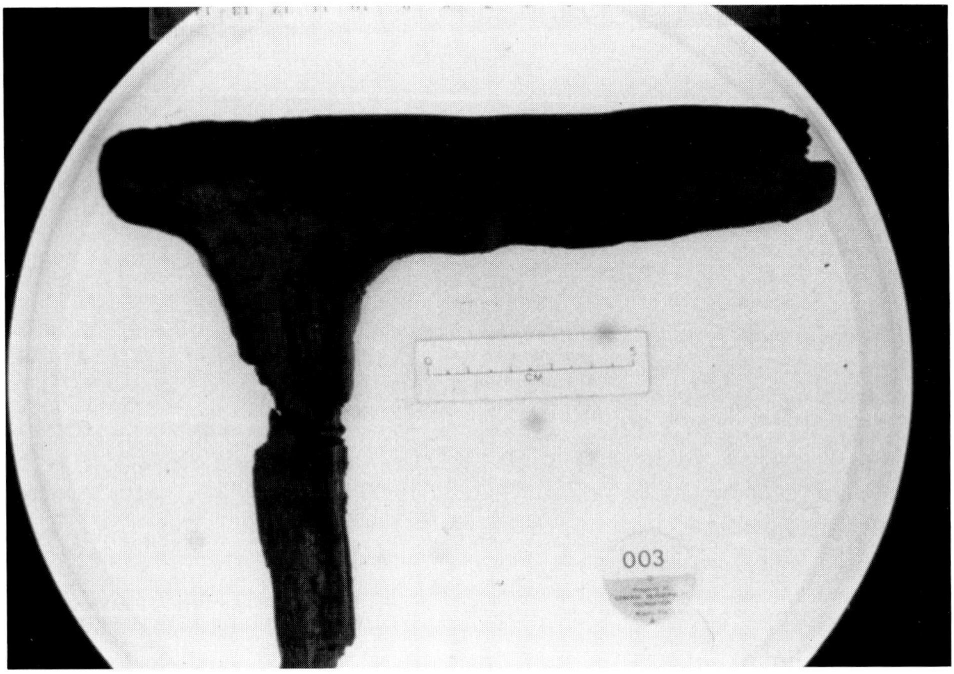

FIGURE 58 Broken nonreturning boomerang from Little Salt Spring probably dating to the late Paleoindian period (picture credit: General Development Corporation).

peat. Wood, bone, shell, and stone artifacts accompanied the dead as grave goods. Test explorations suggest that more than 1000 individuals were interred in the cemetery.

The most recent date received on a human bone from Little Salt Spring is 5220 years. The average water level in the feature apparently began to rise at this time and fresh water again became regionally abundant ending people's dependence on the spring. With the approach of sea level to its present position, heavily mineralized water began to issue from the feature and its usefulness ended. The water today has a salinity of 3.2 per mil, a temperature of 24.4°C, virtually no dissolved oxygen, and flows from the cavity at a rate of 42.8 liters per second (Clausen et al. 1979:609, 613).

Three periods of human habitation are represented at Little Salt Spring. The oldest, corresponding to a water table 26 m lower than present, occurred when several species of now extinct animals still lived in Florida. If the 12,000- to 13,500-year-old dates received on the wooden weapon and the giant land tortoise are correct, this period can best be classified as early Paleoindian or perhaps even pre-Clovis or pre-Projectile Point. The middle period, corresponding to a water table that was still 11–12 m below present, is clearly Late Paleoindian or Early Archaic since it dates to 9000–10,000 years ago, and contains only modern fauna and much of the flora that is recognized elsewhere for sites of this antiquity. It is not clear at present if there is a temporal hiatus separating the early and middle cultural periods or if there is continuous occupation from >12,000 to 9000 years ago with people having to adapt constantly to rapidly changing environmental conditions. The final occupation of the site, from 6800–5200 years ago, falls squarely in the Middle Preceramic Archaic Period. The flora, fauna, environment, artifacts, and hypothesized human behavior of each cultural period are presented in the following paragraphs.

Flora

13,500–12,000 BP—Brown (1981) and Brown and Cohen (1985) conducted palynologic and petrographic analyses of the peat deposits at Little Salt Spring but their study did not include a sample of the stratum containing the 13,500- to 12,000-year-old specimens. According to Watts and Hansen "the paleoecologist who seeks records older than 8500 BP in Florida must study sediments of deep lakes. At Lake Annie [located in southcentral Florida, approximately 60 miles northeast of Little Salt Spring], sedimentation began 27 meters below the present water surface (19 meters of water and 8 meters of lake mud) at about 13,000 BP, below which there is a hiatus . . ." Nearby Lake Tulane contains 18.5 m of sediment under 22.7 m of water. The site has no hiatus and contains a well-dated record of about 50,000 years of sedimentation (Watts and Hansen 1988:313). The pollen record for the period of interest, i.e., 13,500–12,000 years ago reveals "a period of abundant pine with a limited range of associated genera, either of trees or herbs. It appears to have been an unfavorable time for mesic trees, nor did it encourage the strong development of prairie species which followed in the early Holocene. The rather nondescript vegetation makes climatic inference difficult. . . . At the southern sites [i.e., Lakes Annie and Tulane] . . . a simple dominance by pines in vegetation seems assured" (Watts and Hansen 1988:315). At Lake Annie macroscopic remains of sand pine (*Pinus clausa*) are present at this time and Lake Tulane sediments reveal a low incidence of fire during the period of pine dominance around 13,000 BP. Sedimentation is organic indicating that, although surface water was scarce, sedimentary processes in existing lakes were similar to those which followed later in the Holocene. This suggests that there was complete vegetation cover on the upland. Watts and Hansen conclude that from 14,000–12,000 BP southern Florida, as revealed by deep cores taken from Lakes Annie and Tulane, was dominated by a pine forest with some herbs and mesic trees. The water level was about 28 m lower than present, the climate was cooler than today with higher precipitation, and fire frequency was low (Watts and Hansen 1988:315, 317).

It is not known, of course, if the vegetation picture presented above can be applied to Little Salt Spring for the early period. Gifford (1989, personal communication) was funded by the National Geographic Society to conduct a paleoenvironmental study of the sediment deposit that lies 60 m below surface at Little Salt Spring. By vibracoring the sediment to 12 m, he hoped to obtain samples for radiocarbon dating and microfossil analysis that would provide information about climate and vegetation that has been missing for Little Salt

FIGURE 59 An historic moment: the recovery of a 40-foot (12-m) core from the bottom sediments at Little Salt Spring (picture courtesy of John A. Gifford). Note the pristine beauty of Little Salt Spring.

Spring. The project was carried out in March 1990 and was highly successful (Fig. 59). Gifford obtained a total of eight cores, of which six were kept. Four of the cores are 10 m or longer. A large piece of oak (identified by L.A. Newsom) was located at 11 m below the sediment-water interface (1 m above the bottom of a 12-m core) (Fig. 60). A radiocarbon date received on the wood was 9660 ± 160 BP (calibrated to dendrochronology and normalized to stable isotope ^{13}C, Beta 36591). Rather than providing a record extending back to the last full glacial period about 18,000 BP as originally anticipated, this date suggests that deposition of organic materials over the quartz sand core of the spring took place very rapidly beginning during the late Paleoindian Period. When the earliest inhabitants were at the spring around 12,000 years ago, the water was clear and the quartz sand bottom was probably visible to them. At first glance the sediment cores appear to be composed of fairly homogenous organic material but they also contain interesting laminations. Microfossil analyses of the core are underway and additional radiocarbon dates will be available soon. Another piece of wood found at approximately the same depth as the oak sample is probably an artifact since it seems to have been intentionally modified. This project verified that the water depth in the spring is 60 m, but it demonstrated that the deposit was flat, not cone shaped, as previously reported (Gifford, personal communication, May 1990).

The wood of the stake found in association with the extinct species of giant land tortoise (*Geochelone crassiscutata*) (Fig. 61) was identified as red mul-

A

B

FIGURE 60 (A) Location of oak sample taken ca. 37 feet (11 m) below the sediment-water interface at Little Salt Spring. Specimen was radiocarbon dated at 9660 ± 160 BP (Beta 36591; picture courtesy of John A. Gifford). (B) Portions of three 10-foot sections of a 40-foot (12-m) core from Little Salt Spring showing organic sediments and laminations.

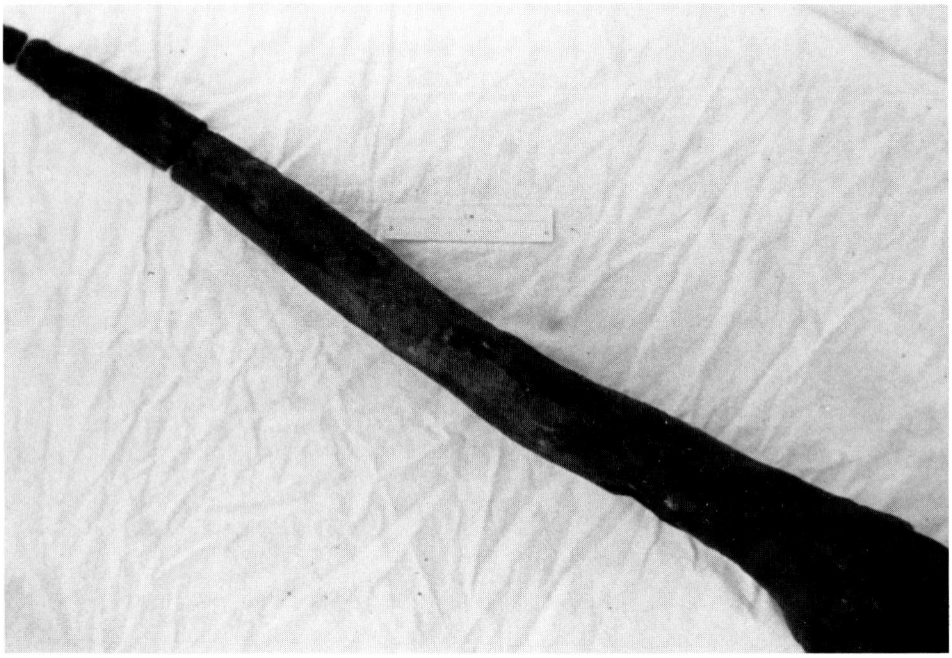

FIGURE 61 Wooden stake excavated from the —26-m ledge at Little Salt Spring in association with *Geochelone crassiscutata,* an extinct giant land tortoise.

berry by the Wisconsin Wood Products Laboratory, Madison, Wisconsin but there seems to be some question about the identification (Clausen 1983, personal communication). No other plant species has been determined for the –26-m ledge at Little Salt Spring.

10,000–9000 BP—Samples of freshwater gastropods (principally *Helisoma* sp.) were taken 12 m below surface from the sloping basinlike depression ringing the lower cavern at Little Salt Spring. They came from the same sediments as hickory nuts that were 9900 years old. The organically rich freshwater calcitic mud trapped within the gastropods was analyzed for its pollen content (Clausen et al. 1979; Brown 1981; Brown and Cohen 1985). Pollen was well preserved and moderately abundant but represented a relatively small number of plant types. The dominant arboreal plant types in the area consisted of *Myrica* (wax myrtle), *Quercus* (oak), *Pinus* (pine), and *Carya* (hickory). The dominant herbaceous plants were grasses, sedges, ferns, composites, and much smaller percentages of cattails, and Chenopodiaceaea and Amaranthus (cheno-ams). The lack or low percentage of local plant types that are typical of pond conditions suggests that the surrounding area (and especially the inflowing slough) was still relatively dry (Clausen et al. 1979:610–611; Brown and Cohen 1985:23–24). The pollen picture is substantiated in part by macroscopic samples of at least two species of wood from which artifacts were fashioned. A nonreturning boomerang (Fig. 58) is made of oak and the stakes lining the opening of the lower cavern are made of pine.

The lowermost portion of a core taken from the sawgrass-vegetated portion of the slough consisted of an organic layer (GDF142/Zone 1) that was progressively less sandy toward the top of the section. Its pollen—low frequency of *Salix* (willow), absence of *Nyssa* (black gum) and emergent aquatics such as *Nymphaea* (waterlily) and *Sagittaria* (arrowhead), along with the high frequency of *Pinus* and *Quercus* pollen—indicates that conditions were slightly drier than those that followed (Brown and Cohen 1985:27–29). This diagram supports the findings of the pollen analysis conducted on the mud-filled gastropods recovered at ca. 12 m below surface that date to approximately the same time.

The greatest influx of charcoal to the sediments of Lake Tulane occurred between 11,000 and 7000 BP and provide a record of upland fires. These were probably natural fires caused by lightening strikes that occur most readily when there is plenty of dry fuel on the ground (Watts and Hansen 1988:311).

6800–5200 BP—A drying trend apparently began around 7600 years ago and by 6800 years ago the dropping local groundwater level and concomitant reduced availability of surface water again made Little Salt Spring attractive to humans, this time to Middle Archaic Period people (Clausen et al. 1979:612). Pollen from a core sample corresponding to this zone is characterized by decreasing *Salix* and increasing *Myrica*, *Quercus*, and ferns (Polypodiacea and *Achrostichum)*.

Utilization of botanical species by the Middle Archaic Period people included grass which was used to wrap bodies of the deceased, wax myrtle (*Myrica cerifera*) used for biers, and oak from which was fashioned a pointed digging stick. Another wooden artifact (species unidentified) from this period is a tablet or plaque bearing what may be the partial profile of a long-necked or long-billed bird (Clausen et al. 1979:612; Fig. 62). This specimen no longer exists.

From 5000 years ago to the present, the area immediately around the spring has supported a bayhead hammock community. The floral picture compiled by Brown and Cohen (1985:29–31) strongly supports the conclusion of Clausen et al. (1979) that increased Indian occupation of the site coincided with periods during which climatic conditions in this area were the driest. It was during these times that the spring became a primary source of fresh water.

Fauna

13,500–12,000 BP—The faunal remains are fairly well known from this time period in contrast to the lack of information about botanical remains. In addition to the giant land tortoise (*Geochelone crassiscutata)*, other animals found on the –26 m ledge at Little Salt Spring include three smaller extinct tortoises of the same species, the extinct large box turtle (*Terrapene carolina putnami)*, the extinct ground sloth (*Melgalonyx* sp.), two extant species of freshwater turtle (*Pseudemys floridana* and *P. nelsoni)*, the land tortoise that presently inhabits Florida (*Gopherus polyphemus)*, the modern diamondback rattlesnake (*Crotalus adamanteus)*, the rabbit (*Sylvilagus* sp.), the wood stork

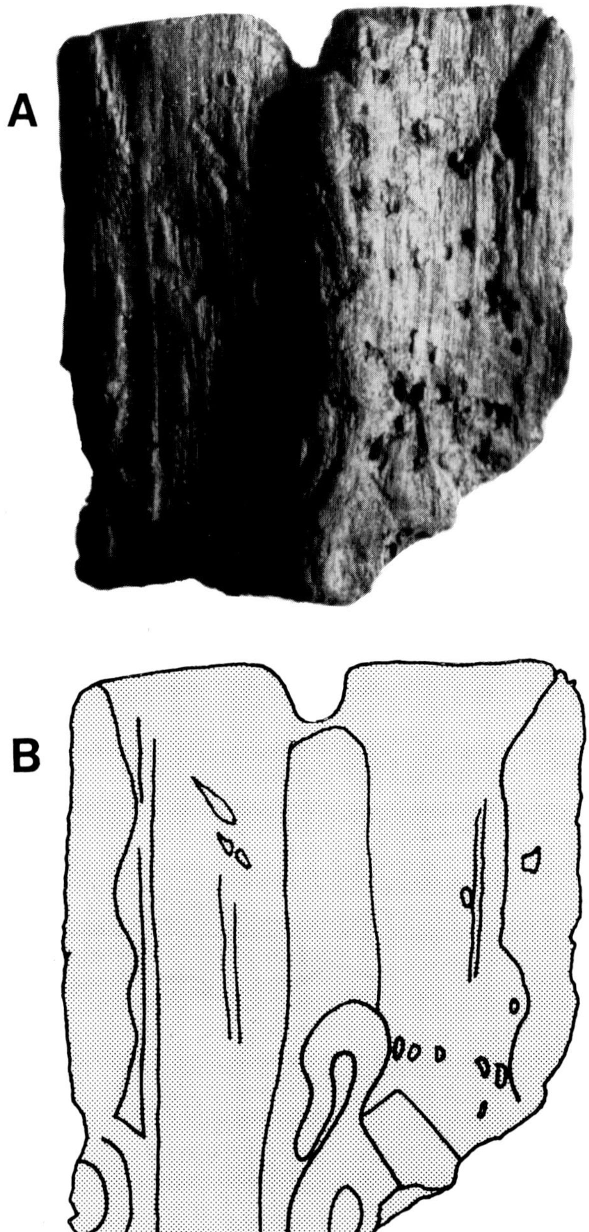

FIGURE 62 Flat wooden plaque with bird carving found with Archaic period burials at Little Salt Spring (picture courtesy of Division of Historical Resources, Tallahassee).

(cf. *Mycteria*), a large mouth bass (*Micropterus salmoides*), and portions of an immature mastodon (*Mammut americanum*). Altogether, 16 individual animals were recovered from the –26 m ledge. The largest number of individuals were turtles and the number of coldblooded animals exceeded that of warmblooded ones (Holman and Clausen 1984:152).

Holman and Clausen (1984) describe the above specimens thoroughly. They furnish measurements of the elements recovered, discuss the possible utilization of the animals by humans, and provide an assessment of the type of habitats required by each species. Almost all of the animals recovered could have been food animals—e.g., the bass, turtles, rattlesnake, rabbit, and ground sloth—but positive evidence for human involvement exists only for the giant land tortoise found in association with a wooden stake.

> The shell was upside down and collapsed. A wooden stake lay between the carapacial and plastral bones. The shell bones were given individual numbers prior to a photo-recording process that preserved their spatial relationships to the artifact. Apparently the stake entered the axilla at an angle of about 50° to the midline and must have penetrated the skin between the right foreleg and the neck. The artifact appears to have been thrust about 28 cm into the body cavity to a position about 10 cm beyond the midline on the opposite (left) side. The stake entered under the second and third right peripheral bones and passed under the first right costal and across the midline under the second neural. The tip of the stake came to rest under the median edge of the third left peripheral bone. The stake must have penetrated the lungs and could have damaged the heart as well. . . . The possibility remains that the animal was killed by some other means, and that the stake was used merely to manipulate the animal over fire.
>
> . . . Both the left femur and the right humerus have the trochanteric ends roughly broken, suggesting they may have been chopped away to free them from their points of articulation with the girdles. These ends are blackened. . . . The distal end of the ulna was also crudely broken, possibly when the front foot was removed. The distal end of the ulna is also black. It is possible that heat at the points of articulation of the femur, humerus, and ulna produced the black coloration. This would have facilitated dismemberment of the tortoise by softening the cartilaginous and ligamental structure of the joints . . .(Holman and Clausen 1984:149–150).

The outer surface of the shell bones had also been altered by fire. The carapace of the specimen was estimated to be 94 cm long and 75 cm wide (Holman and Clausen 1984:150).

The vertebrate fauna from the –26 m ledge at Little Salt Spring indicate two major habitats:

1. A subterranean pond with mesophytic vegetation on the shallow basin made possible by the presence of paleo seep springs that emitted water near the periphery of the basin. The moist basin would have formed a small oasis in a semiarid surrounding. Animals of habitat 1 include the largemouth bass, Florida cooter (*P. floridana*), Florida red-bellied turtle (*P. nelsoni*), extinct box turtle, wood stork, extinct ground sloth, and mastodon.

2. The xerophytic upland. Animals of habitat 2 include the extinct giant land tortoise, the Florida gopher turtle, and the Florida diamondback rattler.

Clausen et al. (1979) and Holman and Clausen (1984) believe that the sinkhole acted as a trap for animals 12,000 years ago when the water was at about the level of the –26 m ledge. Animals stranded there became easy prey for humans but unanswered questions remain as to the way people used the sink as a water and food resource and how they entered and exited the cavern which would have been both difficult and dangerous.

10,000–9000 BP—The only published reference to animal remains from this time period is white-tailed deer (*Odocoileus virginianus*) and other vertebrate food refuse (Clausen et al. 1979:610). Clark (1969) mentions that alligator bones were recovered from the basin slope but no provenience is provided.

6800–5200 BP—The only published reference to animal remains is undrilled shark teeth, bone, and shell. The most common artifacts are tapered points with roughly beveled bases, averaging 10 cm in length and made from the long bones of deer (Clausen et al. 1979:612).

Human Remains

Skeletal material—No human remains have been reported at Little Salt Spring in association with the sediment deposit 60 m below surface or with the vertebrate fauna and wooden stake on the –26 m ledge (13,500–12,000 BP) although Clark (1969:167–169) relates that on March 18, 1959 Royal found a human skeleton there with a huge rock on top of a leg bone as if the person had been pinned under the rock. He had left the skeleton undisturbed so that pictures could be made of it in place but later, when he was attempting to show it to Clark and the rest of the diving party, they were foiled in their efforts to reach that depth because of trouble with their aqualung equipment.

None of the human bones recovered from the basin depression date as old as the nonreturning boomerang (10,000–9000 BP). Some of the bones seem to be more heavily mineralized than others (Clark 1969:167; Goggin 1962:78) and may indicate greater antiquity, but additional chronometric dates are needed.

Skeletal material dating from approximately 7000–5200 BP was removed from the basin depression and the slough area, and large quantities of human bone and accompanying grave goods still remain in the 6000 m^2 cemetery (Clausen et al. 1979:612). Some of the bones have been analyzed and preliminary results were reported by Charles F. Merbs and Carl J. Clausen in a paper entitled "The People of Little Salt Spring" at the 1981 annual meeting of the Society for American Archaeology in San Diego, California.

Most of the human skeletal remains from the deeper portions of the basin were recovered by William R. Royal in 1959 (Fig. 63) but their provenience is not well documented. Well-controlled excavations of three 2-m squares located at varying depths of the basin from 7–12 m provide a better picture of the nature of the deposits containing the human skeletal remains. Minimum counts of individuals found in the three squares were 4, 3, and 5 from the 7–12 meter

A

FIGURE 63 (A) William R. Royal (on left) in the late 1950s with human skeletal remains at Little Salt Spring (picture courtesy of W.R. Royal). In 1990, the area surrounding the spring remains the same as seen in the background. (B) William R. Royal and Eugenie Clark examine human skeletal remains from Little Salt Spring in the late 1950s (picture courtesy of W.R. Royal).

depths, respectively. All of the squares were considerably disturbed and were missing skeletal elements. In general, an abundance of larger bones were found, such as skull, arm, pelvic, and leg parts, but few small bones, such as shoulder, back, hand, and foot parts. Although human skeletal material appears to be present in large quantity in the lower basin, disturbance and missing parts greatly decrease its value for osteological study. One nearly complete skeleton of a 7–8 year-old child was found in the shallow portion of the basin near the edge of the pond. It was removed from the bottom sediments without proper controls, thus no information is available about its orientation or associated burial goods. There is a possibility, however, that additional complete and undisturbed burials exist in the shallow portion of the basin.

At least since 1973 when several skeletons were found during the construc-

FIGURE 63B

tion of Price Blvd. (now McCarthy), it has been known that the slough area contains information about burials dating from 7000–5200 years ago. The skeleton of an infant was encountered just south of the road and human bones were observed along the main transect line about 100 m south of the road near the spring itself. In 1977 a test pit was dug just to the north of Price Blvd. using a backhoe. The backhoe cut through the skeleton of an adult male, but several parts were recovered. The pelvis and femora of the adult male had not been disturbed and these parts were carefully exposed. The skeleton was found to terminate at the knees; the right patella was present but not the left. Near the pelvis were foot bones apparently from the same individual, which had been carefully wrapped in grass. In excavating the male skeleton, the skull of a 30- to 40-year-old female was encountered. This individual was excavated down to her pelvis during the 1977 excavations. She had been extended on her back, her arms at her sides, head turned slightly to the right, with the long axis of her body oriented to the west. The body appeared to have been placed in a shallow, prepared pit. A chert knife lay next to her head and a wooden digging stick had been placed over her head. Next to the skeleton were post molds, perhaps from effigy poles or from some kind of protective housing erected over the burial. The skull contained peat-stained organic material with convolutions resembling a human brain (see below). In 1980, Merbs and Clausen finished excavating this burial working from the pelvis to the feet. In burning the dry vegetation away from the burial area to facilitate access and minimize skirmishes with hostile slough inhabitants, they noted two unusual burial features. One was a network of sticks or narrow poles lying above the skeleton. Longer sticks followed the longitudinal axis of the skeleton while shorter ones ran across the body. These had been placed in a series of layers, each oriented in a particular direction. The authors believe the sticks represent a collapsed structure that had been placed over the body when it was interred. These sticks were later identified as grape vines. The second feature was a plant fiber shroud which completely enclosed the body. The fibers were not woven but may have been hammered into a fabric much like tapa cloth. The shroud appears to have fit the body snugly.

It was difficult to disengage the shroud, vine network, and wooden digging stick from the surrounding peat matrix since they were nearly of the same consistency. The procedure was to tease the peat away using a bamboo tool. The surface of the excavation had to be well drained to allow them to see the material being excavated but moist enough to prevent shrinkage and damage.

The human skeletal remains recovered from the Little Salt Spring site represent a minimum of 35 individuals. The bones were inventoried for evidence of pathology but the value of the data is reduced by the inability to identify age and sex because so few of the skeletons are complete. One condition present in the people of Little Salt Spring is cribra orbitalia which was found in the child from the shallow basin area and in a young adult from the lower basin. Cribra orbitalia is attributed to anemia which could be inherited (e.g., thalassemia), caused by a vitamin deficiency (e.g., scurvy), a mineral deficiency (e.g., iron), or to excessive blood loss caused by intestinal parasites. Growth arrest lines have been noted in the medullary cavities of many long bones indicating seasonal variations in dietary stress. Enamel hypoplasia, also the result of growth insult, appear to be quite common in the teeth from Little Salt Spring. The teeth in general show extensive, smooth, flat wear with occasional apical abcessing and premortem loss. In some instances, the enamel seems unusually thick. Only one example of caries was noted. Evidence of periosteal inflammation was noted on a number of the adult long bones, especially femora, and on several of the long bones from the child found in the shallow basin. There was only one healed fracture observed. Postmortem mineral deposition is often difficult to distinguish from true pathology. An adult man at Little Salt Spring was about 5 feet, 3 inches tall and probably weighed one hundred pounds. Overall, the people of Little Salt Spring were healthy with strong bones.

Brains—As at several other archaeological wet sites in Florida, brains have been preserved in some human crania at Little Salt Spring (see also sections in this chapter on Republic Groves, Warm Mineral Springs, and Windover). Evidently, the child burial found near the edge of the pond basin contained brain material but it was lost because no one suspected that it was there and it simply poured out of the cranium before it was noticed. Samples of the brain-like material from inside the skull of the >6000-year-old shrouded female recovered in the slough in 1980 were submitted for analysis to Michael Zimmerman, physical anthropologist and pathologist at the University of Michigan, and Wilson S. Hendry, forensic pathologist in Sarasota, Florida. Zimmerman's microscopic analysis using various stains showed intertwining fibrillary material consistent with neuronal processes. Hendry noted that the surface of the intracranial mass shows vague convolutions similar to the surface of the cerebrum. He found structures which appear to be cell bodies and their related processes and also structures which appear to be degenerated capillaries. Hendry saw a vague demarcation between cortex and white matter. The conclusion is that the material found in the slough skull is degenerated brain material, and not simply foreign matter that filtered into the skull (Merbs and Clausen 1981).

Another brain came from a relatively intact burial recovered during exploratory excavations in 1986. The cranium of this individual contained macroscopically well-preserved neural tissue. Immediately upon its discovery the cranium was placed in spring water and refrigerated. Samples of neural tissue were removed the same day through the foramen magnum and frozen in liquid nitrogen. An accelerator mass spectroscopy radiocarbon date of 6860 ± 110 years BP (Beta-17208; corrected by ^{13}C for total isotope effect) was obtained from a small sample of the neural tissue. This date corresponds almost exactly to the age of a pointed oak digging stick, 6830 years BP, found in association with burials from the slough in the 1970s (Clausen et al. 1979:612). An oak digging stick was found also with the burial recovered in 1986.

Pääbo et al. (1988) describe the methods and materials used to obtain the following information about the brain tissue recovered in 1986.

The majority of the extracted DNA was degraded to a relatively small molecular size (50–200 bp [base pairs]) while a fraction of the material was of a larger molecular size (2–5 kb [kilo bases]). Such a reduction in size is typical of ancient DNA.

> Aliquots of 2–4% of the extracts were used for enzymatic amplification of various parts of the mitochondrial genome . . . Initially, various sets of primers were used to determine the maximum size of the DNA fragments which could be amplified. It was found that pieces up to 110 bp in length could be successfully and reproducibly amplified whereas no amplification of over 200 bp was ever successful. This was in sharp contrast to contemporary human DNA extracted from brain tissue by the same procedure from which pieces of up to 2 kb could be amplified. . . . [I]t was estimated that the amplification mixtures contained between 100 and 300 fg of human DNA which could be replicated . . . . Based on the estimation that there exists 1000 copies of mtDNA in one mammalian cell, this would indicate that the amplifications started from approximately fifty copies of the relevant mitochondrial sequences. Two anthropologically informative regions of the mitochondrial genome were selected for further study.
>
> Region V . . ., a small intergenic segment of the mitochondrial genome, carries two direct repeats of 9 bp, one of which has been shown to be deleted in some individuals of Asian origin . . . as well as in members of populations thought to originate from Asia such as North American Indians . . . and New Guineans.
>
> An additional informative site is present in the sequence obtained from region V. At position 8251 the sequence from the brain contains an A whereas that position in the published sequence of the human mitochondrial genome . . . carries a G. . . . Among the 241 individuals examined in a worldwide survey of human mtDNA, this restriction site variant occurs in only three individuals originating from different geographical regions. . . . this is a rare sequence. It might thus be of significance that the HaeIII loss in conjunction with the AvaII gain occur in five mtDNA types forming a clade in a phylogenetic tree of 62 mtDNA types in the Japanese population and that in all those cases the same nucleotide change is responsible for the change in restriction patterns (Pääbo et al. 1988).

Another anthropologically informative site occurs at position 13259 in the Little Salt Spring brain; 13259 is absent in 40% of American Indians living in the southwestern United States but is absent in only one out of 55 Orientals analyzed (Pääbo et al. 1988:9782–9783).

Based on the above findings, the authors conclude that the Middle Archaic Period Indian excavated at Little Salt Spring does not belong to the mitochondrial lineages previously described for the New World and thus the prehistoric colonization of the New World involved a minimum of three distinct maternal lineages.

In a discussion of their findings, Pääbo et al. (1988) mention how the study of ancient DNA poses numerous technical problems. The DNA sequences exist in low amounts in the presence of an excess of other DNA, often of fungal or bacterial origin. The vast majority of the DNA molecules extracted from archaeological remains have suffered physical damage that precludes replication. The technique used to study the brain material from Little Salt Spring can be expected to overcome these problems since the technique makes possible the specific *in vitro* amplification of extremely small numbers of a relevant DNA sequence. "However, DNA extracts from ancient tissue remains often contain an unidentified component which inhibits the activity of *Thermus aquaticus* polymerase. Here, we show that this inhibition can be overcome by the addition of bovine serum albumin and increased amounts of enyzme" (Pääbo et al. 1988:9783–9784).

Authenticity of ancient amplified DNA is another problem discussed by Pääbo et al. The most likely source of contamination is DNA from other humans and may stem from a variety of sources such as handling during excavation or removal of samples. Some contamination can be eliminated by careful extraction and laboratory methods. An additional criterion of authenticity is made possible by the finding that for ancient DNA there is a strong inverse dependence of the amplification efficiency on the size of the segment to be amplified. The small molecular size of amplifiable template DNA of ancient origin is a feature that is a consistent characteristic of archaeological retrieved DNA (Pääbo et al. 1988:9783).

Modern science, ancient brains, combining to open new windows to the past!

Discussion

Investigations at Little Salt Spring have helped confirm the existence of a water table many meters lower than present during the Late Pleistocene. Studies of flora, fauna, sediments, and other environmental variables have demonstrated that a freshwater cenote attracted humans to the area during periods when surface water was scarce in Florida. The site had no allure when the water became too salty and when lake bottoms filled and rivers flowed elsewhere providing easier access to water than the deep sink. As with other archaeological wet sites, the most significant and spectacular information has been derived from the organic remains at Little Salt Spring. The upland midden adjacent to the slough, dating to the Middle Preceramic Archaic Period, has yielded a few stone tools, vertebrate refuse, and little more.

The site is very complex. Human utilization alone represents several millenia of occupation alternating with periods of abandonment: (1) the >12,000-year-old remains associated with extinct animals on the –26-m ledge, (2) the 10,000- to 9000-year-old evidence from the basin depression 12 m below surface, and (3) the 7000- to 5200-year-old cemetery in the drainageway. There is also a ledge 18 m below surface that may contain pertinent cultural and environmental information dating to 10,000 years ago. Each of these is a site in itself and each could be a focal point for investigations. Excavation strategies will differ depending upon the location studied. The slough area, for example, necessitates a knowledge of how to draw down the local water table using an efficient pumping system more than it requires diving experience. Recovery of organic materials from any of the components demands a knowledge of preservation, particularly of botanical remains.

Little Salt Spring has been explored since the late 1950s but information about several of its components is not complete. Major reasons for this lack are insufficient funds and the difficulty of working underwater for the time periods required to excavate systematically at great depths; air supplies are used up quickly and nitrogen narcosis can be a problem. Dr. John A. Gifford's recent vibracoring of the sediment deposit that lies 60 m below surface at Little Salt

Spring is yielding significant results and may be the only practical way to recover stratigraphically unmixed deposits from this level. Clark's (1969:149–169) vivid description of her reactions while diving at Little Salt Spring serves to demonstrate some of the problems that still exist today even with improved diving methods and equipment.

> Unlike most Florida springs, Warm Mineral Springs and Little Salt Spring . . . are not clear. It was in these two springs that I made my weirdest and most dangerous dives—into waters where I couldn't read the depth gauge on my wrist with a flashlight directly over it, and had to feel around blindly for the human remains we sought of people whose bodies got there many years before we did; and to depths where nitrogen narcosis hits you unexpectedly. . . the simple wet darkness can be terrifying.
>
> . . . Little Salt Spring is *cold*. I wore a shirt, tights under blue jeans, and woolen socks and sneakers under my flippers to keep my feet warm.
>
> . . . Suddenly I realized that down was up and I was looking at bubbles rising toward the surface. When I thought I was crawling up the vertical wall, I was actually heading down into the central sinkhole.
>
> . . . at 80 feet, my ears cleared easily and I hardly felt like I was making a particularly deep dive. . . . We checked [the] . . . depth gauge again and read 130 feet. We dived deeper . . . we continued down the rope . . . I began to feel that I was breathing exceptionally fresh air. . . . deeper still, I suspected someone had opened the window in our stuffy dark room . . . The air seemed to be loaded with oxygen and I felt very light.
>
> . . . I was sinking very slowly now into what felt like velvet ooze. I saw a hand sticking out of the mud [from the sediment deposit at the bottom of the spring]. Oh, it's Bill Royal's hand. . . . And Bill is dead. It was not a disturbing thought. . . .
>
> I'm having a baby and I want to go to sleep. Suddenly I realized something was wrong with my thinking. I couldn't be having a baby. I was diving . . .!

Jacques Yves-Cousteau, a French naval officer who developed the first efficient device to explore the underwater world, called nitrogen narcosis "rapture of the deep."

In 1959, John M. Goggin was probably the only professional Florida archaeologist who was also a diver when Little Salt Spring was initially investigated. Practically no precedents had been established by anyone at that time for excavating archaeological sites underwater. Nevertheless, professionals in other fields, such as Harold K. Brooks (University of Florida geologist), Eugenie Clark (marine biologist), William M. Stephens (an underwater photographer), a dentist, and a veterinarian, plus retired Lieutenant Colonel William R. Royal and other skin divers were not considered qualified to draw conclusions about the human remains and cultural materials recovered from the spring.

The story of the estrangement that evolved between the archaeologists and nonarchaeologists (translate amateur or pothunter) is part of the history of the site and I believe has been instrumental in creating many of the funding and other problems that have existed at Little Salt Spring for 30 years. The best accounts of the antagonism appear in Goggin (1962) and Clark (1969). The tale is recounted in the section about Warm Mineral Springs.

Deplorable conditions have existed at the Little Salt Spring site since the more or less continuous explorations and funding of the 1970s ceased. Some of the artifacts, flora, and fauna have been poorly curated; others are missing altogether. This is poor publicity for archaeologists. It carries a message to the people of Florida that archaeologists are irresponsible because they will not work without money to support them. This demand is not made of any other profession.

It is instructive once again to consider Coles' (1988) criteria for establishing priorities relating to funding, excavation, and protection: (1) Little Salt Spring is unique and of international significance, (2) the site is virtually intact, and (3) it contains a full range of environmental and artifactual materials; it may even contain works of art of exhibit quality. In addition, there are qualified individuals who can extract, analyze, and preserve the material. Little Salt Spring deserves major funding and support.

Little Salt Spring was formerly owned by the General Development Corporation which also funded exploratory work at the site for several years during the 1970s. The site is now owned by the University of Miami, Coral Gables, Florida. The artifacts, human skeletal remains, specialists' reports, and pictures are not all centrally located at the present time.

References

Brown, Janice G. 1981 Palynologic and Petrographic Analyses of Bayhead Hammock and Marsh Peats at Little Salt Spring Archeological Site (8-SO-18), Florida. Master of Science thesis, Department of Geology, University of South Carolina.

Brown, Janice G. and Arthur D. Cohen 1985 Palynologic and Petrographic Analyses of Peat Deposits. *National Geographic Research* 1(1):21–31.

Clark, Eugenie 1969 *The Lady and the Sharks*. Harper and Row, New York.

Clausen, Carl J., H.K. Brooks, and A.B. Wesolowsky 1975a Florida Spring Confirmed as 10,000 Year Old Early Man Site. *The Florida Anthropologist* 28(3, part 2):1–38.

1975b The Early Man Site at Warm Mineral Springs, Florida. *Journal of Field Archaeology* 2(3):191–213.

Clausen, C.J., A.D. Cohen, Cesare Emiliani, J.A. Holman, and J.J. Stipp 1979 Little Salt Spring, Florida: A Unique Underwater Site. *Science* 203(4381):609–614.

Coles, John M. 1988 A Wetland Perspective. In *Wet Site Archaeology*, pp. 1–14. Edited by Barbara A. Purdy. The Telford Press, Inc. West Caldwell, New Jersey.

Goggin, John M. 1960 Underwater Archeology: Its Nature and Limitations. *American Antiquity* 25(3):34;8–354.

1962 Recent Developments in Underwater Archaeology. *Southeastern Archaeological Conference Newsletter* 8:

Holman, Alan J. and Carl J. Clausen 1984 Fossil Vertebrates Associated with Paleo-Indian Artifact at Little Salt Spring, Florida. *Journal of Vertebrate Paleontology* 4(1):146–154.

Merbs, Charles F. and Carl J. Clausen 1981 The People of Little Salt Spring. Paper presented at the 46th Annual Meeting of the Society for American Archaeology, San Diego.

Pääbo, Svante, John A. Gifford, and Allan C. Wilson 1988 Mitochondrial DNA Sequences from a 7000-Year-Old Brain. *Nucleic Acids Research* 16(20):9775–9787.

Royal, W.R. and R.F. Burgess 1978 *The Man Who Rode Sharks*. Dodd, Mead, New York.

Watts, William A. and Barbara C.S. Hansen 1988 Environments of Florida in the Late Wisconsin and Holocene. In *Wet Site Archaeology*, pp. 307–323. Edited by Barbara A. Purdy. The Telford Press, Caldwell, New Jersey.

Page-Ladson

Introduction

In Florida, the best potential at present for discovering evidence of Paleoindians and their material culture in association with now extinct Pleistocene megafauna, called the PIM Zone by Webb et al. (1990), comes from underwater settings. Underwater excavations are difficult but they can be quite rewarding when undisturbed sediments are encountered. These sediments furnish a sequential story of interactions among flora, fauna, and humans that is more than 10,000 years old; the abundant organic remains are useful for radiocarbon dating and environmental interpretations.

Such a setting is Page-Ladson (8-Je-591) where investigations have been underway since 1983 by the Florida Museum of Natural History, Gainesville and the Department of State's Bureau of Archaeological Research, Tallahassee. This location offers an opportunity, as at Little Salt Spring and Warm Mineral Springs, to expand the knowledge about the early cultures of the Western Hemisphere that is now restricted primarily to single component kill sites containing a few stone spearheads and bones of mammoth.

Page-Ladson is located along the lower Aucilla River, 15 km from the Gulf of Mexico, in the widest section of the area known as Half-Mile Rise (Fig. 1). Prior to 5000 years ago, discrete sinkholes accumulated sediments in quiet water along this half-mile section of the river that now separates Jefferson and Taylor Counties. The late Pleistocene sediments represent a time when the Florida aquifer was many meters below its present levels. Animals and the people who hunted them sought water from the sinkholes. The sinkholes became interconnected when the level of the water rose and began to flow about 5000 years ago. The early fine-grained deposits of marls, clays, and peats can be distinguished readily during excavation from the later fluvial deposits of alternating sands and peats punctuated by limestone rubble representing bank erosion (Dunbar, et al. 1988; Webb et al. 1990).

For several decades, Paleoindian artifacts (Serbousek 1983) and Late Pleistocene/Early Holocene fauna (Webb 1974, 1976) have been known to occur in the Aucilla River downstream from the Page-Ladson site. Unfortunately, most of the Paleoindian artifacts were found in a redeposited context along with younger materials. A positive association between humans and megafaunal remains came from the lower Wacissa River, 3 km from the Page-Ladson site where an 11,000-year BP *Bison antiquus* skull was discovered with a fragment of a spearhead embedded in the frontal bone (Webb et al.1984; Webb et al. 1990). Because of Page-Ladson's proximity to these previous finds, investigations were initiated at the site in anticipation that a PIM Zone would be encountered in undisturbed strata.

Page-Ladson receives its name from the John Ladson family who own the land where the site is situated and from Buddy Page, an expert diver and formerly a "seal" in the U.S. Navy, who shared his knowledge of the river bottom and contributed to the underwater operations.

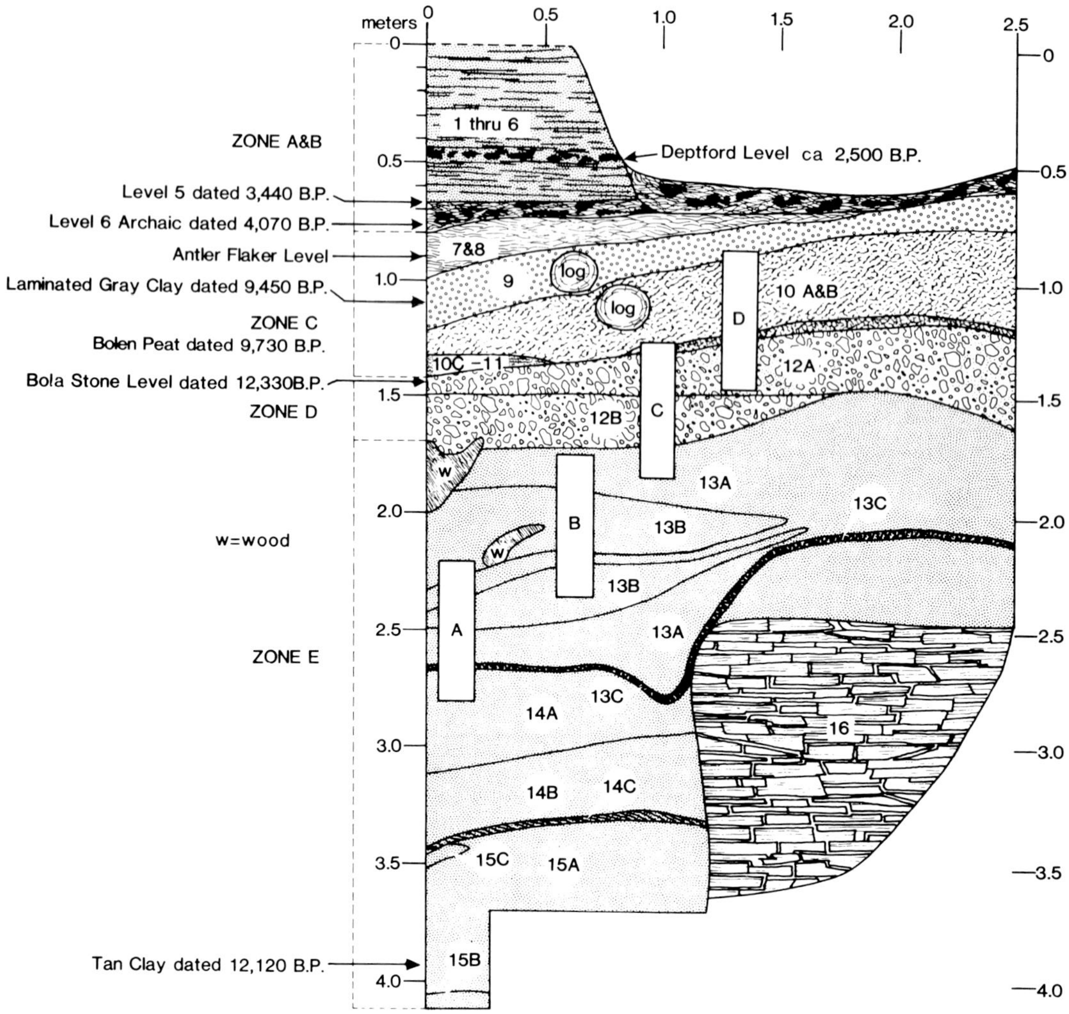

FIGURE 64 North wall profile, Test B, Page-Ladson (photo courtesy of James S. Dunbar).

Excavations and Stratigraphy

Test pits were excavated with the use of 4- and 6-inch water lifts and by hand digging with trowels. Scuba divers worked in pairs within aluminum grid frames or staked out grids. Sediment was removed from the pits via vacuum hoses to a large floating screen on the surface where cultural material that inadvertently may have become associated with the excavation dirt could be collected and sorted. Both regular scuba gear and Kirby Morgan diving hoods with surface supplied air and two-way communication were used. Surface generators supplied 1000-watt lights for needed visibility. Cores, 3 inches in diameter and several meters long, were also taken. Samples from the cores were used for stratigraphic control, carbon dates, pollen, and sediment analyses.

Six test pits, designated A to E, were excavated and six cores, numbered 1 to 6, were taken (Fig. 64). Test B exposed 4.1 m of sediments representing eight distinct strata. The upper three strata were redeposited by flowing water, but the lower five units represent quiet water deposition. Correlating strata between all test sites gives a more general view in which the upper fluvial zones consist of sands and peats while the lower zones usually include a mollusk-rich marl stratum underlain by various strata including calcareous silts, peats, and organic clays. Concentrations of extinct vertebrate bones were found in the lower strata of each test site. In Test E, cypress trunks, branches, and knees were recovered from a highly organic clay beneath a marl zone. This "cypress forest" was dated at 18,400 BP. In Test C, a 20-cm thick dark, clayey soil horizon, referred to as "the dirt" by the investigators, was encountered under 1 m of sediments. (Webb et al.1990). The sediments under "the dirt" consist of gray silt with varying lenses of sandy and shelly material. Expansion of Test C, horizontally and vertically, is planned in future seasons.

Palynology and Paleobotany

Pollen from a core extending 2 m below the bottom of excavations in Test C was identified by Barbara C.S. Hansen and W.A. Watts. The top of the core was dated at 10,600 BP and the bottom at 11,800 BP. The major pollen type represented is oak (*Quercus*) which ranges from 26 to 39%. Other broadleaf trees include hickory (*Carya*), elm (*Ulmus*), American hornbeam (*Ostrya/Carpinus*), sweet gum (*Liquidambra*), hackberry (*Celtis*), and sycamore (*Platanus*). Several juniper and cypress species comprise a total of 4–14% of the pollen, and Pine (*Pinus*) amounted to less than 8%. Other tree pollens made up less than 1% of the pollen sum. Herb pollen comprises 25% of the pollen total, pri-marily ragweed type (Ambrosia). Other compositae, grass, goosefoot, and sedge are also important herbs represented in the pollen assemblage. Pollen of dry dune and scrub species is absent and aquatic plant pollen is scarce. The pollen assemblage suggests a rich deciduous forest indicating a warmer, moister climate than the most recent glacial period which in Florida is defined by high values of pine pollen, few mesic trees, and fairly abundant herbs. It is similar to other early postglacial assemblages documented for Florida (Watts and Hansen 1988).

Many additional species were identified when the macrobotanical remains were examined by Lee A. Newsom (Webb et al. 1990). Macrobotanical specimens came from all test units. The "cypress forest" stratum of Test E included also water hickory, hard maple, button bush, swamp dogwood, beech, ironwood, oak, and plum as well as seeds of aquatic plants and damp ground herbs indicating a bottomland community adapted to frequent submersion. Plant remains from "the dirt" stratum in Test C, which dates to about 10,100 BP, include cypress, willow, American elm, ironwood, grape vine, soft maple, a true hickory, and an oak of the white oak group. A rich bottomland community is indicated. From a Late Paleoindian stratum in Test B, dated at about 9800 BP, additional tree taxa are represented including a member of the red oak group, sweetgum, red mulberry, and pignut hickory. These species are better adapted to well-drained soils. Cypress is absent. In higher strata of Test B, during the Archaic Period, species representative of both the floodplain and mesic bluffs are present as well as a number of other taxa. Charcoal of six different wood types were identified in the Archaic samples.

A total of 86 plant species are represented in the pollen and macrobotanical remains from the Page-Ladson site.

Fauna

Every test at Page-Ladson that penetrated the Late Pleistocene strata contained megafauna especially in marl deposits. These include Columbian mammoth, American mastodon, tapir, long-limbed llama, giant tortoise, Jefferson's ground sloth, horse, and giant armadillo. A heavily abraded radius of *Mammut americanum* (Fig. 65), recovered from eroded surface deposits, was apparently used as a tool (Dunbar et al. 1990). This is an important specimen because it furnishes evidence of interaction between humans and Pleistocene megafauna. Other terrestrial and aquatic species include bowfin, catfish, turtles, alligator, sirens, rattlesnake, and white-tailed deer; many do not differ from modern fauna in the area. Both aquatic and terrestrial environments are indicated by the fauna from the Late Pleistocene and Early Holocene deposits at the Page-Ladson site.

Artifacts and Chronology

The chronometric dates obtained for the strata in which the artifacts were found at the Page-Ladson site are exactly where they would be predicted based on knowledge accumulated at other Late Pleistocene/Early Holocene sites in Florida and elsewhere in North America.

A bone pin fragment and unworked bone, including remains of mammoth, came from the marl zone of Test A which yielded dates of 10,520 ± 90 BP on proboscidean bone and 11,770 ± 90 BP on organic material (Dunbar et al. 1990).

A good example of Late Paleoindian stone tools, including diagnostic Bolen beveled points, came from a peat stratum in Test B dated at 9730 ± 120 BP. The stone assemblage also contained scrapers and adzes. An Early Archaic Period antler flaker (Fig. 66a) was found above the Bolen horizon (just above a level dated at 9450 ± 100 BP), and a bola stone and chert debitage were recovered from a slightly lower stratum that the authors believe is of Paleoindian age.

"The dirt" stratum of Test C yielded an unbeveled Bolen Plain stone point with corner notches (Fig. 67), broken adze bits, and chert fragments. The surface of this stratum dated from 10,000 ±120 BP to 10,280 ± 110 BP. A beveled wooden peg or part of a post (Fig. 66b) made of true hickory was found in an upright position in "the dirt" and the overlying clay. Dunbar et al. (1988) and Webb et al. (1990) presume it dates to the horizon above "the dirt." Many wood

FIGURE 65 Mastadon left radius showing the striations of extensive wear to the outer bone wall. (Photo courtesy of James S. Dunbar.)

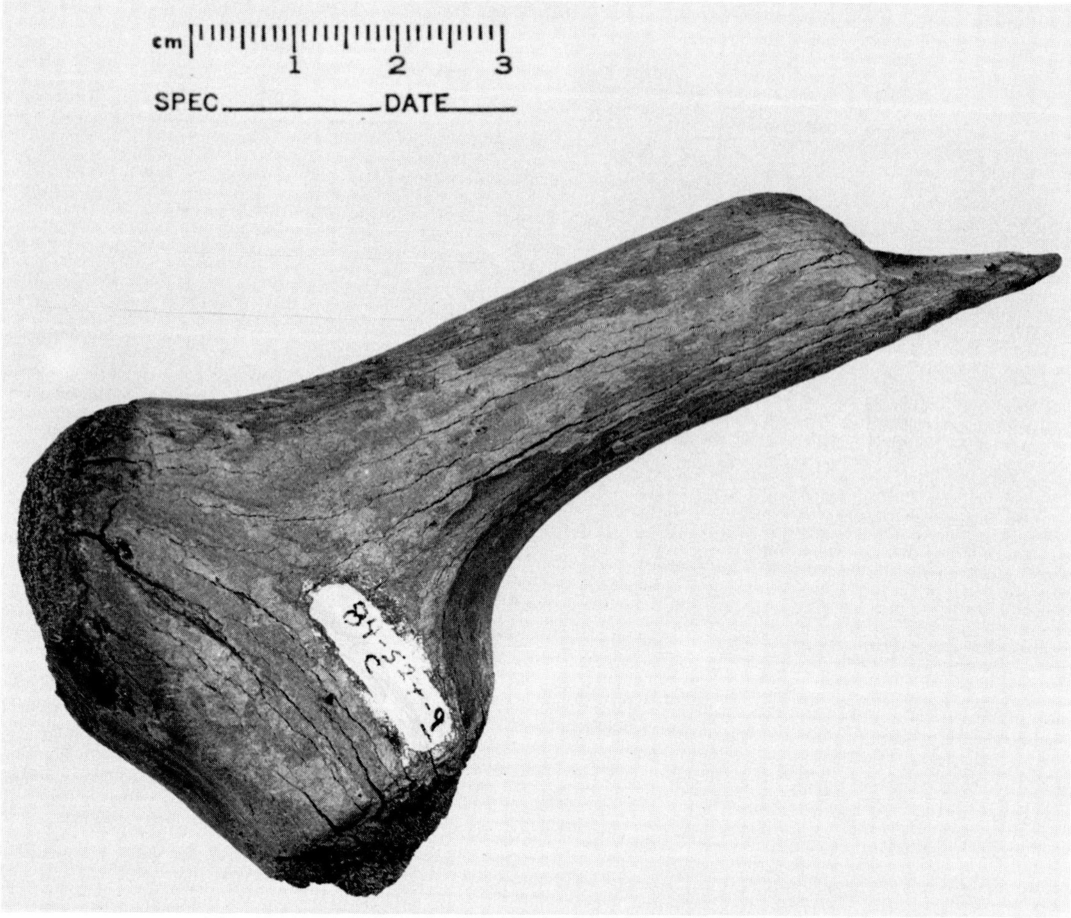

A

B

FIGURE 66 (A) Antler pressure flaker from Test B, level 7/8, Page-Ladson displays desiccation cracks as a major taphonomic alteration of its exterior (photo courtesy of James S. Dunbar). (B) Bevelled wooden peg or post (photo by Stanley Bloneley, Florida Museum of Natural History).

fragments were recovered in "the dirt" zone but they lacked cut marks or other indications of human workmanship.

Discussion

The Page-Ladson site is located far north of any other known wet site in Florida yet it has a number of characteristics that resemble its contemporaries, Little Salt Spring and Warm Mineral Springs. The date for the beveled Bolen component, for example, is almost identical to that received for the stakes recovered from the slope surrounding the cavern entrance at Little Salt Spring.

FIGURE 67 Bone points, and bolen plain spearhead and blade knife of chert from Test B, Level 10c-11. A radiocarbon date of 10,000 BP was received for a bolen plain found in another test unit (photo courtesy of James S. Dunbar).

The Bolen points themselves resemble typologically the Bolen or Green Briar points from Warm Mineral Springs and should date to the same time period.

There are also interesting similarities between the northern and southern sites with respect to the biota. The chronology of the strata at the Page-Ladson site suggests that late Pleistocene megafauna had become extinct by 10,500–10,000 BP. These dates tend to support information from other sites. I am intrigued with the possibility that the Late Paleoindian Period, characterized by the Bolen points, may have been a time when drier conditions prevailed. If there was sustained drought, it may explain why the Bolen points I have observed are very weathered compared to stone artifacts from all other time periods in Florida (Purdy 1981). At the Senator Edwards chert quarry site, I recovered Bolen points from a shallow 30–60 cm sand stratum that also contained Early Archaic and Middle Archaic Period points. The Bolen points were deeply weathered compared to the Archaic styles. The differential weathering and the mixing of artifacts from these time periods indicated to me that the Bolen points were discarded during a time when very little deposition was taking place, i.e., drier conditions (Purdy 1975).

The individuals who are investigating Page-Ladson had the advantage of knowing ahead of time about the two southern springs and they also were aware of the large numbers of Paleoindian artifacts and Pleistocene fauna that had been recovered in the area near Page-Ladson. Using this information to construct a predictive model, they had the added benefit of being able to penetrate strata that were undisturbed by previous activities which would have destroyed much of the potential data base. The disadvantage of excavations at the Page-Ladson site, as at all underwater sites, is that the removal of even small volumes of material is extremely time consuming and expensive.

The archaeological materials and some of the cores from the Page-Ladson site are in the collections of the Bureau of Archaeological Research. Other cores and floral and faunal remains are at the Florida Museum of Natural History, and the pollen samples are in the Palynology Laboratory at the University of Minnesota.

References

Dunbar, James S., Michael K. Faught, and S. David Webb 1988 Page/Ladson (8Je591): An Underwater Paleo-Indian Site in Northwestern Florida. *The Florida Anthropologist* 41(3):442–452.

Dunbar, J.S., S.D. Webb, and D. Cring 1990 Culturally and Naturally Modified Bones from a Paleoindian Site in the Aucilla River, North Florida. In *Bone Modification*. Edited by R. Bonnischsen Center for the Study of Early Man, Orono, Maine (in press).

Purdy, Barbara A. 1975 The Senator Edwards Chipped Stone Workshop Site (8-Mr-122), Marion County, Florida: A Preliminary Report of Investigations. *The Florida Anthropologist* 28:178–89.

1981 *Florida's Prehistoric Stone Technology*. University Presses of Florida, Gainesville.

Serbousek, D. 1983 Explorations of a Paleoindian Site on the Aucilla River. *The Florida Anthropologist* 36:88–97.

Watts, W.A. and Barbara C.S. Hansen 1988 Environments of Florida in the Late Wisconsin and Holocene. In *Wet Site Archaeology*, pp. 307–323. Edited by Barbara A. Purdy. The Telford Press, West Caldwell, New Jersey.

Webb, S.D. 1974 *Pleistocene Mammals of Florida*. University Pressess of Florida Gainesville.

1976 Underwater Paleontology of Florida's Rivers. *National Geographic Society Research Reports*, 1968 Projects, pp. 479–481.

Webb, S.D. et al. 1984 A *Bison antiquus* Kill Site, Wacissa River, Taylor County, Florida *American Antiquity* 49:384–392.

Webb, S.D., J.S. Dunbar, and W.A. Watts 1990 Fauna, Flora, and Culture from the Pleistocene/Holocene Transition at Half-Mile Rise, Florida. Manuscript submitted to *National Geographic Research Reports*.

Republic Groves

Introduction

The Republic Groves site (8-Hr-4) is located in Hardee County (Fig. 1) approximately 5 miles southeast of Zolpho Springs. The area was originally a bayhead swamp that contained three springs, one of which was large enough to swim in. Before 1968, the bayhead covered about 2 acres and was approximately 400 feet long and 250 feet wide. To create land dry enough to plant citrus, an attempt was made to fill in the three springs by bulldozing some of the surrounding surface peat into them. When this did not produce the desired effect, drainage canals were dug. The situation was complicated by the fact that the spoil that was thrown up on both sides of the ditches was later leveled off spreading material from the lower levels across the surface of the site. Since bone was discovered in the spoil and in the sides of the canals, Mr. Mitchell E. Hope was contacted because of his interest in Pleistocene faunal remains.

Hope examined the material and visited the site. He discovered that human remains and cultural materials were present as well as Pleistocene fauna. The soil types consisted of 3–4 feet of surface peat underlain by fine gray clay. Human bone, contemporary faunal remains, and artifacts were found exclusively in the peat layer. The Pleistocene faunal material was found only in the gray clay (Saunders 1972; Wharton et al. 1981).

Excavation Procedures

A datum point and grid system composed of 10-foot squares were established at the site (Fig. 68). Each 10-foot square was excavated in four 5-foot quadrants. Hope spent 85 field days over a 26-month period excavating approximately 400 square feet of the cemetery. He estimated that the area excavated was less than 1% of the cemetery area. All materials recovered were numbered serially although the squares were located in an area of extreme disturbance and the original provenience of most artifacts and bones was lost. Coordinates and elevations were of little value but were useful occasionally in locating missing portions of skeletal material during laboratory analysis. This was often made possible also because Mitchell Hope took very detailed field notes and photographs.

A 2–3 acre upland habitation zone was located adjacent to the bayhead cemetery. Stone artifacts and debitage, worked bone and antler, two ceramic sherds, freshwater mussel shells, and marine clam shell fragments were found on the surface of the upland portion of the site. Newnan and Florida Archaic Stemmed projectile points were represented in the lithic collections. Hope placed a single 3-foot square excavation unit and several shovel tests within the upland zone but failed to find any subsurface materials. The upland portion of the site had either been sparsely populated or previously destroyed by agricultural practices or erosion (Wharton et al. 1981:60).

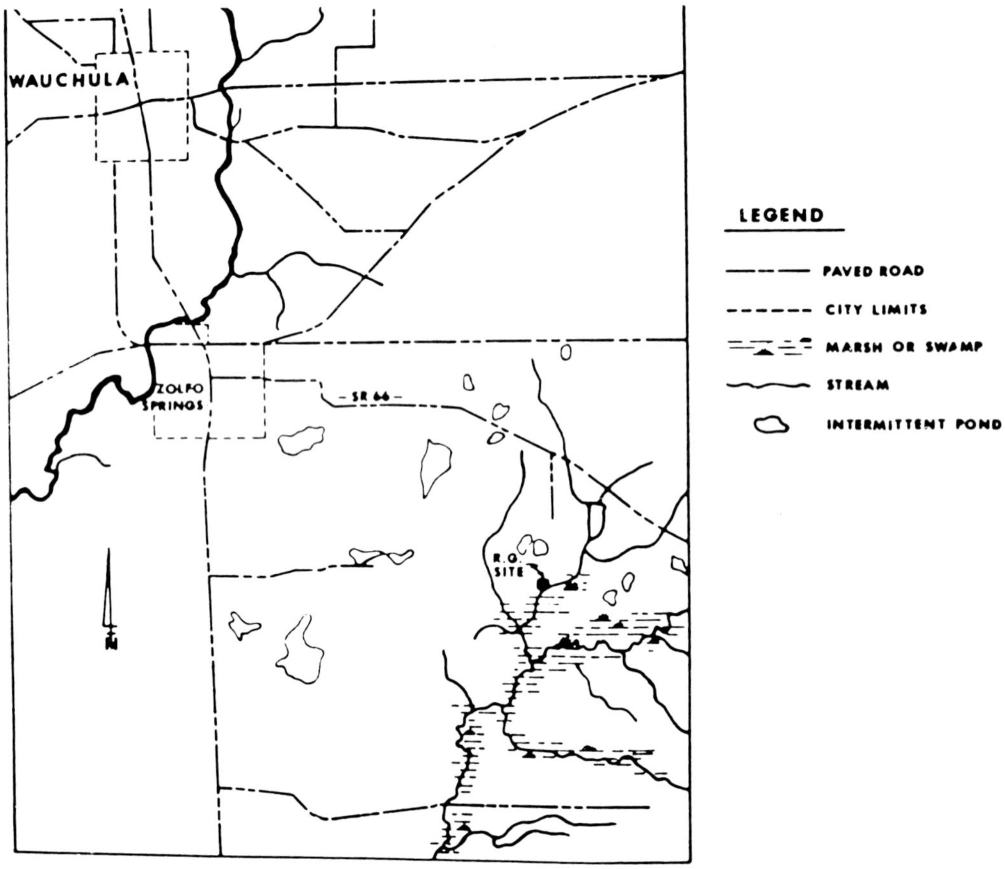

A

FIGURE 68 (A) Location of the Republic Groves Site in Hardee County, (B,C) planview and grid system showing area excavated at Republic Groves site (courtesy of Mitchell E. Hope).

Artifacts

The description of the artifacts recovered at the Republic Groves site is taken from Wharton et al. (1981:64–66) unless otherwise noted.

Stone—The stone artifacts included projectile points or knives, drills, scrapers, a chopper, utilized flakes, a Waller knife, a microcore, a grinding/anvil stone, a honing tool, beads, a pendant made of a non-Florida serpentine-like material, and two miscellaneous pieces of stone. Of these, the projectile points and the Waller Knife are diagnostic in estimating age. Nearly all of the identifiable points are Middle Preceramic Archaic in style including Newnan, Putnam, Marion, and Levy. Thus, these point styles, along with radiocarbon analyses, suggest that the principal occupation of the area occurred about 5000–7000 years ago.

Waller Knives are usually found in association with either Paleoindian or very Early Archaic materials (Purdy 1981).

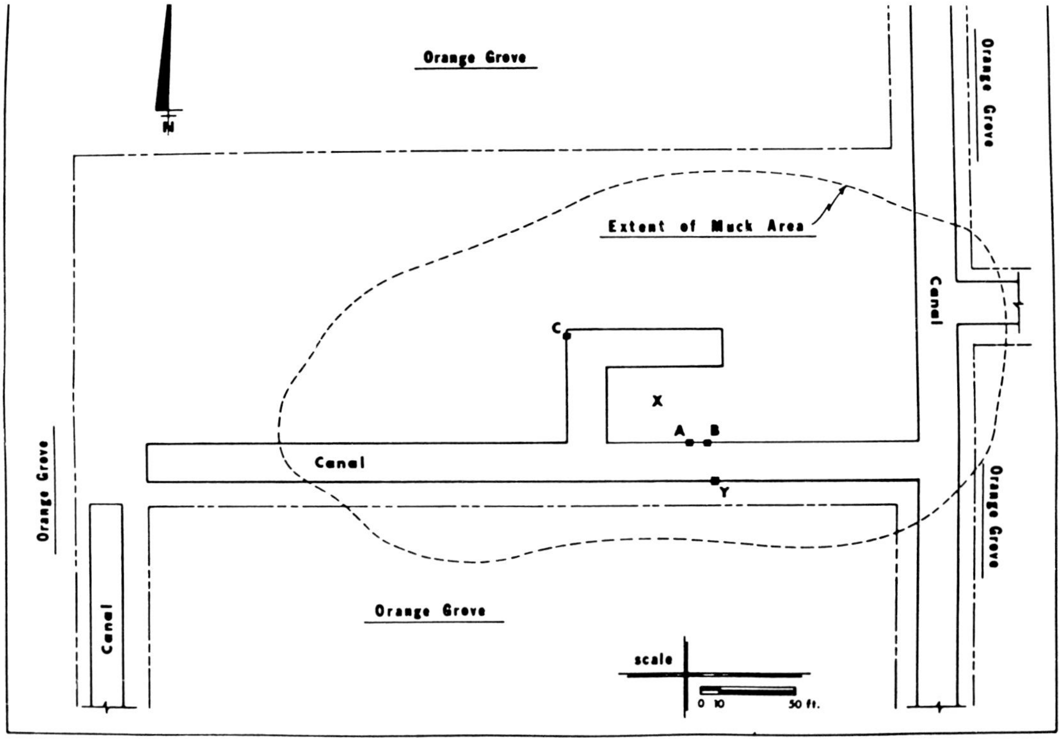

FIGURE 68B

The kind of stone used to manufacture the beads is not given but it is interesting to note that 111 of them were found near the mandible of one of the child interments suggesting they had been strung together on a necklace. Since chert (flint) material is very hard, it is likely that the beads were made of a softer stone, possibly of non-Florida origin. Exotic (imported) stone, such as the pendant made of serpentine, is rarely recovered in Florida and is usually found at sites dating from the latter part of the Middle Preceramic Archaic through the Late Archaic and more recent time periods. Its presence suggests some kind of long distance relationships.

Nearly all of the stone was found in the bayhead cemetery and some of these specimens could be assigned association with a specific burial. Three projectile points, the Waller Knife, the microcore, and a large limestone nodule were surface finds from the upland area.

Wood—A total of 33 tapered wooden stakes made of pine and oak were recovered from the muck in direct or probable association with the burials (Fig. 69). In several instances, groups of three to seven stakes were placed vertically or obliquely in a straight line, sometimes alongside interments. The stakes vary between 10 and 15 inches in length and between 1-1/2 to 2-1/2 inches in widest diameter. They gradually taper toward their distal ends and are bluntly tipped. Whole specimens and proximal fragments show battering on their heads as if hammered upon. Many were charred. Hope immersed six

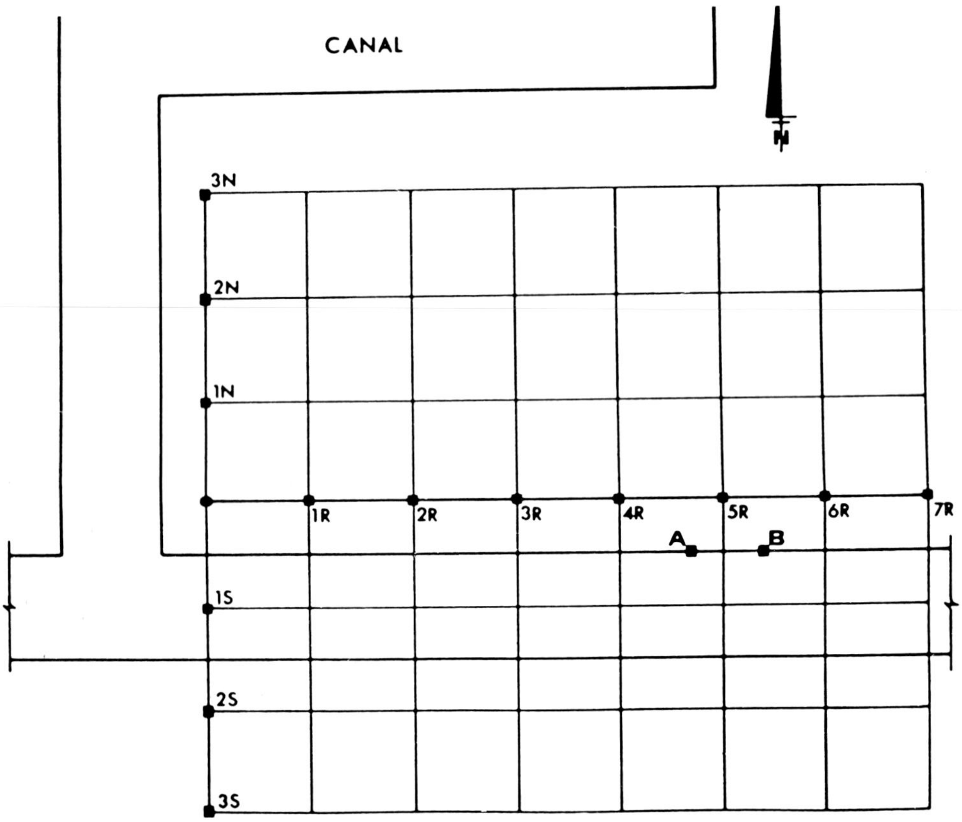

FIGURE 68C

of the burial stakes (R.G. 430–435) in 1 part Elmer's Glue and 3 parts water in an effort to preserve them. Wood stakes were found also at Little Salt Spring, Windover, and the Bay West site.

There may have been a larger quantity of worked wood preserved at the site but the first wood that was found was thought to be roots. After the original stake was found, greater care was taken in examining all of the wood before it was discarded.

Antler—The antler artifacts consist of a grooved plummet, four elaborately incised ornaments or pendants, and what is thought to be an atlatl hook (Figs. 70,71). The plummet is highly polished and is three-quarters grooved. It was found buried 20 inches in the muck. The atlatl hook is a thick section of cut antler with a long spur. It was recovered from a depth of 27–29 inches below the muck surface in apparent association with human skeletal remains. Wharton et al. (1981:65–66) describe the incised ornaments as follows:

> Four artifacts of incised antler were recovered in close co-association within one of the 5 × 5-foot squares. One, possibly a pendant or a hair ornament, had been fashioned from a short section of cut antler, and bore an array of complex design motifs including nested

A

FIGURE 69 (A) Wooden burial stake (RG-409) in situ in square 4R at the Republic Groves site (individual is Ripley P. Bullen), (B) close-up of wooden burial stake in situ (photos courtesy of Mitchell E. Hope).

chevrons, cross-hatched rectilinear bands and triangles, nested triangles, and a combination nested triangle-rectangle design. The ornament is divided into seven zones, set apart by incised lines, and is perforated through the center for suspension.

Another apparent pendant fashioned from a curved antler tine, exhibits a complex geometric design that spirals around the shaft. Two design motifs are present—rectangles and triangles, which have been appended to each other in various combinations. When the designs are unrolled and reproduced in plane view, the combinations of rectangles and squares form a variety of elements, including large squares, triangles, rectangles, and portions thereof. What makes this pendant unique is the manner of design execution. A series of rows and columns of small raised dots fill the individual rectangles and triangles as if the craftsman meticulously carved the background around each dot. Disregarding the manner of executions, the various geometric motifs on this piece are replicated in the incised antler [described above]. [The other two] are curved antler pieces which exhibit two parallel obliquely incised lines. [One pendant] has been worked into a penis or snakelike shape. [Two] show evidence of a circular perforation for suspension.

FIGURE 69B

Wharton et al. (1981) are correct in stating that the ornaments show clear relationships with similar artifacts from sites dating to the Orange period. They also are similar to designs recovered from Florida Archaic Sites and from the Bilbo site in Georgia which is evidently Preceramic. Considering the radiocarbon dates received for the Republic Groves site (discussed below) and the styles of the majority of the projectile points, it appears that these beautifully incised antler objects were made during the Preceramic Archaic Period in Florida. The Windover materials have demonstrated that people decorated bone objects at least by 7000–8000 years ago.

Bone—Bones from deer, unidentified small mammals, and birds were fashioned into a variety of utilitarian and ornamental artifacts. These include beads, a number of awls made from deer metatarsals and ulnae as well as ulnae of small mammals, one bone pin, a flesher, knives, and miscellaneous bone used as tools. Unmodified bone was also present. Several shark teeth (one perforated), representing three genera, were recovered. All of the bone artifacts were from the muck deposits and represented extant species.

Cordage—A few pieces of twisted cordage were recovered and a matted impression underlying one of the burials was observed but not recovered.

A

B

FIGURE 70 (A) Grooved plummet made of polished antler, (B) possible atlatl hook of antler (photos courtesy of Mitchell E. Hope).

Ceramic—Two pottery sherds came from the upland area. The presence of these sherds plus a Sarasota and a Pinellas point establishes that there was some utilization of the area following the site's primary occupation which occurred during the Middle to Late Archaic.

Radiocarbon Analysis

Four of the wooden tapered stakes were dated. The uncorrected ages for specimens R.G. 409, 417, 418B, and 420D, respectively, are 5745 ± 105 BP (UM 2260), 2485 ± 80 BP (UM 2259), 6430 ± 80 BP (DIC-2009), 6520 ± 65 BP (DIC-2010). R.G. 409 had been partly driven into the underlying Pleistocene clay stratum. The date of 2485 BP for R.G. 417 is either in error or the burials span a 4000-year period. The other dates suggest a Middle to Late Preceramic

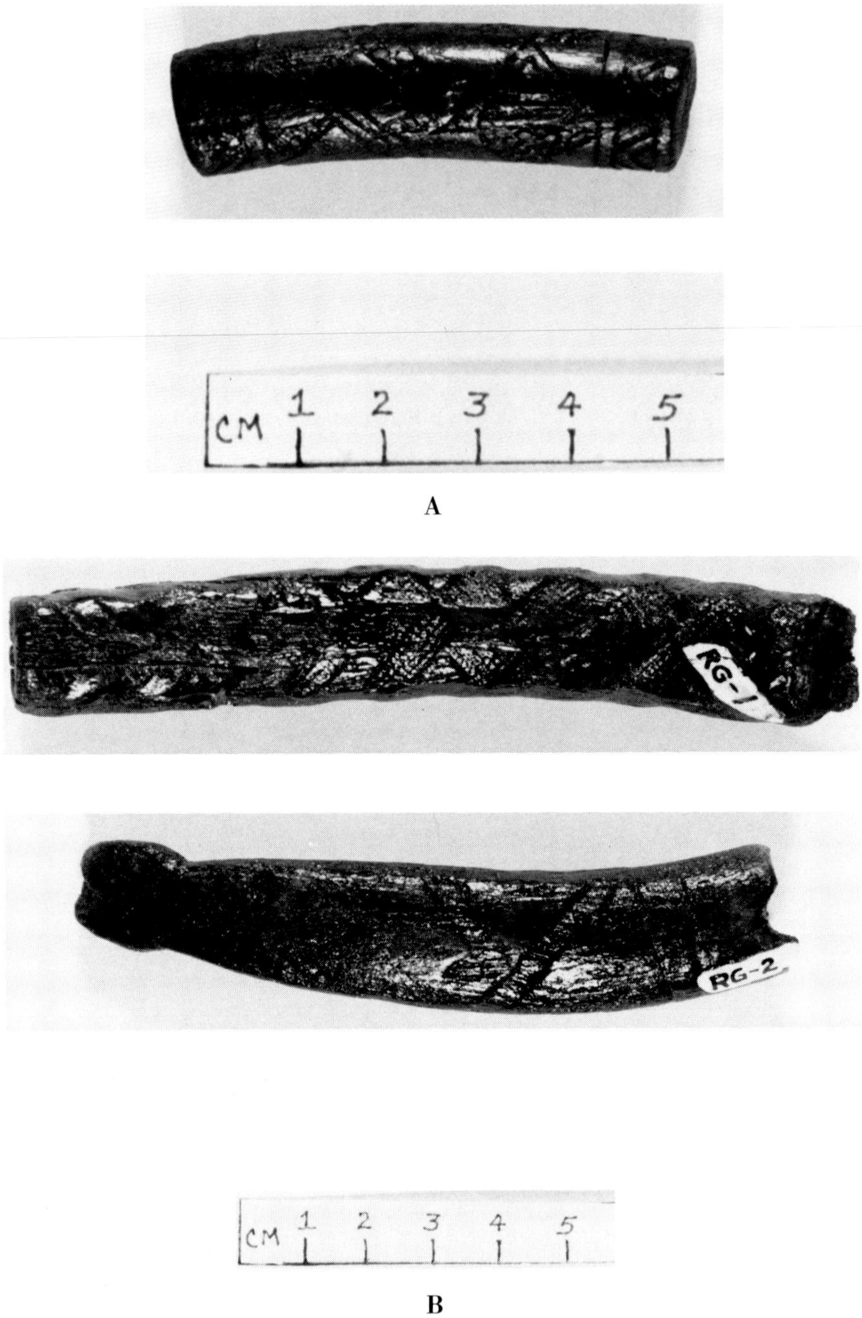

FIGURE 71 Three of the elaborately incised antler artifacts recovered at the Republic Grove site (see text for description). (Photos courtesy of Mitchell E. Hope.)

Period occupation for the site and their accuracy is supported by the stone point styles and by the fact that no ceramics were included with the burials.

Human Remains

The human bones from the Republic Groves site were studied by Saunders

(1972). Because the site had been badly disturbed prior to Mitchell E. Hope's salvage excavations, most of the burials were disarticulated and the skeletal material was fragmentary. Where it was possible to determine, primary flexed interments appeared to be the predominate form of burial (Wharton et al. 1981:76). Saunders explains in detail how she attempted to overcome these difficulties in order to determine minimum number of individuals (MNI), age at death, sex, stature, pathologies, and other attributes of interest and significance.

A minimum of 37 individuals was identified in her sample. Of these, six were children, ages 9 months to 7 years. Saunders believes that the skeletal elements of immature individuals as a percentage of the total population are under-represented because their bones are fragile and not enough was recovered to present an accurate picture of sub-adult mortality rate.

Saunders was able to determine approximate age at death for 20 adults from traits present on the crania which she considered to be the only valid evaluation for this population sample. She provides reasons for excluding post-cranial skeletal elements in her estimation of age. Thirteen (65%) of the adults were early to late middle aged at death; seven (35%) were young adults. The sex of 19 individuals was tentatively determined: 10 males (52.7%) and 9 females (47.3%). This nearly 50/50 ratio of the sexes was also maintained in age at death.

The height was difficult to assess since it is best estimated from the lower limbs and only two femora were complete enough to measure. Saunders used the two femora and ulnae and radii to determine stature for nine indviduals which ranged from 5 feet 2 inches (1.57 m) to 5 feet 8 inches (1.73 m). She could not detect any difference in height due to sexual dimorphism.

Only ten skulls could be restored sufficiently to determine craniometric data. Of these, four (40%) were dolichocranic and six (60%) were mesocranic.

A number of pathologies or abnormalities were identified on the crania. The most frequent was a localized porosity or slight pitting occurring on several bones of the skull (frontal, parietal, and occipital). Out of 14 restored skulls, only two did not exhibit this condition which may have been a result of childhood anemia, possibly caused by hookworm. Arthritis of the temporal-mandibular joint was also common.

Heavy attrition of teeth was noted. Saunders studied this phenomenon on teeth remaining in the dental arch of maxilla and mandibles:

> Attrition on the maxillary dentiton was oblique, at a buccal [cheek] to lingual [tongue] angle, while the oblique angle on mandibular dentition was complementary, slanted in a lingual-buccal dimension. The terms buccal-lingual and lingual-buccal indicate that in the maxillary teeth the buccal side was the higher point on the occlusal surface and the lingual side the lower, and that on mandibular dentition the lingual side was higher with the buccal side lower. . . . The angle of wear was often extreme, with one side worn down to root level while on the other side enamel was still present. . . . attrition is more pronounced on mandibular dentition than on the maxillary teeth, particularly in the molar area (Saunders 1972:19–21).

Dental attrition is caused by cultural habits. Generally, dietary items and the use of teeth as tools account for the majority of tooth wear in a population. The angle of wear on the Republic Groves dentition suggests something being pulled through the teeth in a downward direction placing heavier attrition on the mandible. When this is done, pressure is placed on the teeth of the lower jaw while the upper teeth serve more to complement them. Saunders was able to suggest a high correlation between this cultural practice and the presence of interproximal grooving and abscesses derived from periodontal disease.

The dental trait called interproximal (toothpick) grooving is a feature that consists of a pair of shallow polished grooves just above the enamel line usually occurring between the molars. Peridontal disease may be connected with the wear of the occlusal surfaces of the teeth and the consequent opening up of the interproximal spaces to increasing food impaction. This probably initiated the gingiva irritation. The use of a toothpick would serve to remove the impacted food and thereby relieve the pressure of the sensitive gums. Saunders, citing the work of others also, suggests that the heavy attrition on the teeth of the Republic Groves people led to "food impaction in the opened interproximal spaces and resulted in periodontal disease which is symptomized by irritation of the gums. . . . this sensitivity of the gums resulted in attempts to remove impacted food and relieve the pressure and discomfort. Every adult observed exhibited some evidence of periodontal disease" (Saunder 1972:24). There was a low incidence of dental caries in the sample which is consistent with nonagricultural groups studied from other areas believed to have a low carbohydrate diet.

Osteoarthritis was the most common pathology noted on the infracranial skeletal remains. Several other bone pathologies were present including pyogenic osteomelitis and spina bifida.

Physically, the skeletal remains from the Republic Groves site seem to indicate robust, strongly muscled individuals (Saunders 1972:39).

Several of the crushed skulls contained material other than muck and Hope suspected it might be brain tissue. One skull was sent to the Smithsonian Institution where it was examined by Dr. J. Lawrence Angel:

> . . . the brain tissue inside the skull is fragmented and surrounded by layers of plant material and has changed to amorphous coffee-colored masses. When I broke open one of these, the surface took on the scalloped form of a dissected area of the central part of the cerebral white matter where meshed crossing occurs between fibers of the internal capsule and interhemispheric commissural fibers; one might expect this next to the lateral ventricle and when I cut across the fragment a small hole did indeed seem to show ventricular lining tissue. But all of this is uncertain since decay and denaturation seem to have abolished differences between white (myelinated fibers) and grey (cellular) matter and the masses which I loosened from the cerebellar fossae of the skull were largely compacted black vegetation with very little brain tissue, apparently not cerebellum (Angel, personal communication to M.E. Hope, July 1, 1970).

Angel says that it is unusual for brain to persist for very long in a skull after decay of skin, flesh, dura, and arachnoid. He assumes it happened in the Republic Groves skull because of a chemical change involving denaturation of protein, persistence of fat in myelin, and cholesterol.

Discussion

Saunders did not attempt to compare the Republic Groves people with other populations because, at that time, no analysis had been completed on human skeletal remains that were contemporary and geographically contiguous. She considered both of these conditions essential in order to draw conclusions about relationships. Some observations are possible now, however. There was a full range of ages in the sample from infants to >50 years and both sexes were represented indicating that no status differentiation of individuals existed. Studies of other "cemeteries" have led to the same conclusion. Craniometric data suggest that the majority of the population was mesocranic which appears to be typical of Archaic Period populations. The antiquity of the site is similar to Bay West, Gauthier, the slough at Little Salt Spring, and one component at Warm Mineral Springs. This statement is based on artifact styles and on chronometric dates. Republic Groves shares with Little Salt Spring, Warm Mineral Springs, and the Windover sites the preservation of brain tissue. Brain tissue may have been present in more than one skull but the disturbed nature of the site precluded, for the most part, systematic excavation of intact skeletal remains. The Republic Groves site shares with Windover the recovery of most grave goods in association with infant or child burials. This trait may be more prevalent for hunter-gatherer groups than has been recognized previously.

Wharton et al. (1981:78) projected that several hundred to over a thousand individuals still remain interred at the Republic Groves site. They delineate the kinds of information that might be obtained if investigations of the site could continue. It has now been more than 20 years since Mitchell E. Hope conducted salvage excavations at Republic Groves. Efforts were undertaken locally to help preserve the remaining portion of the cemetery but the drainage canals have dewatered the strata and probably destroyed most of the perishable remains that still lie buried in the muck. The site was also vandalized from time to time (see Hope's field notes). I visited the site in 1980 and took six core samples using a 3-inch diameter soil auger. None of the cores contained organic material more than a few centimeters thick but degraded wood fragments were found in some of them.

Hope mentions in his field notes that he recovered mammoth or mastodon, horse, giant land tortoise, and unidentified animal bones from the clay deposit. He also notes the presence of animal bones from the peat stratum such as deer, rabbit, frog, and alligator teeth. Hope placed all of these into a box containing unmodified animal bones. An analysis of these remains has not been conducted. The greatest emphasis was placed on recovering the human bones and associated artifacts. In addition, no information is available about botanical remains from the site other than the identification of the wooden stakes as being manufactured of oak and pine. Hope's field notes of April 26, 1970 mention a large fernlike object found lying by and under the skull of the child burial that also contained the greatest number of artifacts, including 111 stone beads. He says, "As this dries it will probably shrivel and crack up just as the posts and other organic matter do which are uncovered in this muck."

Mitchell E. Hope should be commended for the way in which he extracted so much information from the Republic Groves site. He contacted archaeologists, physical anthropologists, geologists, surveyors, photographers, and other individuals whose expertise was used to recover and study the disturbed remains in the most systematic way possible. He took extensive and excellent field notes and drew maps which are still available for examination, and he attempted to restore and preserve some of the materials. Most importantly, he made the existence of the site known and shared all of the information with persons he felt could interpret the cultural behavior of the people who lived in this area of Florida several millenia ago. The owners, Republic Groves, Inc., cooperated in permitting excavations on their property.

The human skeletal remains from the Republic Groves site are in storage at Florida Atlantic University in Boca Raton and the other materials are in the possession of Mitchell E. Hope.

References

Angel, J. Lawrence 1970 Report on skull found by Mitchell Hope, Hardee Co., Fl. Unpublished manuscript on file at Smithsonian Institution, Washington, D.C.

Hope, Mitchell E. n.d. Republic Groves Field Notes, 1968–1970. Unpublished manuscript on file, University of South Florida, Tampa.

Purdy, Barbara A. 1981 *Florida's Prehistoric Stone Technology*. University Presses of Florida, Gainesville.

Saunders, Lorraine P. 1972 Osteology of the Republic Groves Site. Unpublished master's thesis, Florida Atlantic University, Boca Raton.

Wharton, Barry R., George R. Ballo, and Mitchell E. Hope 1981 The Republic Groves Site, Hardee County, Florida. *The Florida Anthroplogist* 34(2):59–80.

Warm Mineral Springs

Introduction

In many ways, Warm Mineral Springs (8-SO-19) and Little Salt Spring (8-SO-18), located only 3.12 km apart in Sarasota County (Fig. 1), have had parallel histories involving many of the same personalities: explorations of their deep caverns began in the late 1950s principally by retired Lieutenant Colonel William R. Royal and geologist Dr. H.K. Brooks; both sites have been

A

FIGURE 72 (A,B) Aerial views of Warm Mineral Springs (from Clausen et al. 1975). (C) View of Warm Mineral Springs taken February 1990 showing its use as a health spa (photo by the author).

the subject of extensive publicity and controversy; and both have yielded unprecedented information about flora, fauna, climate, and human presence in Florida that is greater than 10,000 years old. But there are important differences. At Warm Mineral Springs hot connate waters with a maximum temperature of about 37°C (98.6°F) come from deep within the earth. The connate waters combine with cool meteoric water issuing from the Tampa Limestone that is approximately 23°C (73.4°F). The mixing of these two sources produces a year-round uniform temperature about 30.5°C (86.9°F) in the upper two thirds of the water column. These heavily mineralized warm waters are thought to be curative, the Fountain of Youth sought by Ponce DeLeon. People have been swimming in Warm Mineral Springs for many years. In the 1940s the site was known as Salt Springs and was accessible only by a substandard road and cowpath. Since that time the area has been developed as a health spa (Clausen et al. 1975a:196; 1975b; Fig. 72). Clausen et al. (1975a,b) have published the most thorough description and analysis of Warm Mineral Springs. Their article is cited extensively below.

The surface area of Warm Mineral Springs is a round pond approximately 73 m in diameter centered above a deeper shaft. The bottom of the pond is basin shaped sloping from a water depth of 15–20 cm (6 to 8 inches) at its

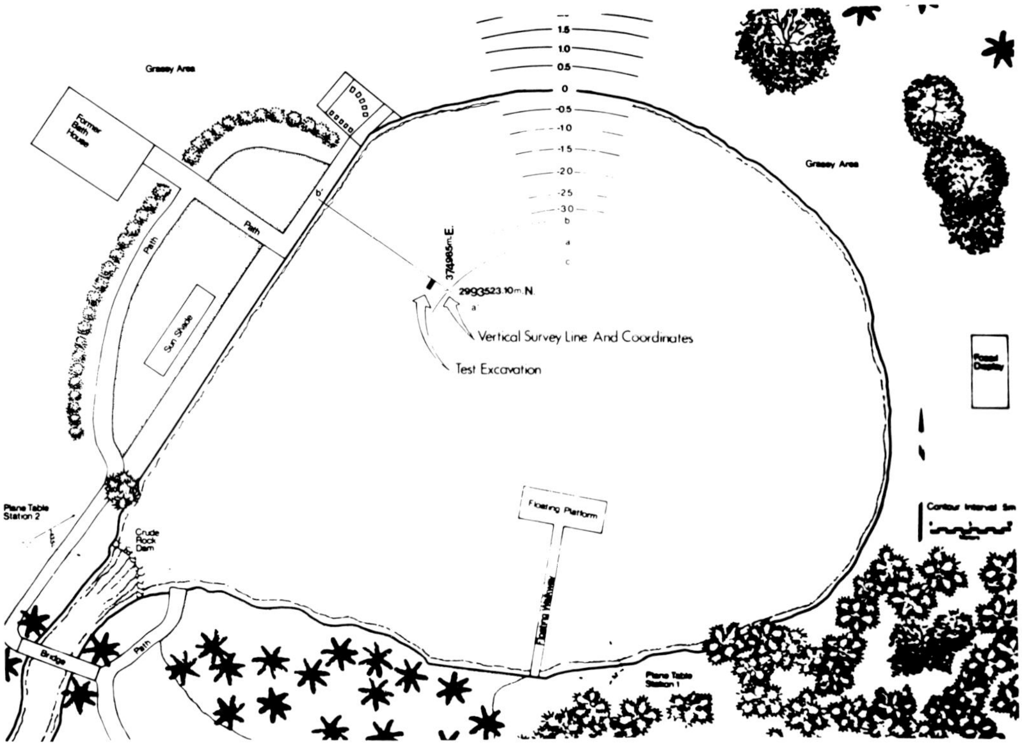

FIGURE 72B

FIGURE 72C

shallow periphery to depths of 3–6 m (ca. 9–18 feet) where the bottom drops away forming a circular opening 46 m in diameter. At a mean depth of approximately 13.40 m (ca. 44 feet), the walls of the spring contract to form a sloping ledge which completely, though not uniformly, rings the cavity. The width of this 13-m ledge varies from 1 m to a maximum of 4–6 m (Clausen et al. 1975a:193–194; 1975b). The diameter of the spring cavity expands rapidly below the ledge to form a large chamber. A debris cone lies directly below the surface pond at an average water depth of 45 m (Fig. 73). The 13-m ledge, in particular, and the debris cone have been the focus of most of the investigations at Warm Mineral Springs.

Stratigraphy and Chronology

From 1958 to 1962, H.K. Brooks mapped, photographed, and procured carefully zoned samples of the sediments in the spring for faunal, floral, petrographic, and radiometric analyses. It was fortunate that Brooks obtained this information because the stratified deposits located on the ledge in the spring were largely destroyed by amateur collectors following the 1959 announcement of the discovery there of 10,000-year-old human remains. The geologic record amassed by Brooks was invaluable when Carl J. Clausen excavated a small but controlled test of the 13-m ledge in January 1972 (Clausen et al. 1975:192).

The more horizontal portions of the ledge nearest the walls caught and held a variety of sediments which settled from above. Three distinct zones were evident when Brooks first obtained samples:

Zone 1—This stratigraphically superior layer is an algal slime, 20–50 cm thick. It is composed of a soft aqueous black algal ooze containing shells of small snails still prevalent in the spring, *Heleobops docima* and *Pyrogophorus platyrhachis*. Bones of alligator, tarpon, and turtle are found occasionally. Zone 1 was already largely destroyed when Clausen worked at Warm Mineral Springs in January 1972.

Zone 2—The middle stratum is composed of unconsolidated gray, calcitic mud, 15–50 cm thick. There is some pine bark, oak leaves, and other plant debris in this layer. Snail shells are common, especially *Physa cubense, Heleobops docima,* and *Helisoma trivolvis*. The few vertebrate remains are those of frogs and mice. Wall *tufa* represent periods of spalling. Radiocarbon dates on charcoal from the upper and lower portions of this zone are 8520 ± 400 years (W1243) and 8600 ± 400 years (W-1241). Since these dates are statistically identical, a period of rapid deposition for Zone 2 is suggested. Zone 2 had been largely destroyed when Clausen excavated at Warm Mineral Springs in January 1972.

Zone 3—The bottom stratum is a leaf bed, 10–80 cm thick. It consists of alternate bands of terrestrial plant debris, predominantly leaves, twigs and

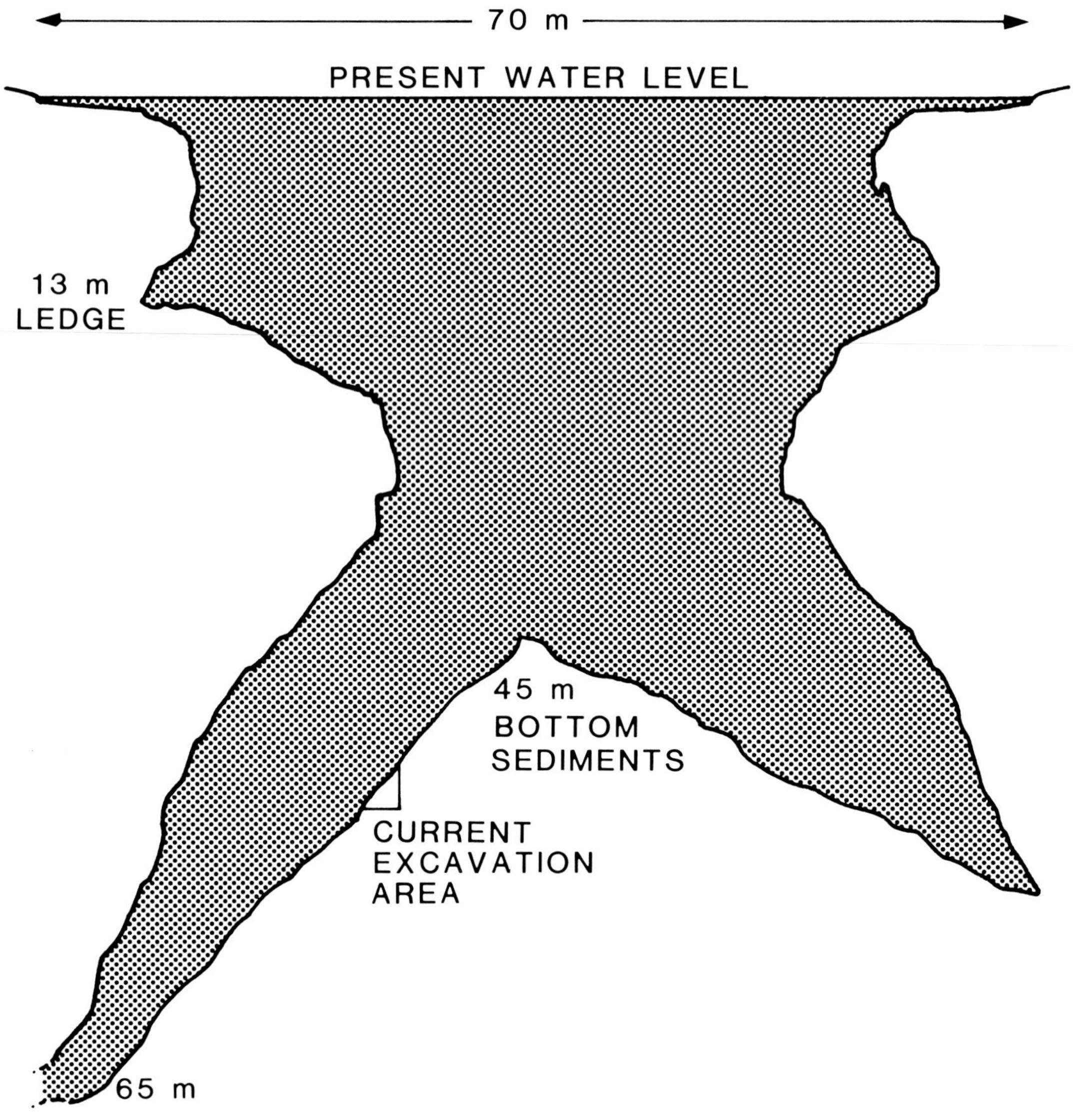

FIGURE 73 Cross sectional view of Warm Mineral Springs showing the present water level, the 45-foot ledge (13m), and the location of recent excavations in the bottom sediments at 155 feet (not to scale).

small logs, seeds, and charcoal intercalated with calcitic mud layers that contain freshwater and terrestrial snail shells as well as fragments of wall *tufa*. Fallen stalactites occur as well as fronds of the sinkhole fern *Thelypteris nomalis* encrusted with a heavy calcitic dripstone layer. Terrestrial snails are more abundant and diverse in this zone. *Helisoma trivolvis* and *Physa cubense* are the most common freshwater species. The calcitic muds contain an abundance of freshwater ostracod shells. Identified plant remains represent the following species: *Pinus elliotii, Sabal palmetto, Quercus virginiana, Q. laurifolia, Ampelopsis arborea, Carya* sp., *Phylocacca rigida,* and *Thelypteris normalis*. The most common polynomorph is a pyospore of a species of *Chlamydomonas* or a related freshwater algae. Vertebrate remains identified are human, deer, opossum, raccoon, rabbit, squirrel, mouse, and frog. Radiocarbon dates on charcoal obtained by Brooks from this zone are 9370 ± 400 years (W1245), 9500 ± 400 years (W-1212), and 9870 ± 370 (W-1153) from top to bottom (Clausen et al. 1975:196–197). In 1972, a similar sequence from Zone 3 was dated at 8920 ± 190 (Gak-3992), 9350 ± 190 (Gak-3993), and 9220 ± 180. Five additional dates were received for materials recovered from Levels 1 through 5 during controlled excavation of Zone 3 in 1972: 9420 ± 150 (Gak 3995), 10,020 ± 180 (Gak 3996), 10,630 ± 210 (Gak 3997), 10,260 ± 190 (Gak 3998), and 9880 ± 230 (Gak 3999). Except for Gak 3997, the close clustering of all these dates suggests a fairly rapid deposition of about 1000 years for Zone 3. It should be noted that Zones 2 and 3 appear to be temporally discrete. Zone 3 sediments had undergone extensive damage by the time Clausen excavated there in 1972.

It is strange that the description of the three zones (layers), principally Zone 2, provided by Royal and Clark (1960) differs rather markedly from the one given above, especially since Royal and Clark explored the spring about the same time as Brooks and acknowledge his contribution to their investigations. There is some variation in species identification, particularly of snail shells, but of special interest is that Royal and Clark (1960:286) mention the recovery of bird, deer, humans bones, and bone artifacts from Zone 2 whereas Clausen et al. (1975:197) state that vertebrate remains are uncommon in Zone 2. Royal and Clark (1960:286) report a radiocarbon date of 10,000 ± 200 (L.J. 120;) from a burned log from Zone 3 which is in fairly close agreement with the dates provided for Zone 3 by Clausen et al. (1975:197,198,200–203). Cockrell and Murphy (1978a; 1978b), however, list 16 radiocarbon dates ranging from about 8700 years BP to 10,500 years BP (1978:3) on samples taken close to the location where a nearly complete human skeleton was recovered. Figure 8 of their article (Cockrell and Murphy (1978a:7) clearly shows this to be Zone 2. Except for the two youngest dates these results are more consistent with those that Clausen et al. (1975a; 1975b) and Royal and Clark (1960) report for Zone 3. It may be important to note also that the 2000-year range in these dates is sizable when one considers that all samples were taken from a restricted area and that ages received by other investigators suggest a fairly narrow range for the deposition of Zone 2. It seems apparent that some sort of sampling error occurred.

Cockrell and Murphy (1978a) describe a lower zone, designated Zone 3, from which they recovered remains of extinct ground sloth (*Megalonyx jeffersonii*) and sabre-tooth cat (*Smilodon floridanus*) as well as extant species already mentioned, and a human mandible. They describe the composition of this stratum as gray-green clay and estimate it to be greater than 11,000 years old (Cockrell and Murphy 1978a; Cockrell 1986). Clausen et al. (1975a:198) state that "Zone 3 sediments rested in turn upon an irregular zone of fragmented gray-green clay from the country rock." Although Royal and Clark (1960:286) report that the burned log, dated at 10,000 ± 200, came from Zone 3, they say that "one end of the log was imbedded in the hard clay floor. . . "

Perhaps the gray-green clay stratum should be labeled Zone 4 to avoid confusion or misunderstanding. Which then of the zones described by other investigators is missing from the stratigraphy presented by Cockrell and Murphy? My guess is that Zone 2 is missing or Zones 2 and 3 have been combined in their scheme because the deposits were so badly disturbed. This suggestion could also account for the excessive range in ages received on the materials recovered in close proximity to the nearly complete human skeleton.

Clausen et al. (1975a; 1975b) concluded from their examination of the various sediments on the 13-m ledge at Warm Mineral Springs that the deposits were all subaqueous in origin. The authors present a number of logical explanations for their interpretation citing, for example, the excellent preservation of botanical materials which surely would have degraded if the deposit had dried out, the lack of rodent gnawing on any of the large bones which usually occurs at terrestrial sites, and the level of the water that was necessary for deposition to take place on the ledge. They believe the latter point is extremely significant because it establishes a general range for the water level in the central cavity 9000–10,000 years ago when Zone 3 materials were deposited and thus provides valuable information about the eustatic rise in sea level following the retreat of Wisconsin glaciation (Clausen et al. 1975a:205). Cockrell and Murphy (1978a:1,6) do not agree that the deposits on the 13-m ledge at Warm Mineral Springs were subaqueous because they found rodent gnaw marks on the fibula of a ground sloth (*Megalonyx jeffersonii*) indicating it had decomposed in a dry environment. Since the extinct species found by Colonel Royal while diving at Warm Mineral Springs over a period of 15 years and by Cockrell and Murphy (1978a) apparently did not come from the same geologic stratum identified by Brooks as Zone 3 which contained all modern fauna, it is quite possible that the water level in the sinkhole was lower at the time the extinct species were deposited than it was at 9000–10,000 years BP. The presence of stalactites in the cavern supports this conclusion and substantiates other evidence already accumulated about extremely low water tables and sea level during the late Pleistocene.

Clausen et al. (1975a:206) believe that Warm Mineral Springs ceased to be a passive cenote and started to flow about 8500 years ago initiating deposition of the Zone 1 algal sludge. A similar history has been documented for Little Salt Spring.

Archaeological Excavations

1972—The 1 m × .5 m test unit placed in Warm Mineral Springs in 1972 represents the first time that a controlled archaeological excavation was undertaken in this type of finely stratified deposit under water. The procedures are described explicitly and are well documented with underwater photographs in Clausen et al. (1975a:199–202; 1975b).

The investigators selected a several-meter-long undisturbed section of Zone 3 on the 13-m ledge along the northwest wall of the spring. Zones 1 and

FIGURE 74 Sloping Zone 3 sediments on 45-foot (13-m ledge at Warm Mineral Springs (from Clausen et al. 1975). Location is shown in Fig. 72B.

2 were missing and had evidently been stripped away. A large portion of the deposit they planned to excavate was overlaid by a massive growth of *tufa* weighing an estimated 200 kilos. The *tufa* may have formed directly on top of one of the uppermost leafy lenses of Zone 3 effectively covering the underlying sediments and protecting them from any disturbance. Under Clausen's supervision the *tufa* formation was floated away. Colonel Royal assisted in its removal.

After establishing the coordinates of the area to be excavated, the decision was made to remove the sediments from the side nearest the center of the spring rather than from above. Negative buoyancy was increased with lead weights so the archaeologists could balance on their knees on the exposed sloping outer portion of the ledge with the dark void of the deep central portion of the spring at their backs. Prior to excavation of the established unit, they faced down the exterior of the deposit to expose the stratigraphy. During this procedure they recovered a fragmentary left *ilium* of a human juvenile and recorded its location precisely.

The unit was excavated vertically (rather than horizontally) by arbitrarily imposed 10-cm levels. This approach was adhered to even after it was determined that the strata within the deposit sloped downward slightly from north to south (Fig. 74). The investigators were unsure of their ability to excavate by natural stratigraphy but in retrospect they determined that it could have been accomplished successfully. Seven 10-cm levels were excavated and were numbered 1 to 7 from top to bottom. The procedure was to loosen the material in each level in small sections using a short stiletto-type diver's knife and carefully separate it in the hand using the fingers. All readily identifiable vertebrate faunal material and major botanical specimens were placed in plastic bags and labeled by level. In addition, large samples of sediment were taken to insure preservation of a representative sample of botanical specimens including pollen and of the minute skeletal remains of smaller vertebrates which were known to be in the deposit but were difficult to see because of reduced visibility. This was accomplished by cutting free a block sample 30 cm × 20 cm × 10 cm thick (6000 cc). These were taken to the surface and quick frozen. The Level 6 sample was incomplete and no sample was taken for Level 7. No macroscopic botanical or faunal remains were encountered in excavating Levels 6 and 7.

No artifacts were found in the test unit but an almost intact first sacral vertebra of a human juvenile was discovered during excavation of Level 4. Its precise coordinates are given. Radiocarbon dates on organic materials in association with these bones are about 10,000 years BP. These are the first and earliest human bones recovered under completely documented and controlled systematic excavations in the Southeast. They confirm the claim of Royal and Clark (1960).

On February 5, 1973, a fairly complete skeleton of an adult male was recovered from the 13-m ledge at Warm Mineral Springs by Wilburn A. Cockrell. The recovery of this individual received extensive and sensationalized attention in the popular media. Clausen et al. have the following to say about this skeleton: "Also insupportable is the contention that a human skull and other materials 'discovered'—regrettably during a staged press conference at the spring—represent a burial, or at least a primary purposeful interment. . . . Clausen was shown the location of these bones by Colonel Royal when at the spring in January 1972. . . . a human mandible lay in full view in an irregular tunnel-like hole dug by Royal. . . . The skull later recovered from this point in early February 1973 . . . was not in evidence, but portions of other human bones were clearly visible in the surrounding sediment. (The mandible and some of the other human bones from this location were brought to the surface in November 1972.) It also appeared that the Zone 3 sediments at this point were irregular or distorted, . . ." (1975:207–208).

Cockrell and Murphy (1978a; 1978b) summarize field work carried out at the site in the 1970s. This work included additional mapping, the establishment of a 3-dimensional grid, and excavations in the grey-green clay stratum underlying the zone where "Warm Mineral Springs Man" was recovered. From this lower zone they found sabre-tooth cat, ground sloth, panther, deer, opossum, raccoon, frog, turtle, and human bones. Their explorations and finds were recorded on video tape using a Hydro Products underwater television camera equipped with a special thalium light source and image intensifier.

Murphy (1978) describes in detail the way datum points were established, how above and below water excavations were mapped, how measurements were taken underwater, excavation tools and procedures, specimen removal, and surface screening. After photographing, samples were packed and sent to the Division of Historic Resources laboratory in Tallahassee. From 1975–1977 over 130 video tapes were made and cataloged. In a brief unpublished report about Warm Mineral Springs, Cockrell mentions that existing archaeological land survey, mapping techniques, and excavation methods have been modified to adapt to the underwater environment. Accounts of Cockrell's work can be found also in magazine articles (e.g., Blount 1978; Skow 1986) and unpublished specialists reports. State-sponsored research at Warm Mineral Springs ceased from 1978–1984. From 1984–1987 investigations at the site were funded by a State of Florida legislative appropriation administered by Manatee Community College; since 1987, affiliation has been with Florida State University.

Clausen et al. (1975:212) note that the unconsolidated but stratified organic sediments in the upper portion of the central debris cone hold great potential as a repository of information relative to the understanding of the climatic record in this area and the reason for the human remains in the spring.

> Material has been continually settling to the bottom of Warm Mineral Springs from the time the cavity first opened to the surface to the present. Human skeletal remains are known to exist in this deposit, which at the bottom undoubtedly antedates the approximately 9-to-10-thousand-year-old span of the Zone 3 sediments on the 13 m ledge by thousands, perhaps tens of thousands, of years. [As noted in the section about Little Salt Spring, a date received on a piece of oak 11 m below the sediment-water interface in a 12-m core taken from the bottom deposit at that spring was only 9700 years old suggesting that the accumulation of organic material was rapid; Warm Mineral Springs may have had a similar history.]
>
> Unfortunately, the depths of 38-50 m present a formidable obstacle to efficient, safe excavation by archeologists operating from the surface. Meaningful, attentive, technical work at depths, utilizing repetitive dives on compressed air from the surface, is difficult due to the impairment of faculties experienced . . . (Clausen et al. 1975:212).

Skow (1986), discusses Cockrell's excavations in the debris cone at 155 feet (ca. 48 m) with only 30 mintues of bottom time a day. The first few inches of his trench produced a 1930s Coca-Cola bottle, a 78-rpm phonograph record, a couple of garfish skeletons and, somewhat deeper, deer-skull fragments and catfish skull bones. Most of his finds have been alligator coprolites. With technical assistance from Florida State University's Academic Diving Programs, mixed-gas diving with surface decompression was initiated in 1987 to increase diving time and diver safety. In early 1990, I visited Warm Mineral Springs and learned that excavations in the trench had been discontinued for more than 2 years because of a diving accident and work had begun again only a few days before my arrival. Cockrell (personal communication 1990) explained his diving procedures and described the excavation being conducted in

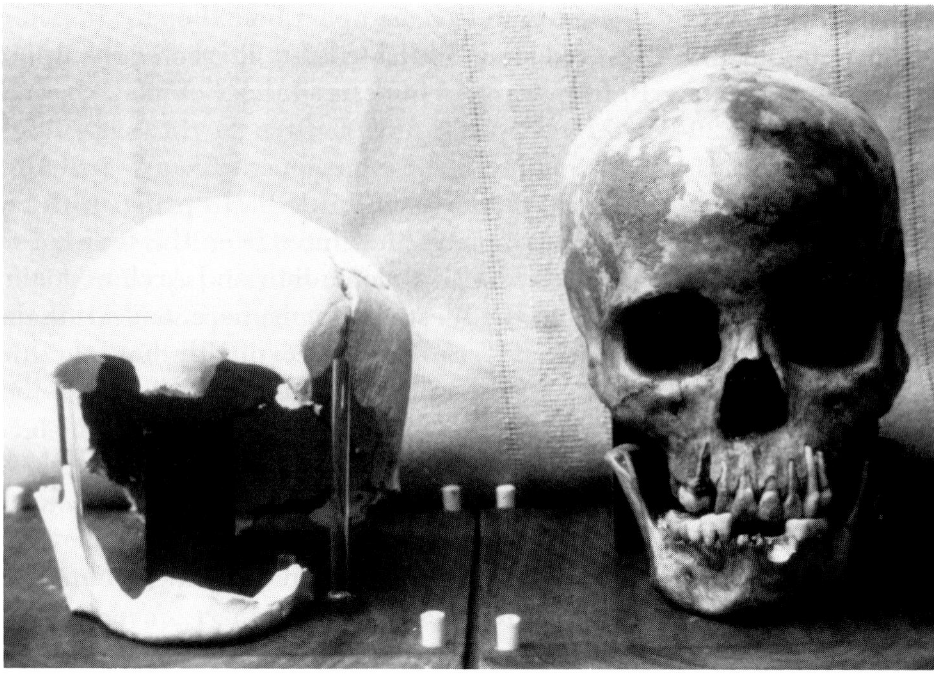

A

B

FIGURE 75 (A) Front view of the Royal and White skulls from Warm Mineral Springs; (B) back view of the skulls (photos by the author). The Royal skull was sawed to remove the brain (see Fig. 77).

ence. On typological grounds the White female and the Warm Mineral Springs male easily fall together and Morris says these two crania represent the earliest aboriginal human inhabitants of Florida yet recovered. He does not mention the vertebra and ilium of the child found in 1972 by Clausen et al. (1975a) that also dated older than 10,000 years.

Morris compares the Warm Mineral Springs male to crania from Vero Beach and Melbourne and concludes the latter two are broader in the temporal region. He questions whether they should be consider Paleoindians from a typological point of view.

Morris mentions the difficulty in generalizing about variability in populations from an examination of only a few crania. But, he says, it may be instructive to consider together the Warm Mineral Springs male, the Royal female, and the Archaic male from Little Salt Spring. The Royal female and Little Salt Spring male may be considered to represent an Archaic population. (It is interesting to note that Morris apparently was not aware that the Royal skull had been dated between approximately 7100 and 7600 years BP and yet he classified it correctly as Archaic.) There is greater temporal fullness in both Archaic crania compared to the Warm Mineral Springs male and White female whom he considers Paleoindians. The Vero Beach and Melbourne skulls are more similar in some respects to Archaic but, he cautions, it is difficult to determine their temporal width because the skulls are fragmentary. Morris concludes, after discussing some of the best authenticated Paleoindian remains from the Western Hemisphere, that Paleoindians were quite varied with respect to the two cranial indices of headedness and vaulting: some are dolichocranic, some mesocranic, and vaulting ranges from low to high. Results also indicate that mesocrany is not necessarily a recent development in American Indians or their ancestral stock. Archaic populations, however, are most often mesocranic rather than dolichocranic and more often high vaulted than low vaulted. The three Warm Mineral Springs crania and those from Melbourne and Vero Beach are all dolichocranic but with variable vaulting. He says that even though there is more variability than thought formerly, the Paleoindians can be separated from later Archaic populations. They are systematically longer headed and lower vaulted than Archaic peoples. Crania become more broad headed and higher vaulted through time and this is probably a plastic response to changing exploitation of natural resources rather than selection; however, this has not been proven.

Morris finishes his study by comparing the Warm Mineral Springs male cranium with Archaic crania and with the White and Royal females. Parietal and temporal fullness characterize all of the Archaic crania in contrast with the Warm Mineral Springs male. The White female is distinguished from Archaic females by its greater length. He summarizes his paper as follows:

> The Warm Mineral Springs finds described and discussed here indicate a group of rather slight, long-headed, medium-vaulted people, affected to some extent by dietary deficiency, who inhabited the area around the spring. Shoveling of the maxillary incisors

> and buccal pitting on the mandibular molars identify the Asian origin of the people. . . . Their congeners on the east coast of Florida at Vero and Melbourne were found associated with extinct animals characteristic of the late Pleistocene.
>
> . . . These old Florida Indians reasonably represent the Paleoindian stock from which the later eastern Archaic peoples derived. The later peoples had increased temporal fullness, broader heads, and higher cranial vaults. There is some evidence that the later group, even though preagricultural, nevertheless had already experienced an increase in stature. The change in head shape and stature can reasonably be related to changes in the environment and its exploitation (Morris 1975:36).

Faunal Remains

Reference has already been made to some of the faunal remains recovered from the 13-m ledge at Warm Mineral Springs. These include shells of several species of snail, tarpon, alligator, turtle, frog, mouse, deer, opossum, racoon, rabbit, squirrel, bird, ground sloth, sabre-tooth cat, and human. While the presence of these animals is mentioned in various papers and Clausen et al. (1975a) systematically recovered and bagged faunal material from each of seven 10-cm levels, the only thorough description of the animal species from the spring and surrounding area is that by McDonald (1975; 1976). Unless otherwise stated, the following information comes from his unpublished papers.

Five classes of animals—fish, amphibians, reptiles, birds, and mammals—represented by 49 genera were identified from the spring and four additional genera were recovered during excavations of the uplands surrounding the spring. Most of the faunal samples in the study were collected by Wilburn A. Cockrell from ten underwater features in 1975 with additional samples added in 1976 from the spring and from terrestrial excavations. (Nineteen human bones were mixed with the animal bones in six of the ten 1975 features.) Other specimens came from the Vertebrate Paleontology Department of the Florida Museum of Natural History, Gainesville. These may be the specimens recovered by Dr. H.K. Brooks and Carl Clausen. Field notes on the museum specimens indicate that they were collected from the same locations as those collected by Cockrell. Additional material, unprovenienced, was donated by William R. Royal. I am listing the faunal species so that this previously unpublished material will be more readily available to interested individuals.

Class Osteichthyes (fishes)
- *Ictalurus* cf *catlus* (catfish)
- *Ictalurus* cf *punctatus* (catfish)
- cf *Pomoxis* (crappie)
- cf *Micropterus* (bass)
- *Negaprion brevirostris* (shark) caudal vertebrae found in terrestrial area

Class Amphibia (amphibians)
- *Siren lacertina* (Greater Siren)
- *Bufo terrestris* (toad)
- cf. *Rana* (frog)
- *Rana catesbieana* (bullfrog)

Class Reptilia (reptiles)

Gopherus sp. (the land tortoise) fragments only found in terrestrial area
Chelydra serpentina osceola (Florida snapping turtle)
Kinosternon sp. (mud turtle)
Sternotherus odoratus (common musk turtle)
Chrysemys scripta (common pond turtle)
Trionyx ferox (softshell turtle)
Deirochelys reticularia (chicken turtle)
Alligator mississippiensis (Alligator)
Elaphe sp. (rat snake)
Natrix sp. (water snake)
Crotalus adamenteus (eastern diamondback rattlesnake)
Sistrurus miliarus (pigmy rattlesnake)

Class Aves (birds)

Phalacrocorax auritus (cormorant)
Ardea herodias (great blue heron)
Ardea (small species great blue heron)
Buteo jamaicensis (red-tailed hawk)
Caracara cheniway (caracara)
Colinus virginianus (bobwhite)
Meleagris gallopavo (turkey)
Gallinula chloropus (gallinule, a duck-like bird)
Tyto alba
Corvus brachyrhynchos (common crow)
Cassidix major (boat-tailed grackle)

Class Mammalia (mammals)

Didelphis virginianus (opossum)
Homo sapiens (humans)
Sylvilagus sp. (rabbit)
Sciurus carolinensis (eastern gray squirrel)
Glaucomys volans (southern flying squirrel)
Geomys pinetis (southeastern pocket gopher)
Peromyscus sp. (white-footed deer mice)
Sigmodon hispidas (cotton rat)
Neotoma floridana (woodrat)
Microtus pennslvanieus
Neofiber alleni (Florida water rat or round tailed muskrat)
Pitymys pinetorum (pine vole)
Urocyon cinereoargenteus (gray fox)
Procyon lotor (raccoon)
Lutra canadensis (river otter)
Odocoileus virginianus (whitetail deer)
Megalonyx jeffersonii (Jefferson's ground sloth)
Smilodon floridanus (Florida sabre-tooth cat)
Felis concolor (cougar or puma)
wolf or dog (Family Canidae)
Equus sp. (horse) protocone of upper molar found in terrestrial area
Small camelid. Non-diagnostic astragalus from terrestrial area.

McDonald (1975) suggests that the habits of the woodrat may be partially responsible for some of the bone accumulation in the spring. None of the bones showed any indication of butchering or skinning marks or of being worked. Most of the species are represented by fragments of only one individual. Catfish, bullfrogs, chicken turtles, and opossum are the most common animals in the sample. Raccoon is fairly common and has been found at eleven other early sites in Florida. All of the alligator material in the sample collected by Cockrell came from the silt cone at the bottom of the springs but Brooks reports having recovered alligator remains from Zone 1 of the 13-m ledge (Clausen et al. 1975a). The majority of the species present in the sample are still extant and live within the proximity of the springs. A few of the forms indicate they had a wider distribution in Florida at one time; others are extinct or no longer occur in Florida, such as ground sloth, sabre-tooth cat, Pleistocene horse, and camelid. The absence of some species at Warm Mineral Springs may be more apparent than real since preservation is often an artifact of an animals' habits and habitat.

Based on the species represented in the sample and their habitat preference, McDonald identifies three distinct but interrelated kinds of environments at or near Warm Mineral Springs.

First are the animals associated with the springs, slough, and stream. The fish, frogs, siren, snake (*Natrix* sp.), turtles, and alligator require a permanent water supply. The cormorant, heron, gallinule, and grackle are all birds that depend on aquatic plants and animals for food. The Florida water rat, cotton rat, and otter depend on permanent bodies of water for shelter as well as food. The raccoon is usually associated with aquatic situations in wooded areas and utilizes them as a food source.

The second environment indicated is a forested one probably consisting of a mixture of deciduous and coniferous trees. Species present usually associated with a forest environment are the red-tailed hawk, turkey, crow, opossum, raccoon, gray squirrel, flying squirrel, pine vole, and deer. The gray fox and rabbit are probably more associated with the ecotonal area on the edge of the wooded area.

The third environment is that of open country between the patches of forest. The eastern diamondback rattlesnake, caracara, bobwhite, woodrat, and gopher are all associated with this type of habitat. It was very likely an open palmetto prairie with sandy soil.

The ground sloth (one nearly whole and parts of another) and sabre-tooth cat were found together at a location designated as Feature 30 on the north side of the 13-m ledge in the spring of 1976. The bones of these extinct species were imbedded in a soft plastic sticky gray-green clay which is sitting immediately on top of a Miocene clay of the same color. The two clays can be distinguished by differences in consistency and texture. Some parts of the ground sloth and sabre cat were not encased in the clay but were recovered from the sand and vegetable matter on top of the soft clay. Some wood material was imbedded in the clay and one sample was removed from underneath the calcaneum (heel) of the sloth. Cockrell (1981:178) reports a date of 10,980 ± 160 for these remains and says that about 30–40 m to the east a human mandible was found from the same strata. Bill Royal provided additional ground sloth bones which represented parts of at least three individuals. He also supplied sabre cat material. McDonald (1976) gives metric data for the ground sloths and sabre cats. Ground sloth has been recorded at 17 localities and sabre cat at 10 localities in Florida.

There has been much debate about how the animal remains, including human, came to rest on the 13-m ledge. Royal and Clark (1960) believed people lived on the ledge, Clausen et al. (1975a) suggested people and animals could not get out once they had gotten in either intentionally or accidentally, Cockrell and Murphy (1978a) thought the humans (at least one) were buried on the ledge. Goggin (1962) did not think they entered the cenote intentionally saying they would have had to be great alpinists. This argument may never be resolved but my guess is that most of the animals fell in and that some of the humans were killed when they tried to obtain water or retrieve a trapped animal. Anyone who has read of the honey gatherers and the nest gatherers in two issues of the 1989 *National Geographic* realizes that people *are* great alpinists and will go to any length to acquire needed or desired resources. Occasionally they have accidents.

Botanical Remains

King (1975) and Sheldon and Cameron (1976) have compiled palynological profiles and plant macrofossil lists, respectively, from an examination of sediment samples taken from the 13-m ledge at Warm Mineral Springs. Sixteenth century historic records from this area describe thick forests of broad-leaved deciduous trees including live oak and other oaks, hickories, bays, mulberries, cherries, and cabbage palms with large grape vines intertwined through the treetops, and with a thick undergrowth of saw palmetto. Sheldon and Cameron conclude from their examination of plant materials preserved in the spring that the environment and climate 9000–10,000 years ago were probably similar to the present, a mixed hardwood forest composed of mesic species which are adapted to moderately wet habitats.

> Major components of the vegetation included oaks, myrtle, hazelnut, and birch. The presence of ash, willow, cypress, cattail, and elm, all rich woods and riverbank genera, are also indicated in the profile. All leaf fragments so far identified in our laboratory belong to live oak, a sandy, damp woods tree (Sheldon and Cameron 1976:68).

King (1975) supervised the collection and study of pollen samples from a depth of 45 feet below water in Warm Mineral Springs:

> The pollen from the trench is dominated by the genus *Quercus* (oak) and the main feature of the diagram is the increase in *Quercus* pollen and the decrease in Myrica (sweet gale) pollen with depth. Similar decreases with depth are seen in *Morus* (mulberry) and *Corylus* (hazel). *Pinus* (pine) is uniformly less than 10% throughout the pollen column.

A

FIGURE 76 (A) Photograph of green briar points found at Warm Mineral Springs by W.R. Royal (picture courtesy of W.R. Royal); (B) sketches of points showing front and back views and cross sections (sketches by James S. Dunbar); (C) atlatl spur made of marine shell recovered at Warm Mineral Springs in 1973 (picture courtesy of W.R. Royal).

Artifacts

As nearly as I can determine, all of the artifacts found at Warm Mineral Springs were reported in Cockrell and Murphy (1978a) and no additional artifacts have been recovered since that time. The majority came from the 13-m ledge and were found by Bill Royal. The two stone points are Green Briars (or Florida Bolen) which places them typologically within the Late Paleoindian Period. They have been dated elsewhere at 9000–10,000 years old. At the time Cockrell removed the male skeleton in 1973 he recovered what he believes is an atlatl spur manufactured of carved shell (Fig. 76). A thumbnail or end scraper that may also date to the Late Paleoindian period was recovered from Feature 44 of the land excavation.

Archaeologist vs. Amateur

Dr. Eugenie Clark, marine biologist, first swam in Warm Mineral Springs in 1956. She had been asked by the Director of the American Museum of Natural History to find a small tarpon, less than 2 feet long, and was told that small tarpon had been seen in Warm Mineral Springs. Clark says: "I looked forward to diving in Florida's only warm spring. People come to drink and bathe in these hard and sulfurous waters. . . . They also take mud baths at the shore of what is called 'The Foundation of Youth'" (1969:149). Clark waded

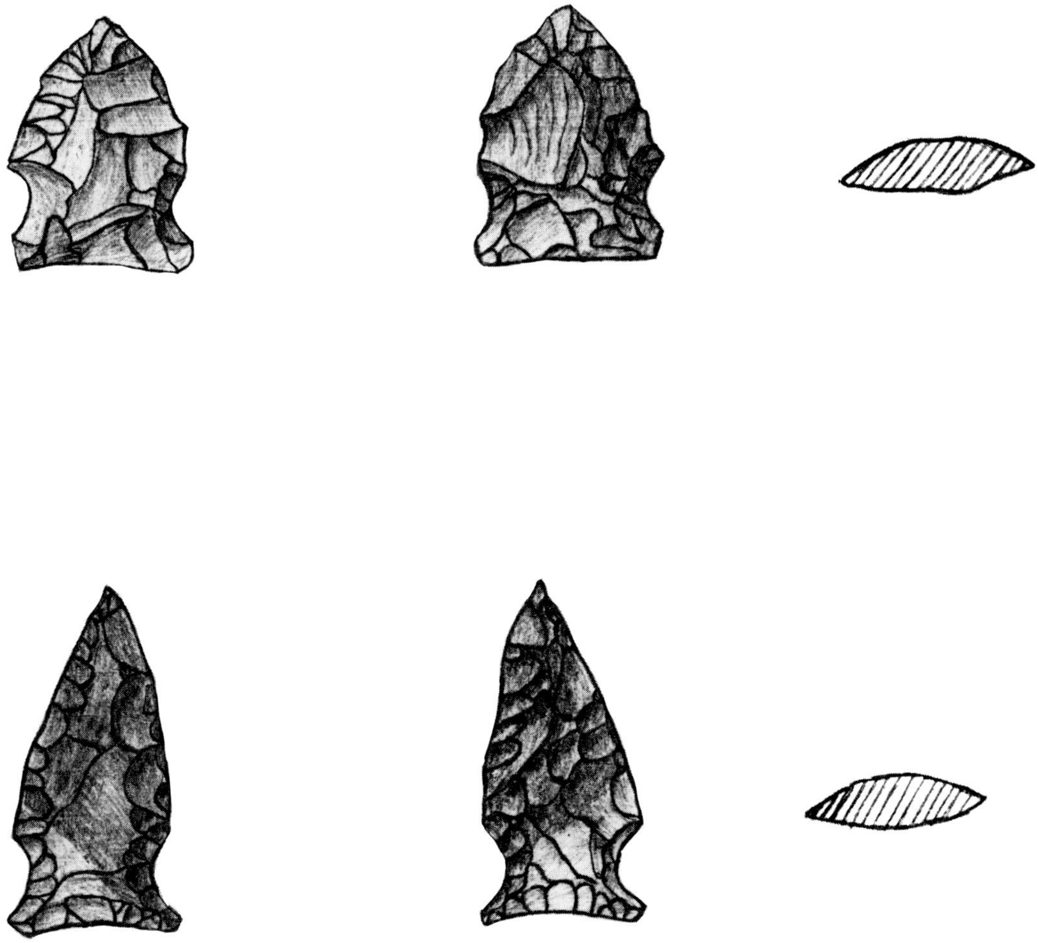

FIGURE 76B

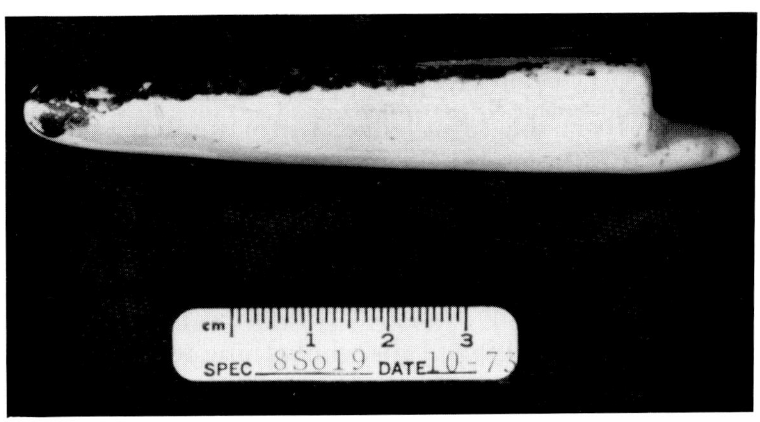

FIGURE 76C

through soft mud, every footstep stirring up a cloud of muck which cut down to almost nothing the already poor visibility in the milky green water. She swam around the circular spring and noted that at the sloping center the bottom dropped down into a huge black hole. She returned to Warm Mineral Springs

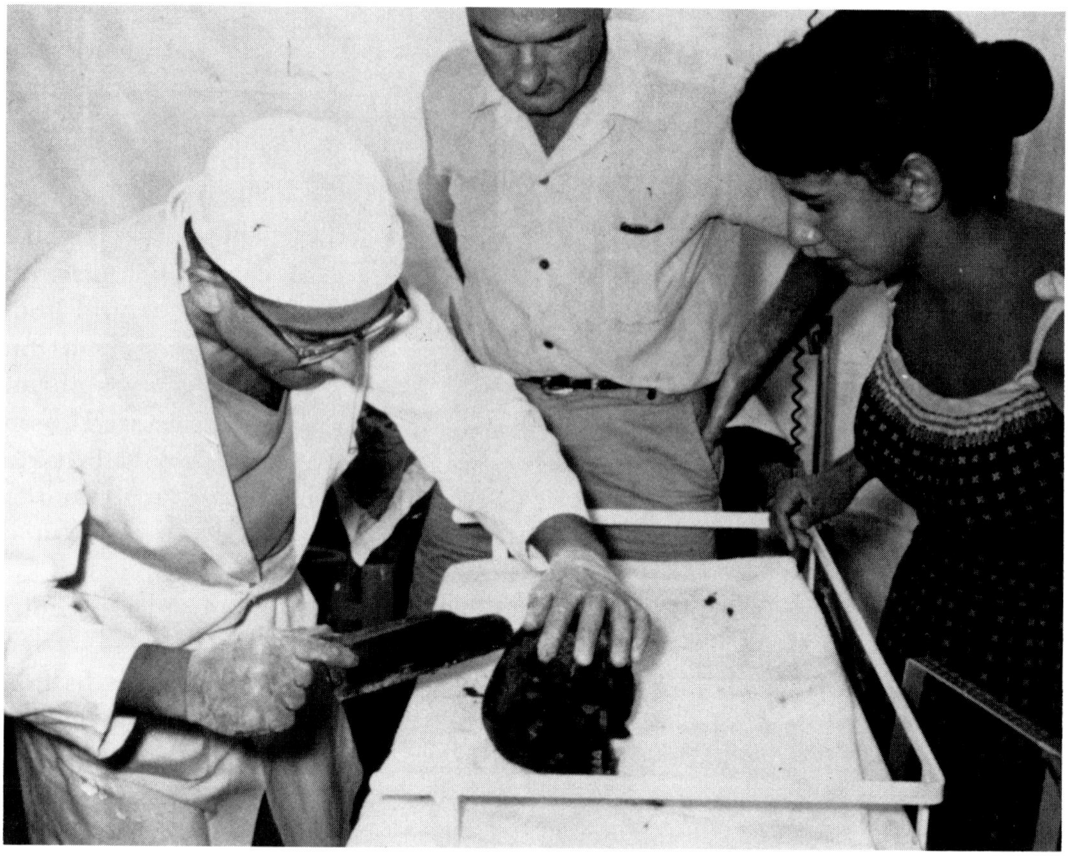

FIGURE 77B

Royal and Clark state that the human skull containing brain tissue was not found in the same zone as the red mulberry log dated 10,000 years old. The skull was found in Zone 2 and Royal and Clark mention that it could be several thousand years younger than the log. A method for radiocarbon dating bone had just been developed and 52 bones associated with the skull were sent to Monaco. Through the help of Captain Jacques-Yves Cousteau, tests were made on them by Dr. J. Thommeret, Chief of the Radiocarbon Laboratory of the Centre Scientifique de Monaco. The date received on the bones, after several years of waiting, ranged between 7140 and 7580 years old (Clark 1969:176) providing a Middle Preceramic Archaic age for this individual. In support of Royal's discovery is the fact that the brain material appeared fresh when it was found but quickly began to disintegrate after it was removed from the deposit in which it had been entombed. If Royal had planted a previously recovered skull to provide spectacular TV coverage, the brain tissue surely would not have been intact. I believe this point has been overlooked. With the hindsight of 30 additional years in which a number of wet sites in Florida have yielded human skulls containing brain tissue dating also to the Middle Preceramic Archaic, archaeologists are eating crow as fast as they can swallow it and we can thank Royal and Clark for their persistence.

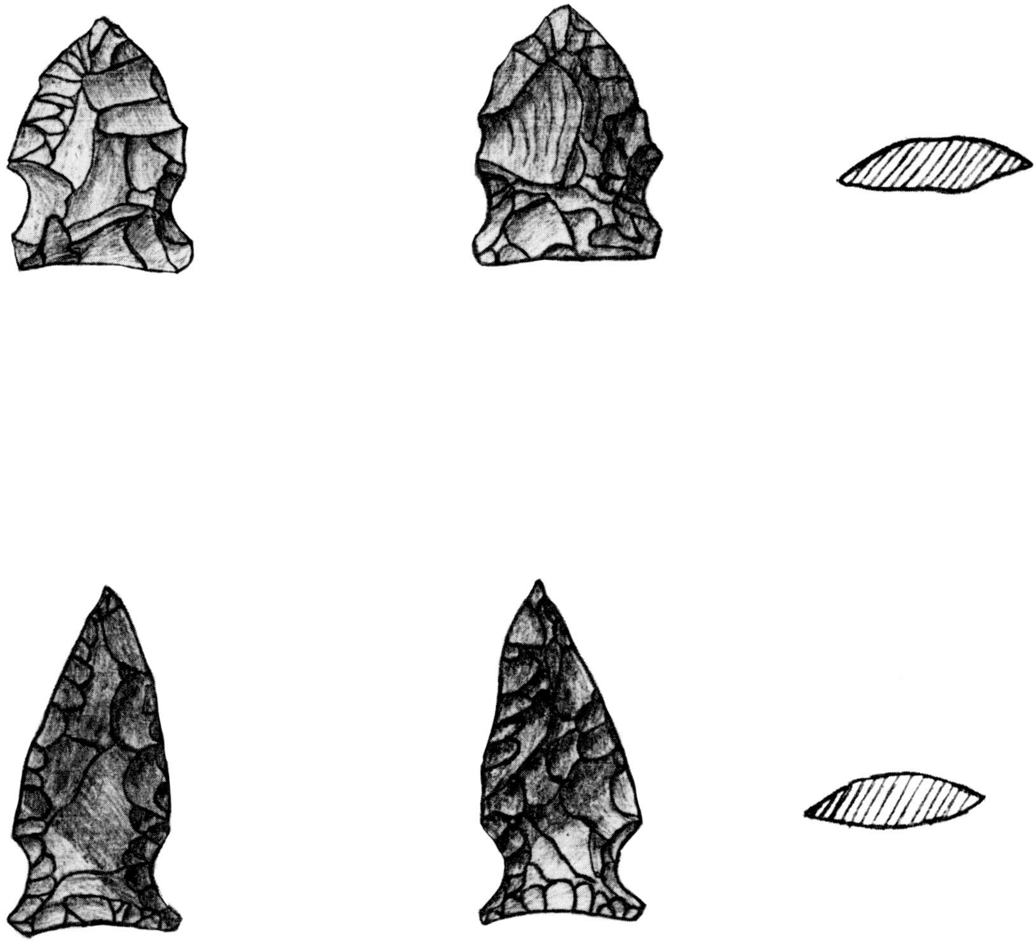

FIGURE 76B

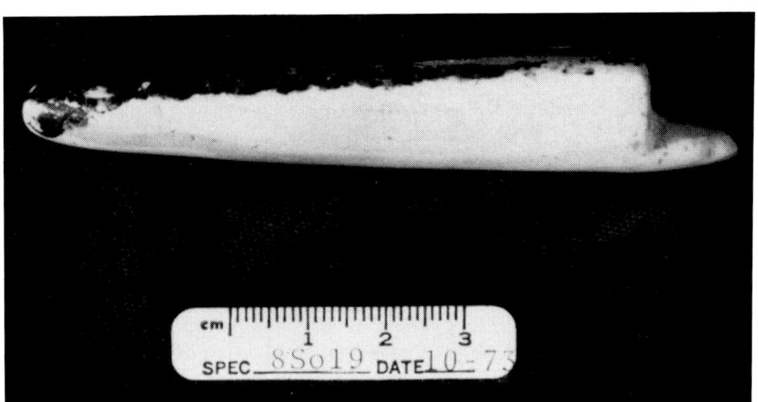

FIGURE 76C

through soft mud, every footstep stirring up a cloud of muck which cut down to almost nothing the already poor visibility in the milky green water. She swam around the circular spring and noted that at the sloping center the bottom dropped down into a huge black hole. She returned to Warm Mineral Springs many times after her initial visit and began to identify the organisms that lived there. Eventually she put on an aqualung and ventured down the vertical drop off into the sinkhole. She discovered that there were no fish or other signs of animal life below 30 feet.

Eugenie Clark had known Bill Royal for a long time. They had dived together in the Gulf of Mexico. It was Royal who talked Clark into diving in Little Salt Spring in 1959 (see section on Little Salt Spring). Shortly afterward, they switched their diving to Warm Mineral Springs and it was at this time they found human skeletal remains and artifacts on the 13-m ledge. Clark relates that they thought the finds were important but they could not convince any Florida archaeologist that it was worth investigating. "Dr. John Goggin—the noted 'Father of Florida Archaeology' at the Florida State Museum and the only archaeologist in Florida who was a scuba diver—disappointed us by being too busy to examine the site and did not think we knew what we were talking about" (Clark 1969:170–171).

But there were interested scientists. Dr. Carl L. Hubbs of Scripps Institution of Oceanography persuaded Dr. H.E. Suess to run a radiocarbon analysis on a burned log of red mulberry (*Morus rubra)* from the ledge. It resulted in a date of 10,000 years old. Royal and Clark thought that this radiocarbon date would bring qualified archaeologists to study the sedimentary layers on the 13-m ledge. But it was difficult. Finally, Dr. Harry Shapiro at the American Museum of Natural History encouraged a scuba-diving colleague to check the sedimentary layers. Unfortunately, he developed a cold on the morning of the dive and could only clear his ears to about 6 m while Royal and Clark waited at the 13-m ledge to show him the evidence. "Word quickly spread that a qualified anthropologist from New York came to dive with us at the springs but could not verify the sedimentary layers" (Clark 1969:171–172).

The plugged ears of the diving archaeologist were minor compared to what happened later. On July 11, 1959, Royal removed an intact human skull containing brain tissue from the 13-m ledge. Unfortunately, the find was made during a televised Huntley-Brinkley news program with a scuba-diving TV photographer, John Light, recording everything on film. Everyone thought that Royal had planted the skeletal material. No one believed that brain tissue could possibly have survived for thousands of years. Even those individuals who had been intrigued by the material recovered previously by Royal and Clark thought that Royal had faked the whole thing to make a spectacular TV story. Some people suggested the brain belonged to a modern-day turkey.

Those who saw the brain tissue and worked on it knew it was real. When Royal first picked up the skull and was displaying it underwater in front of the camera, he turned it upside down and noticed something white and soft inside the foramen magnum. After he surfaced, Royal, still in front of the camera, shook out pieces of the brain. The skull was put in a bucket of spring water and taken to Clark's laboratory where she and a colleague looked at it and thought it appeared as fresh as a brain from a butcher shop (Fig. 77). They could see the convolutions of the cerebrum and cerebellum. After a few hours the brain

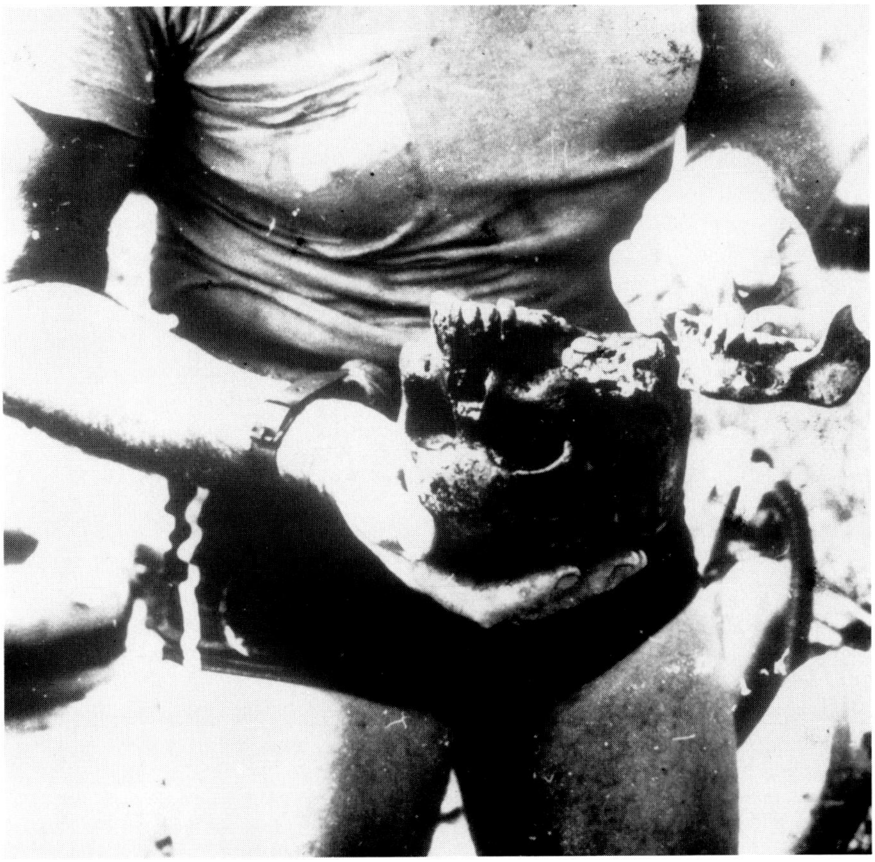

A

FIGURE 77 Historic pictures showing (A) W.R. Royal looking at brain tissue through the foramen magnum of the Royal skull immediately after it was recovered from the 13-m ledge in 1959; (B) surgeon sawing the back of the Royal skull to remove the brain with W.R. Royal and Eugenie Clark looking on; (C) the skull sawed open; (D) Eugenie Clark holding a piece of the brain with opened skull visible in the background; (E) fragments of the brain (all pictures courtesy of W.R. Royal).

started to break into pieces and formalin was added to the water to harden it. The next day, in a doctor's office in Sarasota, the back of the skull was sawed open and the brain fragments were put in a jar of formalin. A neurosurgeon confirmed their observations about the cerebrum and cerebellum, and cholesterol was found in the white matter by a biochemist. No cellular structure was found when a pathologist made microscopic examinations of stained histological sections.

Still no one believed that brain tissue had been preserved and the skull could not be assigned any great age because no archaeologist had checked the site and confirmed the sedimentary layers. Royal and Clark eventually gained some supporters from specialists who told them that human brains had been preserved naturally in peat bogs, in some French cadavers, in a Roman age individual, and in Peruvian and Egyptian mummies. Finally, a report of their findings was published in *American Antiquity* (Royal and Clark 1960:285–287).

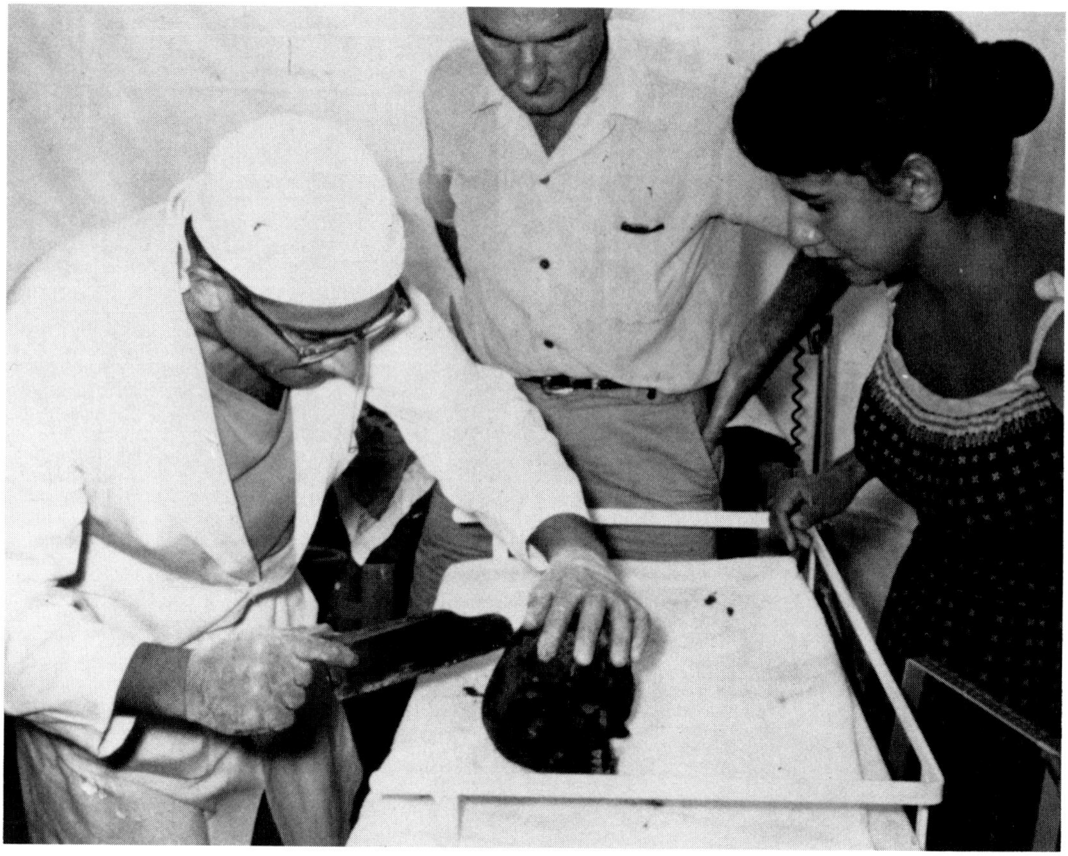

FIGURE 77B

Royal and Clark state that the human skull containing brain tissue was not found in the same zone as the red mulberry log dated 10,000 years old. The skull was found in Zone 2 and Royal and Clark mention that it could be several thousand years younger than the log. A method for radiocarbon dating bone had just been developed and 52 bones associated with the skull were sent to Monaco. Through the help of Captain Jacques-Yves Cousteau, tests were made on them by Dr. J. Thommeret, Chief of the Radiocarbon Laboratory of the Centre Scientifique de Monaco. The date received on the bones, after several years of waiting, ranged between 7140 and 7580 years old (Clark 1969:176) providing a Middle Preceramic Archaic age for this individual. In support of Royal's discovery is the fact that the brain material appeared fresh when it was found but quickly began to disintegrate after it was removed from the deposit in which it had been entombed. If Royal had planted a previously recovered skull to provide spectacular TV coverage, the brain tissue surely would not have been intact. I believe this point has been overlooked. With the hindsight of 30 additional years in which a number of wet sites in Florida have yielded human skulls containing brain tissue dating also to the Middle Preceramic Archaic, archaeologists are eating crow as fast as they can swallow it and we can thank Royal and Clark for their persistence.

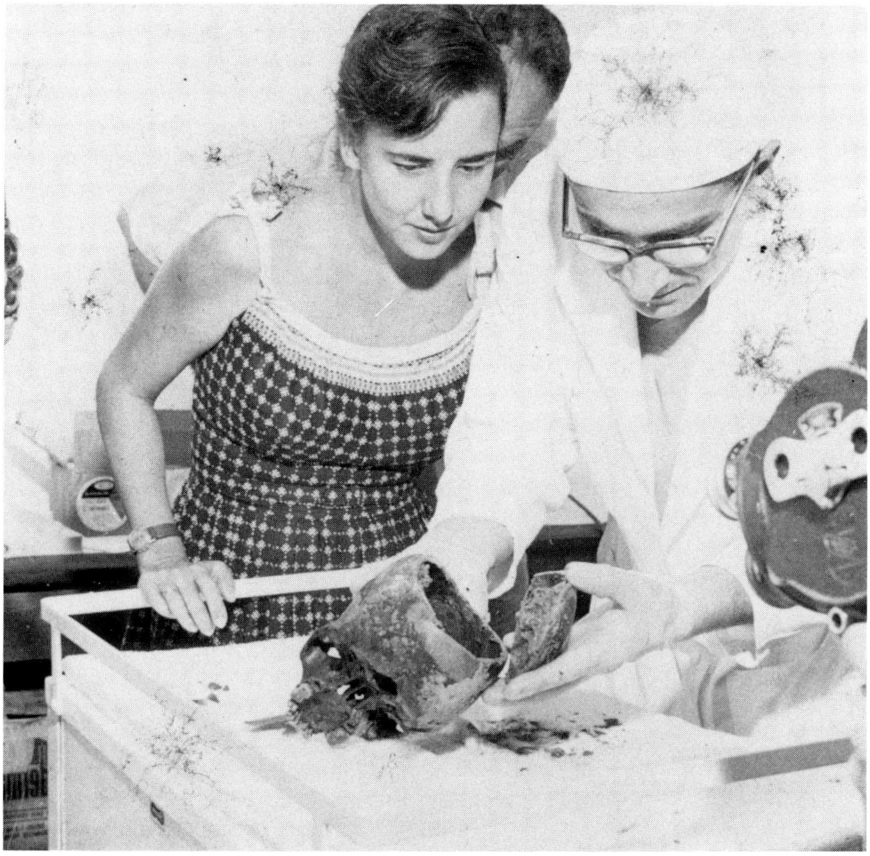

FIGURE 77C

John M. Goggin died in the early 1960s but his reputation as an exemplary archaeologist lingers on. In his discussion of underwater archaeology he made it clear that "underwater archaeology" done by diving enthusiasts does not achieve what "we consider basic coverage in archaeology; that is, the presentation of field methods, adequate description of artifacts both quantitatively and qualitatively, a full discussion of their context, and the drawing of adequate conclusions.... These [accounts] are a matter of much more concern because of their 'professional aura' although an examination of their techniques reveals them to be unacceptable to professional archaeologists" (Goggin 1960:349).

Goggin defined underwater archaeology as "the recovery and interpretation of human remains and cultural materials of the past from underwater by *archaeologists* [emphasis in the original]. This last word should be stressed.... a distinction should be made between underwater *archaeology* and underwater *salvage*. One is a science, the other is collecting [which he equates with pothunting]. It is far easier to teach diving to an archaeologist than archaeology to a diver!" (Goggin 1960:350).

Goggin had been diving since 1947 and had initiated several underwater archaeological projects in Florida. In 1959, he dove in Little Salt Spring and

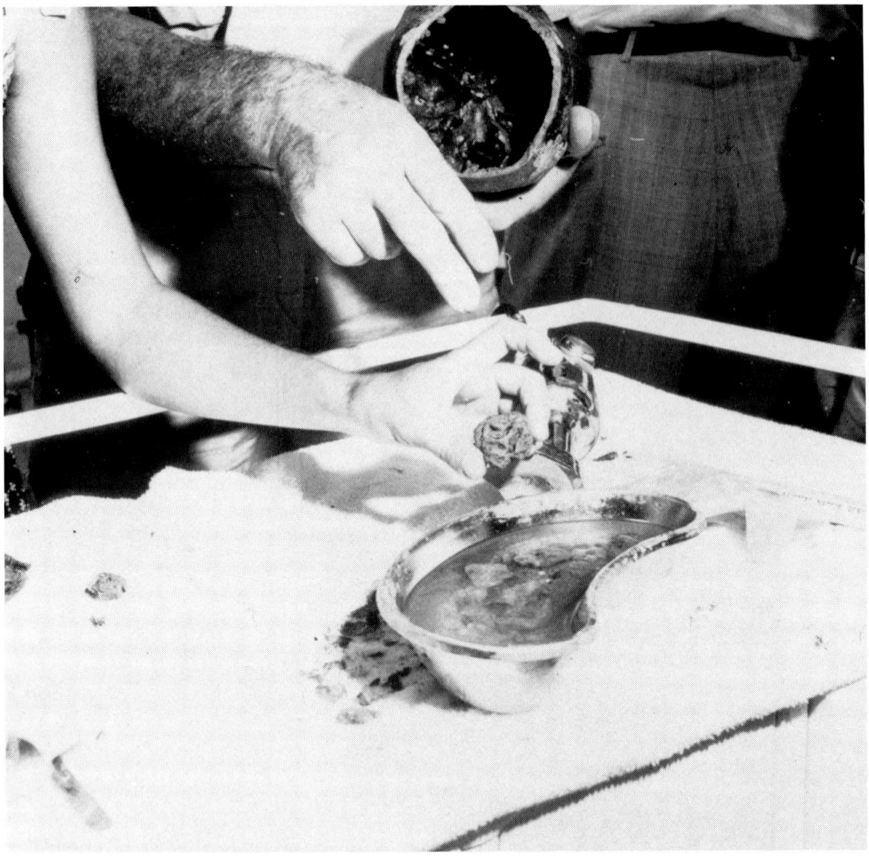

FIGURE 77D

"made one brief dive [in Warm Mineral Springs] down to about a hundred feet with Lieutenant Colonel Royal, the man who had found the first bones.... At the time I went down with Colonel Royal... they had found [previously] one human bone from this ledge [the 13-m ledge]. Subsequently, they went back diving. And as you probably know, if you have been looking at television, they have made some very sensational discoveries all by themselves without any consultation with any archaeologists..." (Goggin 1962:80). Goggin's 1962 article is a published version of a paper he presented at the Southeastern Archaeological Conference. At that time, he was asked by several colleagues why he had not been part of the team that recovered the skull containing the brain. Goggin was quoted "by a newspaper man's excuse for a scientist" as saying "Tragic that this work was done by such irresponsible people" (Goggin 1962:81). In speaking of the log that yielded a radiocarbon date of 10,000 years, Goggin said: "The observations, as far as I know, were made in the water by Colonel Royal. I dove with him and before I went down he explained to me what I would find in these springs. What I found was certainly different than he described. I have no faith in his power of observation.... It is a very fascinating problem, but this business of ichthyologists and retired Air Force officers setting themselves up as archaeologists is a little discouraging" (Goggin 1962:82–83). One

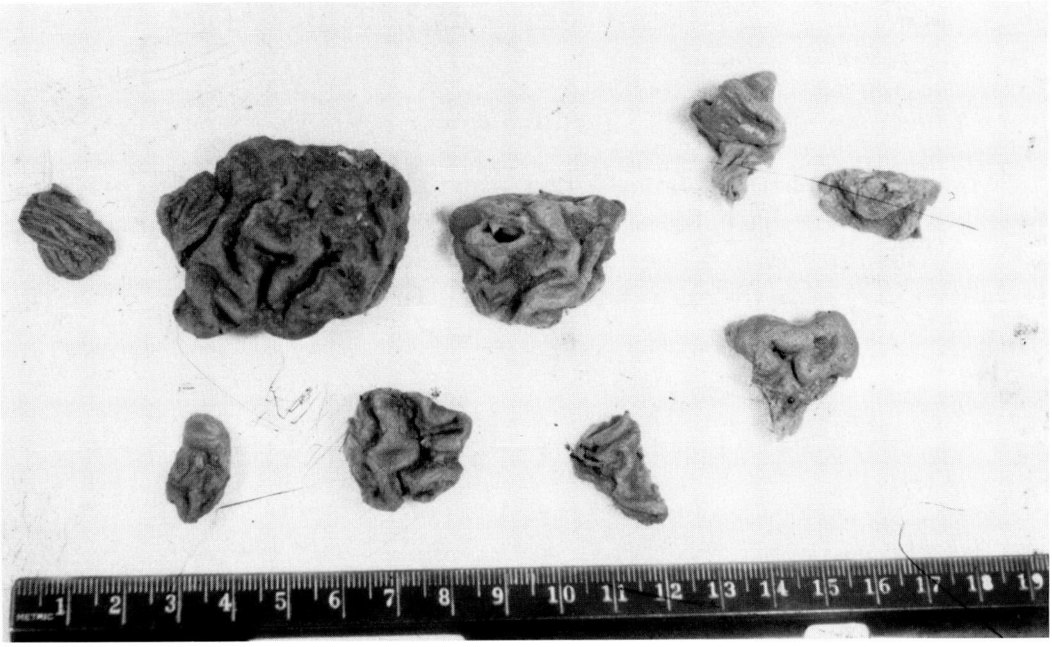

FIGURE 77E

of Goggin's colleague's "felt that the American Anthroplogical Association should protest the publicity on a thing like this, without contacting any professional people.... they were represented as archaeologists....over and over again during the film." And Goggin responded: "This is the discouraging thing; that this is called archaeology.... that someone who is a responsible scholar in one discipline seems to carry no responsibility over into another" (Goggin 1962:83).

It is lamentable that we will never know the exact provenience of many of the finds made in Warm Mineral Springs and that Royal has cemented many of the items into his fireplace (Skow 1986). On the other hand, if Royal had not been adventurous enough to dive in the springs would the springs, and all of the information they have provided over the years, still be undiscovered? In 1973, former Florida Secretary of State Richard Stone presented Bill Royal with a special commendation for his part in identifying and preserving the deposits at Warm Mineral Springs. It was felt that cooperation between scientists and sport divers would lead to important underwater research (Blount 1978). This attitude contrasts dramatically with that of the 1960s which appears to prevail again today (Skow 1986).

In addition to the damage caused by collectors, the spring area has been modified to accomodate bathers. For more than 30 years, sand has been dumped in the shallow water of the spring to provide a wading beach. Some of this sand has moved into the cavern where it erodes the walls and debris cone. In 1985 Warm Mineral Springs was nominated for acquistion through the Conservation and Recreational Lands (CARL) Program. Under state ownership the site would be protected from future destruction, but Warm Mineral

Springs was sold recently to a foreign investment group expected to begin management sometime in 1990.

Present Location of Materials Taken from Warm Mineral Springs

Materials and records from Warm Mineral Springs are at the Division of Historical Resources in Tallahassee, at the Florida Museum of Natural History in Gainesville, at Warm Mineral Springs, and in the possession of William R. Royal who lives near the spring.

References

Blount, Steve 1978 Warm Mineral Springs. *Sport Diver:* May issue.

Clark, Eugenie 1969 *The Lady and the Sharks.* Harper & Row, Publishers, New York.

Clausen, Carl J., H.K. Brooks, and Al B. Wesolowsky 1975a The Early Man Site at Warm Mineral Springs, Florida. *Journal of Field Archaeology* 2(3):191–213.

1975b Florida Spring Confirmed as 10,000 Year Old Early Man Site. *The Florida Anthropologist* 28(Part 2):1–38.

Cockrell, Wilburn A. 1981 Pleistocene Man in Florida. In: *In the Realms of Gold.* Edited by Wilburn A. Cockrell, pp. 176–181. *Fathom Eight Special Publication* No. 1. San Marino, California.

1986 The Warm Mineral Springs Archaeological Research Project. Manuscript on file at the Division of Historical Resources, Department of State, Tallahassee.

1990 Personal Communication during author's visit to Warm Mineral Springs.

Cockrell, W.A. and Larry Murphy 1978a Pleistocene Man in Florida. *Archaeology of Eastern North America* 6: 1–13.

1978b 8-SL-17: Methodological Approaches to a Dual Component Marine Site on the Florida Atlantic Coast In: *Beneath the Waters of Time.* Edited by J. Barto Arnold III, pp. 175-180. *Texas Antiquities Committee Publication* No.6.

Goggin, John M. 1960 Underwater Archaeology: Its Nature and Limitations. *American Antiquity* 25(3):348–354.

1962 Recent Developments in Underwater Archaeology. *Southeastern Archaeological Conference Newsletter* 8:77–88.

King, James 1975 Analysis of Pollen from Warm Mineral Springs. Unpublished manuscript on file, Underwater Archaeological Research, Department of State, Tallahassee.

McDonald, H. Gregory 1975 The Warm Mineral Springs Fauna. Unpublished manuscript on file, Underwater Archaeological Research, Department of State, Tallahassee.

1976 Additions to the Warm Mineral Springs Fauna. Unpublished manuscript on file, Underwater Archaeological Research, Department of State, Tallahassee.

Morris, Donald H. 1975 Warm Mineral Springs Man. Unpublished manuscript on file, Division of Historical Research, Department of State, Tallahassee.

Murphy, Larry 1978 8-SO-19: Specialized Methodological, Technological, and Physiological Approaches to Deep Water Excavation of a Prehistoric Site at Warm Mineral Springs, Florida. In: *Beneath the Water of Time.* Edited by J. Barto Arnold III, pp. 123-128. *Texas Antiquities Committee Publication* No. 6.

Royal, William and Eugenie Clark 1960 Natural Preservation of Human Brain, Warm Mineral Springs, Florida. *American Antiquity* 26(2):285–287.

Sheldon, Elizabeth and Marguerita L. Cameron 1976 Reconstruction of Prehistoric Environments: The Warm Mineral Springs Project. *Southeastern Archaeological Conference Bulletin* No. 19:68–69.

Skow, John 1986 This Florida Spa Holds a Surprising Lode of Prehistory. *Smithsonian* 17(9):73-83.

Windover

Introduction

By archaeological standards established anywhere in the world, the Windover Site (8-BR-246) is one of the most significant and informative sites of the 20th century. The full potential of the Windover site is being realized thanks to the generosity of the landowners and because the principal investigators, Glen H. Doran and David N. Dickel, did a superb job of directing the fieldwork and guiding the studies of enormous quantities of diverse materials.

The Windover site is located in east-central Florida near Titusville about 8 km west of Cape Canaveral (Fig. 1). It was discovered in 1982 during construction within the Windover Farms housing development. Peat was being removed from a small pond prior to road construction when an observant backhoe operator recognized skeletal material in the peat matrix. Construction was delayed until archaeologists were brought in to evaluate the materials. Windover Farms, Inc., the Florida Legislature, and a number of other organizations supported archaeological investigations during three field seasons totaling 17 months in fall-winter 1984, 1985, and 1986 (Doran and Dickel 1988a:356).

Human bones representing a minimum number of 168 individuals were recovered from the 5400 m^2 (1.7 acres) Windover mortuary pond. Ninety-one of these had brain tissue remaining in the crania. This discovery led to immediate international attention. There are many other features of the Windover site, however, that are equally spectacular. Its antiquity, 7000–8000 years old, provides previously unknown glimpses of early utilization of organic materials that usually do not survive. These include wooden artifacts, textiles, incised bone objects, deer antler tools, faunal remains, and plants including the oldest known bottle gourd (*Lagenaria siceraria)* in North America (Doran, Dickel, and Newsom 1990). Marine shell tools were also recovered. Windover furnishes insights about burial practices, the use of grave goods, cause and age of death, and the environment in this area of Florida during the early Holocene. All of these components form an integral picture of aboriginal life in Florida many millennia ago. They are described below along with the innovative excavation, preservation, and analytical techniques employed by the principal investigators and their multidisciplinary team of collaborators.

Excavations

Every archaeological wet site requires the development of creative methods of excavation whereby provenience can be maintained without harming the fragile materials. At Windover, the challenge was particularly great because the objects of interest were buried in a peat deposit more than 2 m below the pond bottom which contained 1–3 m of standing water. This fundamental problem was solved eventually by installing a dewatering (wellpoint) system capable of capturing both the surface and subsurface waters and discharging them away from the excavation area. Doran and Dickel (1988b:268–270) describe thoroughly the system and the adjustments made in it over the 3 years of fieldwork. Their objective was to use the dewatering system to dry the pond but still maintain the integrity of the peat deposit. After successful experimentation in 1984, the entire pond was encircled in 1985 with 158 wellpoints. There was concern that the center of the peat deposit would be so far from the wellpoint perimeter (ca. 30 m) that it would not dry out, but this was not a problem. The expense of reinstalling the wellpoints in 1986 was avoided by leaving the wellpoints in the ground from January to August 1986 when no field excavations were in progress. The points survived the long contact with heavily mineralized water and the system was pumping an estimated 700 gallons of water per minute shortly after it was put in operation on August 16, 1986. The amount of rust and minerals deposited in this 8-month period, however, indicated the points could not have been left in the ground indefinitely. A conservative calculation of the total cost of the dewatering system during the 3 years of excavation, including all peripheral expenses, was $120,000 (Doran and Dickel 1988b:270).

A total of 163 2-m squares was excavated at Windover: 31 in 1984 and 132 in 1985 and 1986. Preceding systematic excavations in all three field seasons, a backhoe removed the uppermost noncultural zone under the supervision of Drs. Doran and Dickel (Nabergall-Luis 1990).

At the end of each field season the excavated area was backfilled and the rising waters gradually refilled the pond. Two thirds of the burial area was excavated. The pond is now on the National Register of Historic Sites and preserved from further disturbance.

Several methods of collection were employed to obtain faunal and macrofloral remains and environmental information. Controlled volumes of peat from all stratigraphic levels were fine screened and all identifiable material was retained. The matrix surrounding each human burial or feature was water screened and materials encountered were bagged and retained for subsequent analysis. Sixteen column samples were removed and water screened through nested screens ranging in size from 1/4 to 1/16 inch mesh (Doran and Dickel 1988b; Nabergall-Luis 1990). Continuous core samples were taken for palynological and petrographic analyses.

Stratigraphy and Chronology

Doran and Dickel (1988b) describe four distinct peat strata identified palynologically and petrographically (Fig. 78). From bottom to top, the oldest and deepest is a water lily peat which is underlain by gray Pleistocene sand. The water lily peat grades into a colloidal peat (called rubber peat by the authors). It is replaced by a red-brown peat stratum which is divided equally into a lower and upper zone. The lower zone of this stratum contains most of

MAMASL	Excavation Zones	Description	Radiocarbon Dates[1]
2.77			
	ONE	BLACK MUCK	
2.27			6,070 ± 90 B.P.
	TWO B	UPPER RED/BROWN PEAT	
1.67			7,360 ± 70 B.P.
	TWO A	LOWER RED/BROWN PEAT	
1.07			7,950 ± 140 B.P.
	THREE	RUBBER	
0.47			8,990 ± 90 B.P.
	FOUR	WATER-LILY FIBEROUS PEAT	
-.33			10,750 ± 190 B.P.
-.53	FIVE	SAND	

Vertical Scale= 10 cm.
MAMASL: Meters Above Mean Average Sea Level

[1] uncorrected radiocarbon dates. Klein et al. (1982) reports radiocarbon dates between 7,000 and 8,000 B.P. are approximately 800 years too young. Thus, the 7,210 B.P. date (Beta-7186) is corrected to 8,010 B.P. or 6,060 B.C.

FIGURE 78 Diagram of Windover stratigraphy (taken from Nabergall-Luis 1990).

the human bone and artifacts. The uppermost stratum is a black, sawgrass peat which began accumulating about 6000 years ago and represents the most recent deposition to the pond bottom.

Twenty-seven radiocarbon dates on a variety of materials are provided by Doran and Dickel (1988a). The water lily peat evidently began forming shortly after the end of the Pleistocene and the macrofloral component of this stratum is dominated by enormous quantities of water lily (*Nymphaea*). The base of the water lily peat in contact with the underlying sand is greater than 10,000 years. Water lily requires water depths of 45–91 cm. There are occasional buttonwood roots found and ash levels are highest in this stratum. Sulfates and pyritics are relatively low (Spackman, cited in Doran and Dickel 1988b; Stout 1986).

The rubber peat dates from approximately 9000–8000 BP and contains a high density of freshwater snail shells (formerly called *Heliosoma*, now *Planorbella* sp., *P. duryi*, *P. scalaris*). Except for a few small twigs, this stratum has almost no identifiable plant material. The precise petrographic and chemical source for this zone is not clear. It may have resulted from subsurface salt water intrusion coupled with high algal/bacterial activity. It is a clay-rich sediment similar to gyttjae peat and Stout (1986) believes that the rubber peat sediments were organic, oozy muds derived from the pond margin. This stratum is chemically distinct from the one below and those above. Pine pollen drops dramatically and Cheno-Am pollen climbs. Sulfates, calcium, and pyrites also rise. Water depths were probably at 1–2 m during the deposition of this stratum.

The *Planorbella* shells were almost entirely restricted to the rubber peat stratum although they were found occasionally in the lower red-brown peat zone which began to form around 8000 BP. Naturally occurring and culturally modified wood is prevalent in the lower red-brown peat. Three Archaic stemmed stone bifaces, one of a type called Kirk Serrated, were recovered from the lower red-brown peat. They have been dated elsewhere at ca. 8000 BP and thus their presence supports the radiocarbon dates for this zone. The lowest levels of the red-brown peat began around 8000 BP and by about 7000 BP water and deposition conditions were sufficiently different to result in visible changes in the peat layers. Cheno-Am and oak pollens decline slightly, sulfates remain relatively high, pyritics decline, and there is an enrichment of phosphorus (Spackman in Doran and Dickel 1988b; Stout 1986). The entire stratum represents a prolonged drying trend. The floral, palynological, and petrographic components are very distinct. The pond was probably no more than 20 cm deep. The area during this time was a woody marsh with water at or slightly above the marsh surface. The peat that accumulated is best described as Cephalanthus-Myrica-Salix peat (Stout 1986). The upper red-brown peat zone ends about 6000 BP with wood debris at its highest which suggests maximum drying and maximum pond margin encroachment. Cheno-Am pollen continues to decline and pine pollen rises slightly. The distinctions between the lower and upper red-brown peat stratum were made originally in the field based on texture and visual observations. The differences between the substrata were later confirmed through fauna, flora, and peat analyses (Nabergall-Luis 1990; Stout and Spackman 1988).

The black sawgrass peat began to form about 6000 BP. The black color comes from seasonal burning and natural fires which are a common feature today of central and south Florida. Sulfates and pyritics virtually disappear. Wood is scarce except for charred material. Pine pollen gradually rises while oak pollens are at their lowest levels. Essentialy modern climatic features prevailed during the deposition of the sawgrass peat stratum.

Floral Remains

Palynology—A total of 272 cm of material was removed from a cleaned, exposed profile at the Windover site in November 1984. In addition, 63 cm of a core that penetrated to the underlying sands was obtained from the floor of an excavation unit. The measured depth of the core was 1.26 m but the peat strata were compacted during extrusion. Five soil samples were collected during the 1986 field season, three from burials and two from a diagnostic feature. All of these samples were analyzed for pollen and microfossils. Holloway (1987) describes in detail the procedures he used to extract and examine the samples. Pollen influx calculations of sediments were accomplished by the addition of a known quantity of exotic marker grains.

Sedimentation at the Windover site was initiated at a time when the late glacial mixed deciduous forests were retreating northward (Zone 5). Pine (23–34%) and oak (12–25%) dominate, and hickory is 3–5%. Pollen reflect a change from an early Holocene coniferous woodland to one dominated by oak (40%) marking the beginning of a warm trend (Zone 4). Pine pollen decreases to an average of 8% and aquatic taxa peak during Zone 4. There was a concomitant reduction in water levels at this time. Cheno-Ams (53%) become the dominate pollen in the rubber peat zone (Zone 3) and oak decreases significantly to 15%. In the red-brown peat zone (Zone 2), Cheno-Ams increase (62%) and willow (*Salix*) peaks but only in the lower and middle portions of the zone. Zone 1 shows a return to more forested conditions with additional taxa present during the recent past. When Holloway (1987) conducted a secondary calculation of pollen frequencies by removing the Cheno-Ams from the pollen sum, the results revealed more subtle changes in the environment than previously thought. Pine and oak are more consistently present and do not show the large decreases reflected in the primary counts during Zones 2 and 3. This supports an interpretation of more stability within these zones than reflected using only the primary counts. Holloway (1987) concludes that the vegetational changes he observed took place gradually. He infers that a constant, uninterrupted sedimentary record is present at Windover rather than erosional episodes at the zonal contacts.

The coprolite (fecal) samples examined by Holloway, originally hoped to be human, were probably from large reptiles (alligator?). Their pollen assemblages were primarily Cheno-Ams. The macro remains consisted of finely masticated bone fragments, sandstone pellets, undeflocculated fecal materials, and eggs of parasites not specific to humans. These nonhuman coprolites were recovered from the same levels as the human osteological material. They were apparently deposited at the same time as the burials, rather than being intrusive from a younger deposit, because the large amounts of Cheno-Ams present suggest the same environmental conditions as those that prevailed at 7000 BP. The standing water and loose sediments would be ideal for reptile habitation.

Other samples analyzed by Holloway (1987) were the suspected colon contents of a burial, a grass mat, and a special feature. The colon sample and the mat were dominated by Cheno-Am pollen although the colon sample had a concentration value approximately one half that of the control sample. The special feature was composed almost entirely of Cucurbitaceae phytoliths which probably represent native vegetation. Seeds of native Cucurbitaceae were identified in the macrobotanical remains (Newsom 1988).

Macrobotanical remains—Approximately 10,000 plant remains were identified by Alexander (1986) from one third of each column sample and general level material to determine the number and kinds of botanical species present at the Windover site. The other two thirds have been saved for specialized analyses. Alexander describes in detail the methods utilized to process the samples and quantify the results. Plants are listed by categories of trees and shrubs, herbaceous plants, sedges and rushes, algae, vines, and water plants. Nine thousand seeds and spores were identified from the column samples. Wood and leaves were counted but not identified. The sawgrass peat (Zone 1) is dominated by >71% *Ambrosia* (ragweed). The upper red-brown peat contains 87% *Amaranthus* sp. (Cheno-Ams) with *Ambrosia* and *Myrica cerifera* (wax myrtle) the next most abundant species each composing about 5% of the count. The lower red-brown peat zone is still dominated by 84% *Amaranthus*. *Ambrosia* sp. makes up almost 10% of the total and *Chara/Nitella* (algae spores) 3.75%. All other species identified are present in quantities <1%. *Amaranthus* still dominates the rubber peat zone with 61% of the total. *Chara/Nitella* makes up 24% and a number of other species are present in nearly equal amounts (2–4%).

From the general level samples, almost all of which came from the lower red-brown peat zone, hickory (*Carya* sp.) represents 79% of the total identified, persimmon (*Dyospyros virginiana)* 10%, and swamp bay (*Persea palustris)*, 5%. Alexander mentions that these samples are biased toward larger specimens.

Alexander's (1986) primary objective in identifying the taxa from the Windover site was to use the information for environmental reconstruction. It was not intended for subsistence studies but she noted a number of plant remains that were potentially edible: hickory nuts, acorns, persimmon, elderberry, saw palmetto, grape, maypop, prickly pear, cherries or plums, roots of cat-tails and arrowheads, water shield, pond lily, water lilies, knotweed, and amaranth. Alexander's (1986) findings confirm other analyses in concluding that the red-brown peat zone represents a drier period from ca. 8000–6000 BP with a return to wetter conditions beginning with the development of the sawgrass peat. She mentions also that these conclusions agree with studies conducted at Little Salt Spring and other areas.

Direct associations of edible plants and human remains are rare at archaeological sites. The paleoethnobotanical component of the Windover site is largely without comparison in Florida. Newsom's (1988) study of the plant remains from Windover provides rare insights into the use of plants for food, medicine, and technology by people who lived in this area of Florida 7000–8000 years ago. Newsom's study focuses on the ethnobotanical aspect including subsistence remains found with human burials and identification of wood species used to manufacture utilitarian and other items. Samples were taken in direct association with human burials with control samples coming from the matrix surrounding the burials. One third of each sample was analyzed. Samples also came from the water-screened material. Newsom describes thoroughly the procedures she used to identify the macrobotanical remains. Unless noted otherwise, all of the information contained in the following paragraphs is taken from Newsom (1988).

Thirty species of wood were identified. These were representative of hardwood hammock, freshwater swamp, pine woodland, and other habitats. The total is increased to 56 when herbaceous taxa, vines, and fungi are added. The taxa closely parallel the present vegetation of the area except for the lack of tropical species in the assemblage. Habitat diversity and resource availability existed within the immediate environment of the Windover site.

A number of the human burials were found fully articulated and undisturbed. The abdominal region of these burials furnish the greatest potential for recovering subsistence remains because no midden or habitation area was found at Windover. Of the six samples that have been analyzed from the abdominal area of burials, only one contained a dense concentration of seeds that differed significantly from the matrix control samples. A total of 3324 seeds came from this burial of a 55-year-old female. Of this total, 83% (2753 seeds) were elderberry representing at least 550 individual elderberry fruits. Since the sample analyzed represented only one third of the total taken, 1650 individual fruits may have been present. In addition, 124 grape seeds (about 40 grapes), 19 prickly-pear seeds, acorn, and black nightshade were identified. Other seeds present are those of species that are also abundant in the matrix control samples. The specimens from this burial differed significantly from the other abdominal samples which closely mirror the control samples for seed content. One of the abdominal samples, however, was interesting because of the relatively high concentration of mostly pulverized and unidentifiable bone it contained.

Additional samples came from burial pits and from various locations of human interments. Grape, hackberry, persimmon, prickly pear, maypop, and other edible seeds/fruit were identified. Hackberry was the only seed type observed that was carbonized. A 40-year-old female was buried with a cactus pad which lay on her leg, over the right femur and beneath the right hand. A shelf fungus was recovered with another burial. One of the most interesting and informative specimens was that of a nearly whole bottle gourd *(Lagenaria siceraria)* found in association with a subadult burial. The gourd was found beneath the skeleton but above the fabric on which the body rested or was wrapped. A portion of the rind was accelerator dated at 7290 ± 120 BP and is the earliest firmly dated occurrence of bottle gourd in North America. Bottle gourds have a worldwide distribution and are a favorite plant for manufacturing containers of various sorts because they are durable and lightweight (Newsom 1988; Doran, Dickel, and Newsom 1990).

Other edible plants were identified at Windover but were not found in direct association with the human remains. These include plum, hickory, oak, red mulberry, and *Polygonum* sp. Creeping cucumber *(Melothria pendulata)* was identified in levels postdating human activities and could have been the source of the cucurbit phyoliths discussed by Holloway (1987).

Seed beads were recovered with two burials. Of the five beads submitted for identification, four were drilled *Sabal* sp. (probably cabbage palm) and one was made from the large stone of an unidentified fruit. Plant fiber was utilized to make cordage, bags, blankets, and baskets (Andrews et al. 1988).

Wooden artifacts from Windover can be grouped into three categories: sharpened stakes, tool handles and shafts, and miscellaneous. Stakes are most common and appear to have framed or marked burials. Eight different wood species were used to make the stakes: pine and seven locally available hardwoods. Of 36 stakes sampled, 19 were pine, 4 live oak, 4 privet, 2 persimmon, 2 indeterminate oak group, 1 white oak group, 1 dogwood, 1 ash, 1 buttonbush, and 1 unidentified hardwood. Wood species selection for this artifact type appears to have been random, a function of convenience rather than specific properties of the wood. The seven tool handles or shafts provide a good example of deliberate selection. Two shafts which were inserted in plumb-bob shaped bone artifacts were fashioned of highly resilient woods (dogwood and holly). A shaft associated with an atlatl cup and three others which were inserted in drilled manatee rib were made of hickory, a wood prized today for its resistance to impact and shock, great strength, and high elasticity.

The miscellaneous wood group includes a paddle-like object, a large pestle (Fig. 79), a small bowl, an incised stick, and two artifacts of unknown function (Fig. 80). The pestle and bowl are made of live oak which is a very hard, strong wood. The two artifacts of unknown function were made of compression wood of pine. Compression wood is anatomically distinct and exhibits unique chemical, physical, and mechanical properties including an increase in density over normal wood of the same species. The Windover woodworkers probably understood the benefits of this attribute.

Newsom (1988) suggests that at least some of the plant species identified from the lower red-brown peat zone are present because of disturbance associated with human activities. Thirty-one potential food/medicinal plants were identified. At least 13 edible species were found directly associated with human remains and 3 of these were found exclusively with burials (hackberry, black gum, and cabbage palm). As a whole the species recovered in association with

FIGURE 79 Oak pestle from the Windover site (picture credit, Richard Brunck).

the burials are predominately those of seeds of edible, fleshy fruits and were probably ingested. There is a possibility that some or all of them could have served as grave offerings instead. Even so, these plants were undoubtedly a valued part of the prehistoric economy.

Based on the season that most of these plants are usable, Newsom (1988) was able to suggest that human activity at the site probably occurred primarily from summer to fall.

> In further support of a mid-summer/fall period for site activities, wooden stakes (pine) associated with at least six burials had their outermost growth rings preserved so that growth cessation, that is, time of cutting, could be determined. In five of the six cases, the summer or "latewood" increment was fully formed, indicating that the branches/stems were cut around September. In the sixth sample latewood formation had just begun, indicating that the piece was harvested around late June-July (Newsom 1988).

Faunal Remains

The faunal remains were recovered only in samples clustered around the rubber peat stratum and in the lower red-brown peat stratum. These zones were so osteologically abundant that their presence was even noted in the peat cores as was the virtual absence of fauna from the other zones, including the upper red-brown stratum. Nabergall-Luis (1990) supplies possible explanations for the differential preservation citing neutral to basic pH, high levels of calcium carbonate, and the presence of pyritic sulfur as evidence for sulfate-reducing bacteria which can only live in anaerobic situations optimum for preservation.

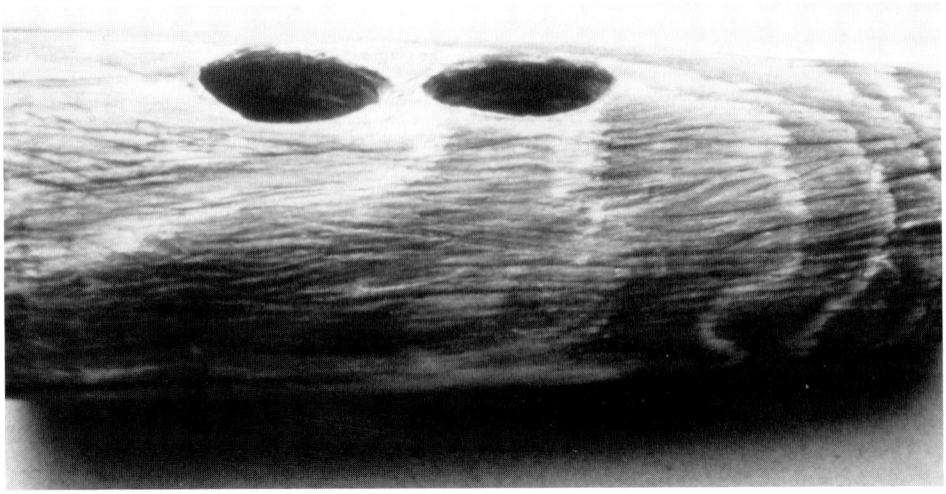

A

B

FIGURE 80 (A,B) Wooden artifact of unknown function from the Windover site; note the excellent condition of the wood after more than 7000 years (picture credit, Richard Brunck).

There is no indication that any of the animal bones were deposited as a result of human subsistence activities. Of the 53 taxa identified at least to genus, taxa of 14 birds, 13 mammals, 12 fish, 10 reptiles, and 4 amphibians are represented. All species are indigenous to this region of Florida and the southeast in general. Fish constitute 74% of the MNI, with the remainder being 14% reptiles, 7% amphibians, and less than 5% mammals and birds. All those represented are either pond dwellers or frequent pond margin visitors. Of the 13 species of invertebrates identified, there are 9 freshwater, 2 marine, and 2 terrestrial species. The most abundant fish are the minnow-like species such as the Florida flagfish and topminnows, followed by the freshwater catfish and the large mouth bass (Nabergall-Luis 1990:84). Although alligator bone was scarce, coprolites of a large reptile, probably alligator, were examined and are discussed in the section on pollen analysis.

Human Remains

The human skeletal material from the Windover site is the largest, most demographically intact skeletal sample from this time period in the New World (Doran and Dickel 1988b:275). A minimum number of 168 people were buried in the pond and 91 of the crania retained brain tissue. Age groups from infants to 65+ years old are present and there are approximately equal numbers of both sexes. The average adult female was 5 feet, 2–3 inches tall (ca. 157–160 cm) and the average adult male was 5 feet, 6–9 inches tall (ca.168–175 cm) (Doran and Dickel 1988a; Dickel 1988). Skeletal remains of subadults (individuals under 18 years of age), often under-represented in prehistoric populations, comprise 52% of the sample from Windover. Textiles were found with 37 of the burials and grave goods were placed with some individuals, particularly children. This large, well-preserved sample of human bones is making it possible to examine burial practices, health and disease patterns, metric and nonmetric variations, growth and development of subadults, fractures, remnant mitochondrial DNA, genetic relationships, and postmortem changes (Dickel 1986, 1988, 1990; Doran and Dickel 1988b:276). Many of these studies are still underway, and may be for years, but some interesting observations are available and summarized below.

Skeletal material—Because only a few bones out of thousands exhibit evidence of rodent gnawing, Doran and Dickel believe that the bodies were placed in the pond within 48 hours after death. Rapid burial and burial below the water table in an anaerobic environment also ensured excellent chemical preservation. The presence of brain tissue with identifiable internal and external structures argues strongly for rapid burial because generally within 48 hours of death the brain begins to break down (essentially it begins to "liquify"). Disarticulation of the bodies, however, is common in the sloping pond margins of the southwestern part of the pond, excavated in 1984 and 1985, because the bodies gradually moved downslope toward the deeper center of the pond. Bodies interred in flat locations on the northeast side of the pond, excavated in 1986, remained relatively stationary and articulated. They were buried flexed on their left side predominantly with heads oriented to the west and face to the north (Doran and Dickel 1988a:366; 1988b:273).

Adults commonly show minor osteophytic (bone outgrowth) lipping and articular surface roughness and pitting of observable joints—metatarsals, carpals, metacarpals, and phalanges, the superior/medial border of the proximal radius head, the inferior margin of scapular glenoids, and all recovered vertebral zygapophyses (Dickel 1988). Anterior teeth are heavily worn possibly due to non-dietary attrition with advanced alveolar resorption, especially right maxillary incisors. Pubic symphyseal and auricular (joint) surfaces of the pelves are badly eroded in older adults (Dickel 1988; 1990; Dickel et al. 1989:166).

No ulnar fractures have been identified in subadults but they are present in 10.1% of all adult ulnae examined (7 right and 9 left of 169). Of the 74 adult individuals for which one or more ulnae are available, 15% have one or more fractured ulnae (6/35 males and 5/28 females) including a male and female with bilateral midshaft "parry" fractures. As these tabulations indicate, there is no side or gender associations (Dickel et al. 1989:168).

One of the adult males, age about 47, who had a healed midshaft "parry" fracture of the left ulna, also had a well-healed right orbital floor blowout fracture accompanied by a superior orbit puncture scar and other periorbital changes. This is not an uncommon type of traumatic injury but it appears to be the first reported case regarding an archaeological specimen probably because the orbital floor is fragile and not often preserved. This type of fracture is caused by a sudden increase of intraorbital pressure resulting from a hard blow to the soft tissue of the globe imparted by something like a rock or a stick. Fractures like these may heal in 3 weeks but healing is dependent upon the age and health of the individual. The injury may have resulted in restricted right eye movement and double vision (diplopia). With time, double vision may have become less noticeable through behavior modification which in turn may explain some of the vertebral degeneration zygapophyseal joint changes observed in this individual. Both the ulnar fracture and the orbital blowout fracture exhibit about equal appearance of healing and remodeling, and any functional accommodations the body was going to make probably occurred a long time before death. Dickel et al. (1989:169) offer clues about the cause of the injury to the orbital floor. There is a small puncture lesion on the orbital roof and it appears that a pointed object entered the orbit and punctured the bone. The angle of the blow would have depressed the orbit toward the floor. The authors note that many bluntly pointed, hollowed-base antler and bone points were found at Windover and sometimes these were used for interpersonal violence because one was found deeply embedded in the auricular rim of a male pelvis. Their suggestion of the possibility of violent trauma is strengthened somewhat by the fact that three adults and two subadults had cranial depression fractures (i.e., blunt instrument trauma indicating interpersonal violence). This is good detective work but the authors caution that both the ulnar and orbital floor fractures could have occurred at the same time if, for example, the individual fell from a tree. This burial was placed in the peat matrix partially flexed and on his left side. He may have been held down by stakes, similar to most of the intact burials. He was found with a number of grave goods including a shark tooth, chert biface, butchered deer bone, deer antler tine, and two ulna awls made from deer and canid. His cranium contained saponified brain tissue (Dickel et al. 1989:167).

Spina bifida was found in 21% (5/24) of a small series of sacra from the Windover site. The term spina bifida encompasses a continuum of increasingly severe defects involving a failure of the vertebral neural arches to completely enclose the posterior spinal cord. Dickel and Doran (1989:325–334) describe the multiple pathologies related to spina bifida that affected a 15-year-old individual, probably male, from the Windover site. These include curvature of the spine in the lumbar region, a severe infection of the right tibia and fibula, disuse atrophy of long bones, and other anomalies which led to a loss of mobility, ulceration, and risk of serious infections. The right tibia was actively infected at time of death (one fourth of the distal portion was missing and the stump was draining). This individual had probably been handicapped for at least 3 years prior to death. These defects provide insights about the long-term health care afforded some members of this community 7500 years ago.

A study of 258 bones from the Windover site was undertaken to evaluate measurable skeletal indicators of physiological stress in juveniles and adults and to use the information to develop a broader picture of the occupants' health (Estes 1988:131). The sample included 75 tibiae, 58 femora, and 125 crania that were relatively intact and free of extensive pathology. The results of the study show that there were episodes of growth arrest as determined by the presence of Harris lines in the proximal tibiae of 92% of the sample. The average line count of five and the range between 0–14 was considered moderate compared with some populations (e.g., Indian Knoll) but greater than others (e.g., Grasshopper). There were no significant differences between juveniles and adults, or males and females. The highest frequency of line formation occurred between the ages of 7 and 11 (Estes 1988:132; Estes and Dickel 1988).

In an evaluation of cortical thickness at the midshaft of the tibia, Estes established that 16% of the sample showed abnormal results. A statistically significant difference was observed between two female age sets (18–35 and over 36). As measured by total cortical thickness, females lost 12% of bone mass after 36 years of age whereas males lost only 0.8% (Estes 1988:132).

Iron-deficiency anemia was present in 41.6% of the Windover sample and was manifest in the orbits and on the cranial vault (Estes 1988:133).

A subsample of 64 was selected for the purpose of analyzing the co-occurrence of stress markers. The sample included individuals whose recovered remains exhibited at least two out of three elements used for stress analyses (Dickel 1988). The association of each stress marker with each of the other two markers was statistically significant suggesting a relationship between all three. There was no significant difference between the frequency of multiple stress markers in adults and subadults. Sex difference in the occurrence of multiple markers was significant with males under-represented (Estes 1988:133).

Episodes of physiological stress during the growing years were common in the Windover population but the fact that children survived these episodes is a sign of good basic biological health. The inhabitants appear to have had adequate nutrition and enough activity to build and maintain relatively normal bone mass during their lives. The osseous changes accompanying iron-deficiency anemia are not severe. Juveniles between the ages of 5 and 10 were most in jeopardy, followed by females of all ages. Children less than 3 years old had few stress markers and died of causes that left no observable osseous changes. Skeletal markers indicate that increasing stress began at the age of 4 and juvenile mortality peaked around age 6 (Estes 1988:133–137).

A study was initiated by Smith et al. (1988) to evaluate the Windover skeletal material for the presence of albumin polymorphism and draw inferences about the relationship of the Windover population to living Native American groups which carry an albumin variant. Albumin is the most abundant protein in human serum and thus the most likely to survive in detectable quantities. Bone fragments (5 g each) of 83 individuals were recovered. Seventy-four of these had been frozen immediately after excavation and shipped in dry ice to the University of California at Davis. The remaining nine samples had been soaked in preservatives and stored at room temperature for up to 3 years. Smith et al. (1988) utilized the most recent technological developments designed to amplify electrophoretic phenotype patterns in dilute solutions. They discuss thoroughly the methods they used to prepare the samples which yielded approximately 1 mg soluble protein per gram of dry bone weight. Extracts from most of the bone samples, including those that had been stored, yielded detectable albumin phenotypes. The method used is suitable also for resolving phenotypes for serum proteins less abundant than albumin. Studies of serum albumin polymorphisms can be used to reconstruct phylogenetic relationships among living and prehistoric groups. From the present analysis, it appears unlikely that the Windover population is closely related to prehistoric ancestors of any group yet studied in North America.

Bone collagen is the most resistant of the serum proteins. Tuross (1988) conducted a stable isotope analysis of carbon and nitrogen of bone collagen from Windover. Based on the isotopic analysis, trace elements, artifact inventories, and proximity to the Atlantic Ocean, it seems likely that marine resources were exploited to some extent (Hancock 1987; Doran 1987:404).

Brain tissue—The Windover site skyrocketed to fame in December 1984 when Dickel noted "greasy tan stuff" inside a child's skull which he knew was not peat. Eventually brain tissue from a 45-year-old woman was positively identified by Drs. Philip J. Laipis and William W. Hauswirth, medical microbiologists at Shands Hospital, University of Florida (Stephenson 1985). When the brain was removed from this individual, the immediate visual impression was of a human brain. It was extremely fragile and difficult to handle. The crania of four other adults and four subadults recovered during the first season of excavation at Windover also contained soft tissue remnants. The skulls of three adult males, about 25–35 years old, "were appraised using conventional X-ray imaging, computerized axial tomography, and proton magnetic resonance imaging. Chemically different components within the intracranial mass could be recognized and differentiated . . . . When these brains were sectioned, peat and brain tissue were clearly distinguishable" (Doran et al. 1986:2). The brains had shrunk to about one-fourth of the original size, were missing a number of features, and were altered in consistency, but gross anatomical characteristics of contemporary brains were present (Doran et al. 1986:2; Hauswirth et al. 1988).

Rapid advances in molecular biology have occurred in recent years and have led to an enhanced understanding of gene activity on a biological level. When these new techniques are applied to nontraditional materials such as the human brain tissue from Windover, they provide new avenues for understanding prehistoric human population dynamics, human evolution, and colonization of the New World which are usually addressed through skeletal remains, surviving material culture, and linguistics (Hauswirth and C. Dickel 1988).

Nucleic acids were extracted from 15 g of relatively peat-free cortex and identified as DNA. The DNA was determined to be of human origin and mitochrondrial DNA (mtDNA) was present. The DNA was only 1% of that normally isolated from fresh tissue and only about 0.05% of the old DNA was mtDNA. DNA isolated from brain tissue would normally yield 0.5–1% mtDNA (Doran et al. 1986; Dickel et al. 1988).

The characteristics that make mtDNA studies attractive are its high mutation rate, estimated at ten times that of nuclear DNA, and its uniparental and haploid pattern of inheritance. Prehistoric mtDNA, if intact enough, could be used for investigating gene flow and population divergence. It is still uncertain if the Windover material provides the preservation needed for these types of studies (Hauswirth and C. Dickel: 1988), but about 40–50 generations are represented by the 1000 years that the Windover site was utilized as a cemetery. The mtDNA materials could be used to establish lineages for this nonliving human population, trace DNA sequenced diseases, and discover new versions of genes or genes not found or not sequenced in modern human populations (Hauswirth 1989, personal communication).

The brain tissue from Windover is the oldest human soft tissue yet analyzed at a molecular level. The only New World parallels in the preservation of soft tissue in other wet sites are preserved brain tissue in Florida (see other sections of this chapter). Genetic material may not always survive even when soft tissue is preserved. Efforts to extract DNA from an English bog body failed (Doran 1988:402). Hauswirth and C. Dickel (1988) offer an explanation for the long-term preservation of the brains from Windover.

> Materials most similar to Windover came from sites where adipocere formation occurred. Adipocere, also known as "grave wax" is a mixture of free fatty acids, primarily palmitic acid, and soaps resulting from the postmortem hydrolysis and hydrogenation of normal body fats. Damp conditions, fatty tissue, electrolytes (which may come from body fluids) and some putrification (to initiate hydrolysis) must be present for adipocere formation to take place. . . .
>
> The conditions at Windover which most likely inhibit bacterial growth and therefore

contribute to enhanced tissue preservation are: the high sulfur levels in water and peat, the highly mineralized waters, and the anaerobic conditions which generally exist 30 cm below the peat surface.

Oxidative damage to the DNA was minimized because of the anaerobic property of the water and DNA within the tissue was preserved due to its near neutrality (pH 5.3–6.8) (Hauswirth and C. Dickel 1988:26).

To ensure the survival of the brain material after the skeletal remains were removed from their burial matrix, tissue was removed from the cranium in the field, sealed in nitrogen-flooded containers and frozen at –20°C. Within 24 hours, they were transferred to the University of Florida, School of Medicine where they are maintained at –70°C until needed for further analysis (Doran 1987:405; Doran and Dickel 1988b:279–280).

Artifacts

Nonperishable materials—Shell and stone artifacts were not abundant at the Windover site but those that were present are significant. Of eight marine shells recovered, six were whole shells that accompanied burials and two were probably tools. The occurrence of these specimens demonstrates that shell technology, so prevalent during and following the Middle Preceramic Archaic Period, was at least in its infancy in the Early Archaic (Doran, personal communication 1990).

Five chipped stone artifacts were found at Windover. Three of these are Archaic stemmed points and one is typologically similar to Kirk-Serrated, an Early Archaic style, which supports the chronometric dates of ca. 8000 BP for the site.

Perishable materials—Wood, plant fiber, bone, teeth, and antler were used to manufacture tools, weapons, textiles, and decorative items that accompanied the dead as grave offerings in the Windover mortuary pond.

The wooden artifacts, discussed above in the section about macrobotanical remains, consisted of debarked, fire-hardened stakes, tool handles and shafts, and miscellaneous items such as a paddle-like object, a large pestle (Fig. 79), a small bowl, an incised stick, and two artifacts of unknown function (Fig. 80). Conservation of 29 wooden artifacts from the site has been completed at the University of Pittsburgh, Department of Anthropology. Preservation included a multistep procedure which involved desalinization to remove minerals by soaking in deionized water; treatment with alcohol, ethulose 100, and PEG 4000; freezing; and finally freeze drying (Doran and Dickel 1988a, 1988b; Gardner 1988).

A large number of technologically sophisticated textiles and other perishables were recovered from 37 burials. This assemblage currently represents the oldest textiles from the Southeast. The plainweave and twined textiles from Windover are presently the oldest in the hemisphere. Close simple twining and close diagonal twining, both with S twist wefts (paired and trebled); open twining with paired, Z twist wefts; and balanced plainweave specimens of nonheddle loom woven cloth are present. Represented forms appear to include circular or globular bags, hoods, blankets, and clothing. Cordage, braid, and fiber construction materials were also present (Andrews and Adovasio 1988; Andrews et al. 1988). Andrews et al. (1988) comment

> Whatever the vagaries concerning production and use of the Windover fabrics in life, the function of these items in death is certain. Specifically these accoutrements were probably specially selected (or possibly, in the case of subadults, manufactured) by family and/or friends to garb, hold the possessions of, as well as to enwrap the deceased, and when transfixed with cordage-enlaced stakes, to secure him or her beneath the quiet surface of the bog forever.

The specimens were treated with PEG/Ethulose at the University of Pittsburgh and are still undergoing conservation. At this time (Doran, personal communication 1990) a process involving parylene conformal coating technology appears the most effective way to conserve the fabric remains (Grattan 1989; Humphrey 1984). Their continued care should include climate control, restricted access, and extremely careful handling (Gardner 1988). The plant fibers used to manufacture the textiles probably came from the Sabal palm (Doran and Dickel 1988b).

Bone artifacts, both utilitarian and decorative, were manufactured from a variety of animals including manatee, deer, canids, felids, and birds. These include awls, bone pins and points, and incised objects (Fig. 81). The bone was removed from the field in plastic bags and treated with Rhoplex AC-33, an acrylic emulsion to prevent cracking, splitting, and contortion (Stone 1988; Doran and Dickel 1988a; Stone, Doran, and Dickel 1990).

Antler tools made from deer *(Odocoileus virginianus)* are numerous. Drilled and shaped antler artifacts may have been used with atlatls (Doran and Dickel 1988b).

Dentary tools were manufactured from shark teeth (some drilled) and pitch-coated dog canines (Doran and Dickel 1988b; Fig. 82).

The bottle gourd and cactus pad, discussed in the section about macrobotanical remains, can be considered artifacts also and should be added to the long list of usually perishable items that were recovered from the Windover site.

Discussion

The significance of the Windover site cannot be overstated. Its antiquity, superb preservation, diversity, abundance, and quality of materials are unparalled on the North American continent and probably in the Western Hemisphere. This is the first time a controlled, multiyear, multidisciplinary investigation of such a site has been possible. Taken one by one, most of the

A

B

FIGURE 81 (A,B,C) Decorative and/or utilitarian items of bone from the Windover site; (D) close-up of *C* (picture credit, Richard Brunck).

FIGURE 81C

FIGURE 81D

objects recovered at Windover have been seen before and some have come from sites even earlier than 8000 years. But collectively, the information from the Windover site has provided the oldest, most thorough view of aboriginal life yet known.

Laboratory investigations are ongoing and provide an exciting opportunity to expand our understanding of prehistoric populations, human biology, and early biocultural adaptation. The analysis of skeletal collections provides the only way to understand the interactions of environment, culture and human biology in a prehistoric setting. The importance of skeletal material, even material excavated many years ago, has exciting new potentials given the dramatic advances in protein and mtDNA isolation, identification, and sequencing (Doran, personal communication 1990).

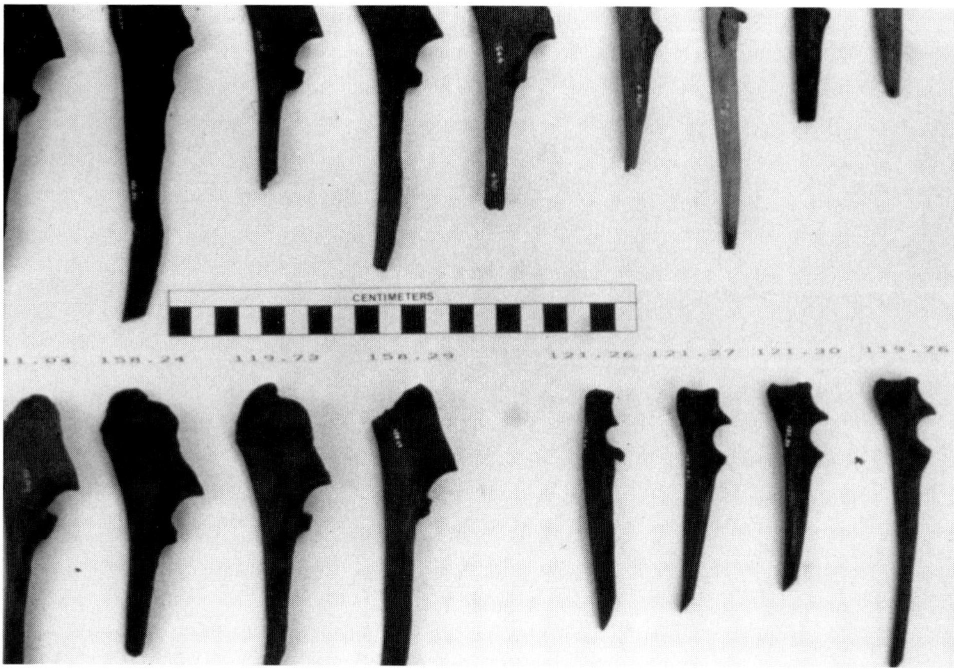

A

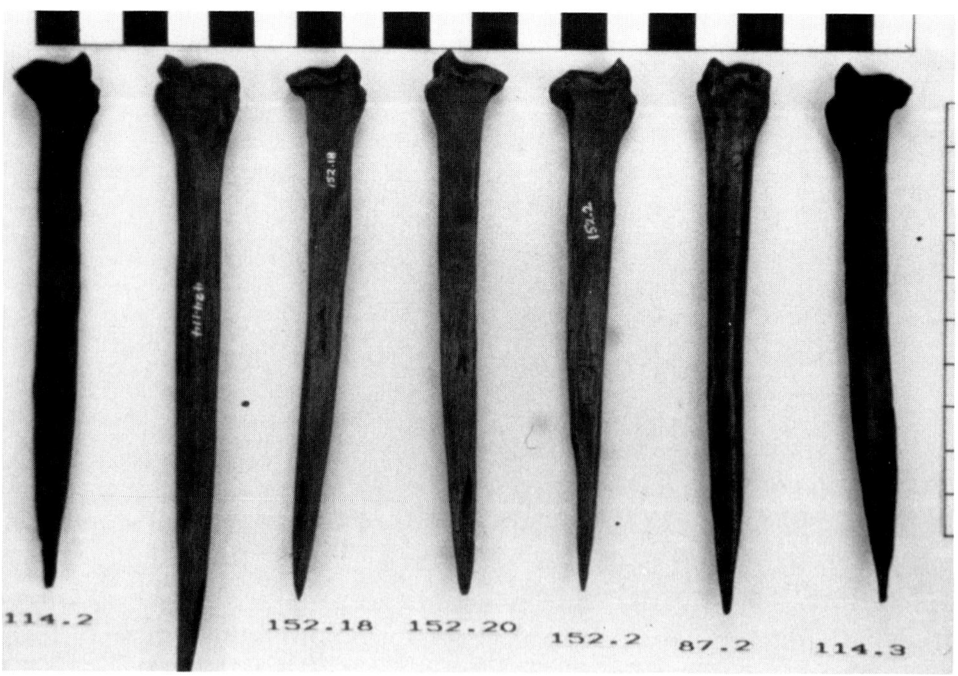

B

FIGURE 82 Bone and tooth artifacts from the Windover site: (A) ulna awls, (B) other bone tools, (C) shark teeth, (D) animal tooth artifacts. In addition, there were many drilled and shaped artifacts produced from antler (picture credit, Richard Brunck, Florida State University.

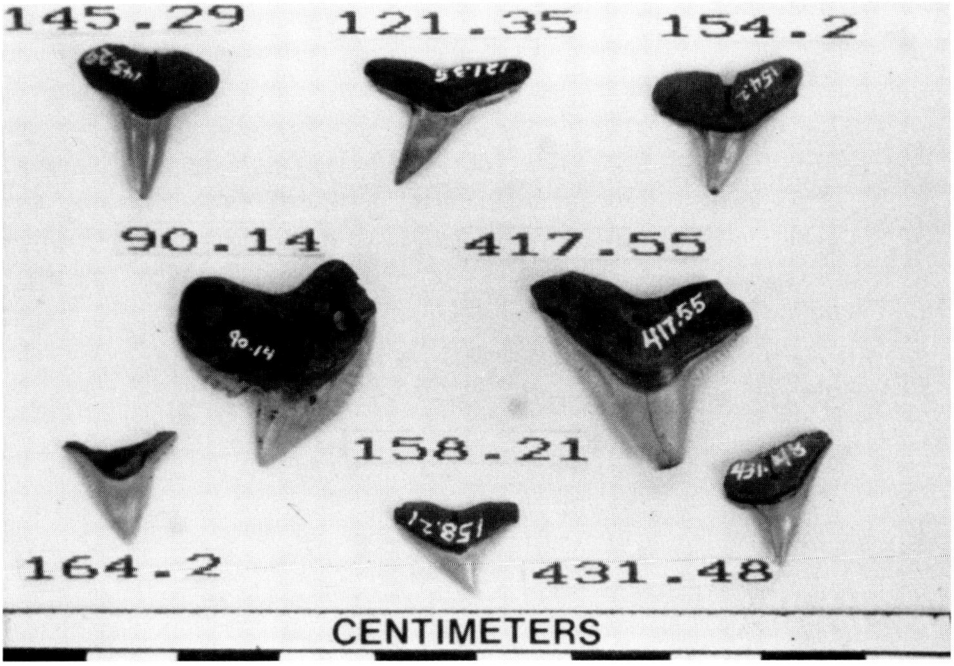

FIGURE 82C

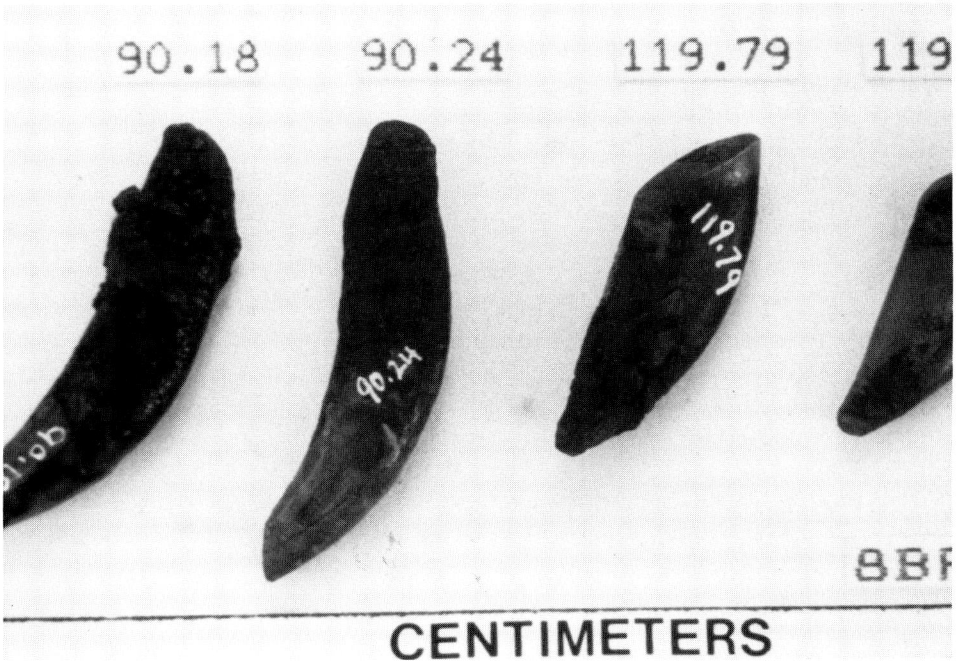

FIGURE 82D

The cooperation between developer, government, and archaeologist has been excellent. The usual problems exist with funding adequate enough to carry each analysis to completion, however, because it is difficult for most people to realize how long it takes to segregate, quantify, and interpret wet-site materials. Using Coles' (1988) criteria to establish priorities, Windover would head his list: it is of national and international significance, it is unique, the archaeological remains are intact and important, excavation and preservation have been carried out, and many of the materials are of exhibit quality.

The majority of the materials from the Windover site are in the Department of Anthropology, Florida State University except those that are currently undergoing analysis and preservation at other institutions.

References

Alexander, Michelle M. 1986 Analysis of Plant Remains from the Windover Archaeological Project (8Br246): 1985 Field Season. Report submitted to the Windover Archaeological Project, Florida State University, Tallahassee.

Andrews, R.L., J.M. Adovasio, and D.G. Harding 1988 Textile and Related Perishable Remains from the Windover Site (8BR246). Paper presented in a symposium, Archaic Adaptations and Biological Diversity at the Windover Site, at the 53rd Annual Meeting of the Society for American Archaeology, Phoenix.

Dickel, Cynthia D., Mary A. Ashley, Philip J. Laipis, and William H. Hauswirth 1988 Isolation of Human DNA from Archaeological Materials. Paper presented in a symposium, Archaic Adaptation and Biological Diversity at the Windover Site, at the 53rd Annual Meeting of the Society for American Archaeology, Phoenix.

Dickel, David N. 1986 Osteological Remains from the Windover Archaeological Site: Interim Report. Paper presented at the 54th annual meeting of the American Association of Physical Anthropology, Albuquerque, New Mexico.

1988 Human Osteology and Adaptation at the Windover Site, Florida (8-BR-246). Paper presented at the 56th annual meeting of the American Association of Physical Anthropology, Phoenix, Arizona.

1990 Maxillary Anterior Attrition (AMA) and Molar Lingual Root Polish (LRP) from the Florida Early Archaic. Paper presented at the 58th annual meeting of the American Association of Physical Anthropology, Miami, Florida.

Dickel, David N., C. Gregory Aker, Billie K. Barton, and Glen H. Doran 1989 An Orbital Floor and Ulna Fracture from the Early Archaic of Florida. *Journal of Paleopathology* 2(3):165–170.

Dickel, D.N. and G.H. Doran 1989 Severe Neural Tube Defect Syndrome from the Early Archaic of Florida. *American Journal of Physical Anthropology* 80:325–334.

Doran, Glen H. 1988a Archaeological Overview and Preliminary Paleodemographic Implications from Windover. Paper presented in a symposium, Archaic Adaptation and Biological Diversity at the Windover Site, at the 53rd Annual Meeting of the Society for American Archaeology, Phoenix.

Doran, Glen H. 1988b A Consideration of Archaeological Wetsites. *Proceedings of the Eighth Annual Gulf of Mexico Information Transfer Meeting*, pp. 402–412. U.S. Department of the Interior.

Doran Glen H. and David N. Dickel 1988a Radiometric Chronology of the Archaic Windover Archaeological Site (8BR246). *The Florida Anthropologist* 41(3):365–380.

1988b Multidisciplinary Investigations at the Windover Site. In *Wet Site Archaeology*, pp. 263–289. Edited by Barbara A. Purdy. The Telford Press, West Caldwell, N.J.

Doran, Glen H., David N. Dickel, William E. Ballinger Jr., O. Frank Agee, Philip J. Laipis, and William W. Hauswirth 1986 Anatomical, Cellular, and Molecular Analysis of 8,000-yr-old Human Brain Tissue from the Windover Archaeological Site. *Nature* 323(6091):803–806.

Doran, Glen H., David N. Dickel, and Lee A. Newsom 1990 A 7,290-year-old Bottle Gourd from the Windover Site, Florida. *American Antiquity* 55(2): 354–360.

Estes, Anna M. 1988 Non-Invasive Investigation of Skeletal Stress Markers in an Archaic Population from Central Florida (Br246). Unpublished Master of Science thesis, Florida State University, Tallahassee.

Estes, Anna M. and D.N. Dickel 1989 Computed Tomography and Cortical Thickness in an Early Archaic Archaeological Sample from Windover. Paper presented at the 57th Annual Meeting of the American Association of Physical Anthropology, San Diego, California.

Gardner, J.S. 1988 Conservation of the Windover Fabrics and Wood. Paper presented in a symposium, Archaic Adaptations and Biological Diversity, at the 53rd Annual Meeting of the Society for American Archaeology, Phoenix.

Grattan, David W. 1989 Paralyene at the Canadian Conservation Institute. *Canadian Chemical News*, October, pp. 25–26.

Hauswirth, William and Cynthia Dickel 1988 Investigations of DNA isolated from Windover Brain Tissue: Methods and Implications. Ms. on file, College of Medicine, University of Florida.

Hauswirth, William, Cynthia D. Dickel, Glen H. Doran, Philip J. Laipis, and David N. Dickel 1988 8,000-year-old Brain Tissue from the Windover Site: Anatomical, Cellular, and Molecular Analysis. Paper presented at the International Congress of Anthropological and Ethnological Sciences, Zagreb, Yugoslavia, August 1988. (In press, anticipate publication summer 1990.)

Holloway, Richard G. 1987 Pollen Analysis of Five Features from Windover Archaeological Project, Florida. Report submitted to Windover Archaeological Project, Florida State University, Department of Anthropology, Tallahassee.

1988 Palynology of Windover Peats. Paper presented in a symposium, Archaic Adaptation and Biological Diversity at the Windover Site, at the 53rd Annual Meeting of the Society for American Archaeology, Phoenix.

Humphrey, Bruce J. 1984 The Application of Parylene Conformal Coating Technology to Archival and Artifact Conservation. *Studies in Conservation* 29:117–123.

Nabergall-Luis, Lee A. 1987 The Archaeological Evidence of a Faunal Community During the Early Holocene/Archaic Period. Paper Presented at the Tenth Annual Meeting of the Society of Ethnobiology, Gainesville.

1990 Faunal Studies from an Early Archaic Wetsite: the Windover Archaeological Site, Brevard County, Florida. Unpublished Master's thesis, Department of Anthropology, Florida State University, Tallahassee.

Newsom, Lee A. 1988 The Paleoethnobotany of Windover (8BR246): An Archaic Period Mortuary Site in Florida. Paper presented in a symposium, Archaic Adaptation and Biological Diversity at the Florida Windover Site, at the 53rd Annual Meeting of the Society for American Archaeology, Phoenix.

Smith, David G., Frederick W. Lorey, and Becky K. Rolfs 1988 Serum Albumin Phenotypes for the Prehistoric Population of Windover and Their Anthropological Significance. Ms. on file Department of Anthropology, University of California, Davis.

Smith, David G. 1988 Windover Serum Protein Phenotypes. Paper presented in a symposium, Archaic Adaptations and Biological Diversity at the Windover Site, at the 53rd Annual Meeting of the Society for American Archaeology, Phoenix.

Spackman, W. 1985 Personal communication. Department of Geology, Pennsylvania State University, University Park.

Stephenson, Frank 1985 The Windover Archaeological Project. Florida State University Bulletin *Research in Review* 79(4):3–10.
Stone, Tammy T., David N. Dickel, and Glen H. Doran 1990 The Preservation and Conservation of Waterlogged Bone from the Windover Site, Florida: A Comparison of Methods. *Journal of Field Archaeology* 17(2):177–186.
Stout, S.A. 1986 Petrography of the Peat from the Windover Archaeological Site, Titusville, Florida: A Preliminary Report. Ms. on file, Department of Geology, Pennsylvania State University, University Park.
Stout, S.A. and W. Spackman 1988 Paleoecology of the Windover Farms Archaeological Site as Determined by the Petrology and Chemistry of the Peats. Paper presented in a symposium, Archaic Adaptation and Biological Diversity at the Windover Site, at the 53rd Annual Meeting of the Society for American Archaeology, Phoenix.

Other Sites and Wooden Artifacts

Tick Island

The Tick Island site (8-VO-24) is located on the southeastern side of the island beside the mouth of Harris Creek (Figs. 1 and 83). Although C.B. Moore did limited excavations in this location of Tick Island in the 1890s (Moore 1893), most archaeological work was of a salvage nature because the midden was badly disturbed by dragline operations. Some shell was removed from the island for road beds as early as the 1920s and commercial shell mining activities nearly destroyed the site in the 1950s. The shell was transported by barge to a large washing machine located near DeLeon Springs where it was washed by pressurized water to remove sand and dirt. The washed shell was sold to be used in the drainage fields of septic tank systems. The sand and dirt washed from the shell was sold for driveways and fill. Artifacts were found as the shell emerged from the revolving washer onto a pile (Stackhouse, E.J. 1966; Stackhouse J.R. 1990, personal communication).

In 1961 Ripley P. Bullen excavated 175 burials of both sexes and all ages at Tick Island (Jahn and Bullen 1978:20–22).

Bone, stone, shell, and ceramic artifacts from Tick Island (Jahn and Bullen 1978) are very diverse and date from the Middle Preceramic Archaic Period or earlier to at least the St. Johns II Period. In addition to the freshwater snails and mussels composing the mound matrix, 18 species of vertebrates were identified including opposum, beaver, black bear, raccoon, otter, domestic dog, deer, alligator, several types of turtles, rattlesnake, cormorant, turkey, duck, garfish, and catfish.

Wooden artifacts were recovered at the site (Benson 1967; Bushnell 1960; Jahn and Bullen 1978: Fig. 49c and h). Their Fig. 49c is a tool handle similar to several from the Key Marco site and their Fig. 49h is described as carved wood. Another specimen, described by Stackhouse (1966), Benson (1967), and Schwehm (1983), bears a strong resemblance to the Florida turkey buzzard (Fig. 84). The skin on its head is even wrinkled. Its beak is opened in anguish

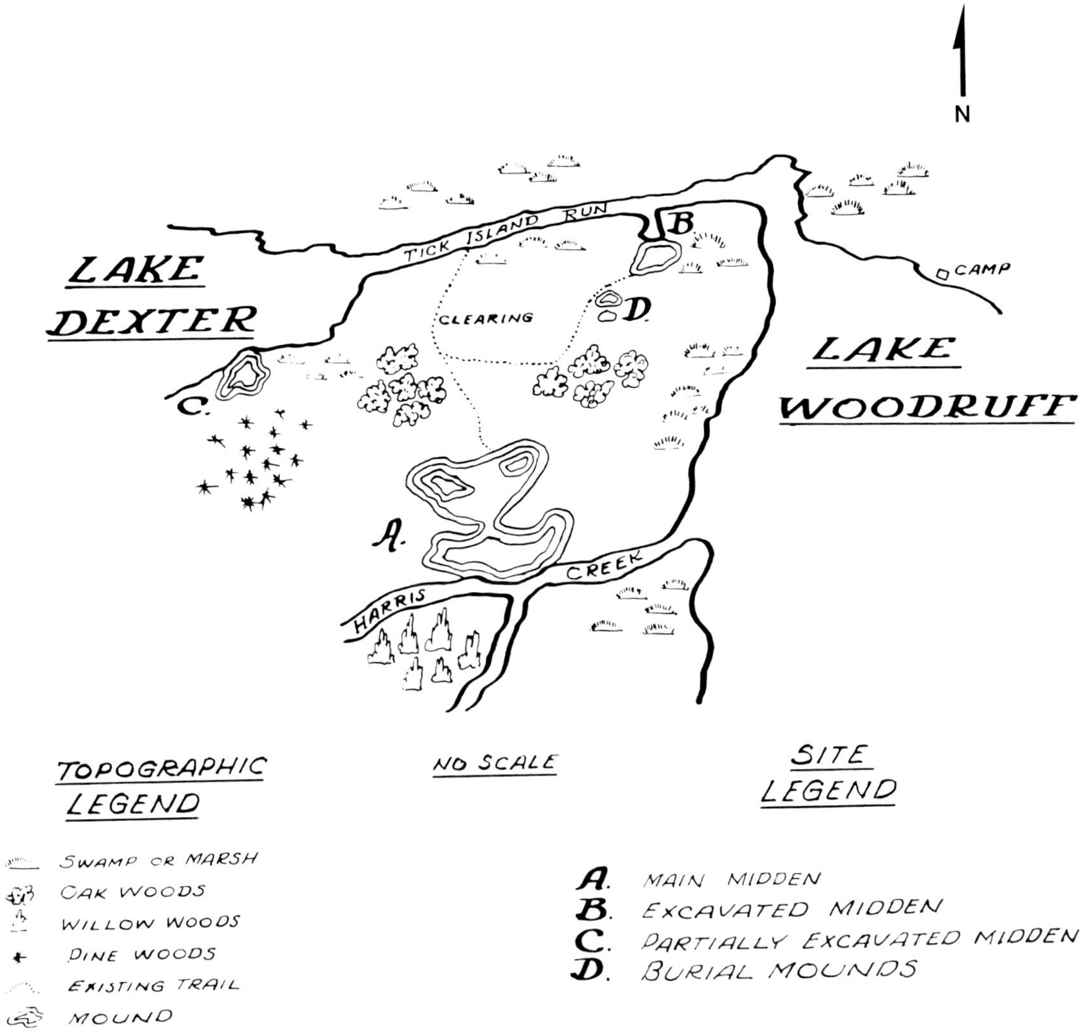

FIGURE 83 Map of Tick Island and location of archaeological sites (courtesy of Carl A. Benson and John R. Stackhouse).

because the talons of a large raptorial bird, possibly an eagle, are clutching it by the throat. The exposed tongue of the distressed buzzard provides further realism. A naturalistic round eye was incised on the head of the buzzard. A dowel pin located on the bottom of the small, 5-inch sculpture indicates that it may have functioned as the decorative end of a ceremonial staff. Stackhouse (1966) estimates its age between 200 BC and AD 500 based on associated artifacts that emerged from the shell washer at the same time. This specimen was preserved with alum and then coated with linseed oil. It remains in good condition (Benson, personal communication 1990). In 1989, this wooden artifact was sold by the owner, R. Stackhouse, to G. Marwick. Mr. Marwick plans to exhibit it in a museum that is being constructed at Silver Springs.

Despite the fact that the Tick Island site is badly disturbed, there are undisturbed deposits on the fringes of the midden. Cores taken in 1982 and

A

FIGURE 84 Three views of the wood carving of a turkey buzzard from Tick Island (courtesy of John R. Stackhouse).

1986 in submerged locations adjacent to the mined area contained well-preserved fauna and flora (including gourd seeds) as well as pottery sherds known as Orange Plain and Orange Incised that are typical of some of the oldest ceramics in North America. It would be informative to excavate a portion of the water-saturated area. At Hontoon Island the earliest dates from the well-preserved strata are only about 2000 years. The materials recovered at Tick Island could extend the archaeological record of human adaptation to the

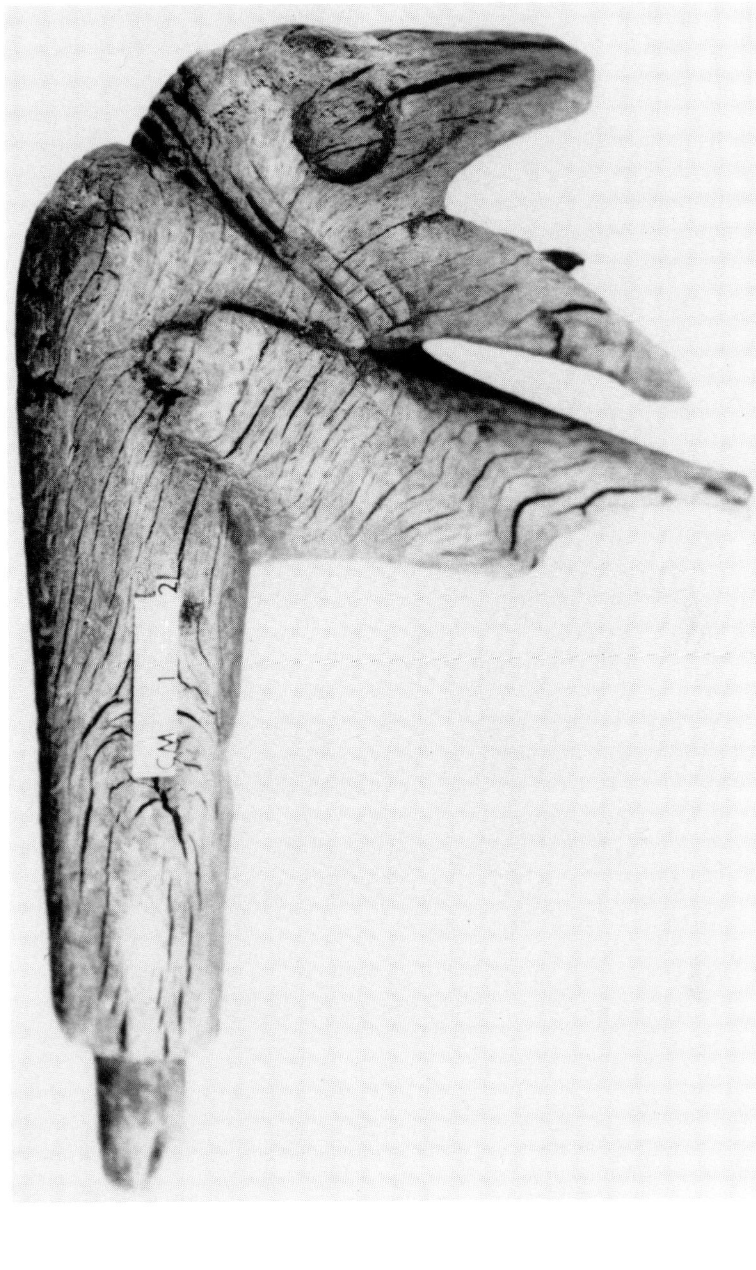

FIGURE 84B

environment in this area back several thousand years into the Preceramic Archaic and furnish a broader picture of the initial utilization of the aquatic resources by the Indians.

In 1891, C.B. Moore dug into a burial mound (8-VO-25) on the northeast edge of the island (Moore 1892a,b). Tick Island is protected now because it is part of Lake Woodruff National Wildlife Refuge (Miller and Griffin 1978). In 1986, 17 boxes of artifacts recovered from Tick Island during shell mining activities by Mr. Wester W. Branton of DeLeon Springs were sold to private collectors following his death. Many of these artifacts are pictured in the publication by Jahn and Bullen (1978).

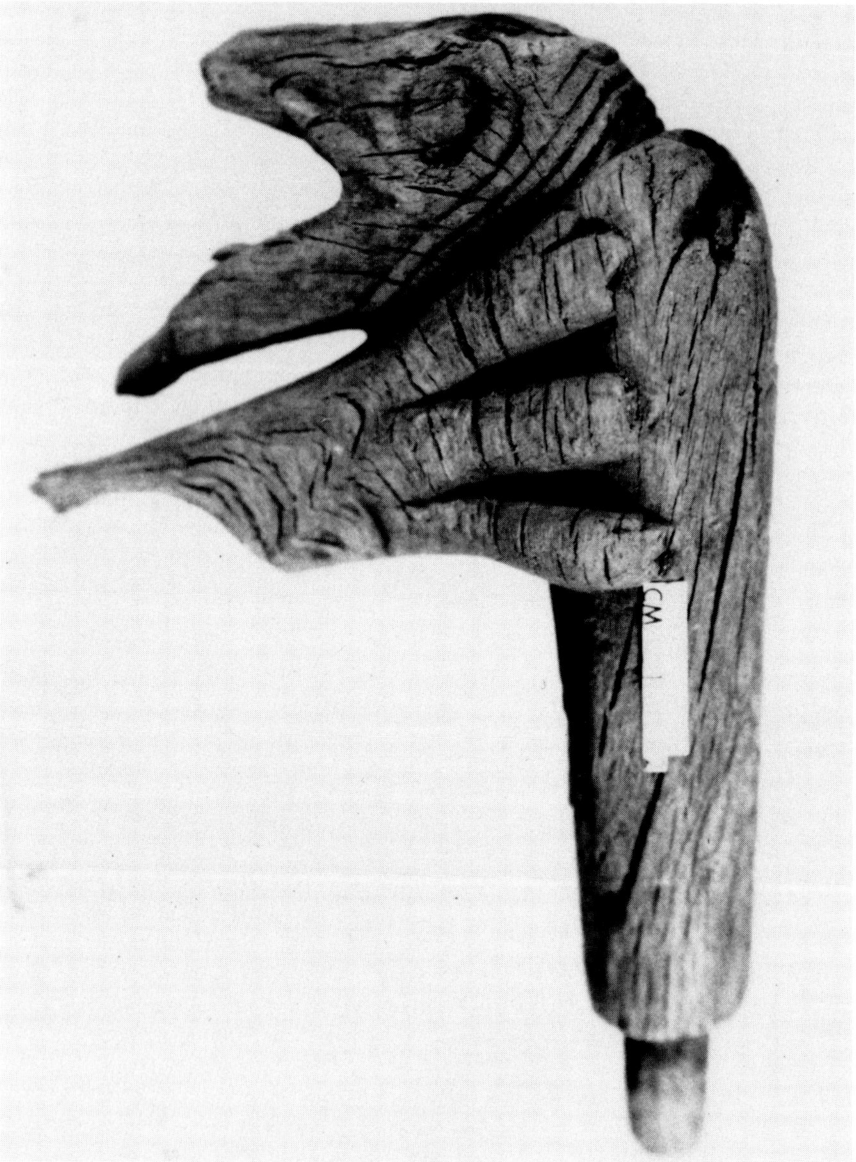

FIGURE 84C

Lake Monroe

In February 1988, a small test excavation was conducted below the water level at the Groves Orange Midden site (8-VO-2601) on the north shore of Lake Monroe (Russo et al. n.d.; Figs. 1 and 85). The site has great potential for the recovery of a full assemblage of environmental and cultural information transending the Preceramic Archaic to the Orange Period. It contains the oldest dated fiber-tempered pottery and the earliest known examples of *Cucurbita pepo* varieties yet found in Florida. The fauna include 20 species of snails and bivalves, 13 species of fish, 3 species of snake, salamander, alligator, 5 species of turtle, 3 species of birds, and 5 species of mammals including *Homo sapiens*. In addition, there were a few unidentified bones representing nearly all classes of animals. Thirty-one botanical species were identified including cane, several species of hickory and oak, dogwood, persimmon, ash, sweetgum, magnolia, mulberry, palm, pine, cherry/maple, willow, elderberry, cypress, elm, grape, toothache tree, fringe tree, bottlegourd, and the cucurbits mentioned above. These identifications were made from seeds, rind, nuts, woods, charcoal, and other preserved plant remains. Artifacts consist of several types of ceramics (primarily fiber-tempered), marine shell tools, nondiagnostic chert flakes, and several bone pin fragments. Dr. J.J. Stipp (University of Miami and Beta Analytic, Inc.) provided radiocarbon dates that ranged from ca. 3254 ± 65 BP (1265 ± 65 BC) for Level 2 to ca. 6032 ± 85 (4843 ± 85 BC) for Level 8. No wooden artifacts were recovered during our limited excavations in 1988 but Grant Groves, for whom the site was named, found an interesting object (Fig. 86) when he discovered the site in the early 1980s. This specimen has been preserved successfully in PEG. Future excavations at this site would furnish valuable information about this little known time period in Florida's prehistory. All specimens from the Groves Orange Midden are in the Department of Anthropology or the Florida Museum of Natural History, University of Florida, Gainesville.

Similar Sites

Plant remains were not preserved at the Cutler Fossil site (8-DA-2001) in Dade County (Carr 1986a; 1986b; Fig. 1) or the Devil's Den site (8-LV-44) in Levy County (Martin and Webb 1974; Clausen et al. 1975:208–209; Fig. 1) but they resemble other early Florida wet sites because the sedimentary, biological, and archaeological evidence they contain indicates that humans were associated with them when both sea level and/or local ground water levels were lower than present (Clausen et al. 1975:209). These sites are Late Paleo/Early Archaic in age. They both contained a variety of extinct and modern animal species, remains of *Homo sapiens*, and artifacts.

The Gauthier site (8-BR-193) in Brevard County (*Florida Anthropologist* 1981; Fig. 1) is a Middle Archaic Period cemetery (Fig. 87) that has many features resembling Bay West, Republic Groves, and the slough at Little Salt Spring but it lacks the botanical component.

Single Wooden Artifacts

Broward County—The margate-Blount site (8-Bd-41) consisted of a burial mound and village site that were greatly disturbed by development activities and the construction of a drainage canal in 1959. Members of the Broward County Archaeological Society conducted excavations at the site periodically from 1959–1961. Several wooden artifacts were recovered including a canoe

A

FIGURE 85 (A) Coring in Lake Monroe in 1987; water is less than knee deep even 100 feet offshore. (B) Coring in 1987 near the area excavated in 1989. (C) 1989 excavation unit.

paddle and a double-ended pestle. The double-ended pestle was made from a heavy unidentified wood species. Its shape is very similar to those recovered at the Belle Glade, Ft. Center, Windover, and Key Marco sites. Of interest also was the recovery of a carved turtle effigy pendant made of bone, 2 historic period glass beads, and 44 human skeletons (Williams 1983; Isçan 1983).

Clay County—Two large wooden bowls made of pine (Fig. 88) were recovered in the same approximate area as 19 canoes (see Chapter 4) on the property of Stricklin Peat Co. (Fig. 1). Their dates are 2420 ± 110 BP (470 ± 110 BC; Beta 17847) and 3020 ± 70 BP (1070 ± 70 BC; Beta 20207). It is interesting that there is a 600-year difference in the ages of these bowls but they are nearly identical in size and shape. It is interesting also that the older bowl appears to be finished more carefully. (Information on file, Department of Anthroplogy, University of Florida, Gainesville.) See also Fig. 125.

FIGURE 85B

FIGURE 85C

FIGURE 86 Wooden artifact recovered at Lake Monroe by Grant Groves and preserved in polyethylene glycol by the author. The specimen is approximately 10 cm (4 in) in diameter and 1.5 cm thick (photo by the author).

Dade County—Goggin (1942:327–328) describes a wooden club (Fig. 89) that was recovered during the dredging of Collins Canal across Miami Beach in 1913 (Fig. 1). The dredge was of the suction type that threw the club out of its spoil from a depth of 8 feet. Goggin says the specimen is made of black mangrove wood (*Avicennia nitida*). It is 9 cm (3.5 inches) wide, 29.2 cm (11.5 inches) long with a 9.5-cm (3.75-inch) long pierced handle. As of 1990, the object was still in good condition and was on exhibit at the Historical Museum of Southern Florida, Miami. (Information on file, Department of Anthropology, University of Florida, Gainesville.)

Two other wooden artifacts (Fig. 90) are on exhibit at the Historical Museum of Southern Florida and are reported to be in good condition. One is a club with a carved handle (11 inches long, 5 inches widest point) and the other is a horn-shaped implement (18 inches long, 2 inches wide). They were found while digging a boat slip on the north bank of the Miami River (8-DA-11; Fig. 1). Information on file Department of Anthropology, University of Florida, Gainesville.)

De Soto County—In 1978, Donald T. Yeats recovered a large wooden bowl made from a cypress burl (Fig. 91) from a swamp along the Peace River at Hull, 9 miles south of Arcadia. The bowl measures ca. 56 cm (21.5–22.5 inches) in diameter at the rim and ca. 28 cm (10.5–11 inches) in diameter at the base, it is ca. 24 cm (9.5 inches) high, the body is ca. 11.5 cm (4.5 inches) thick, and

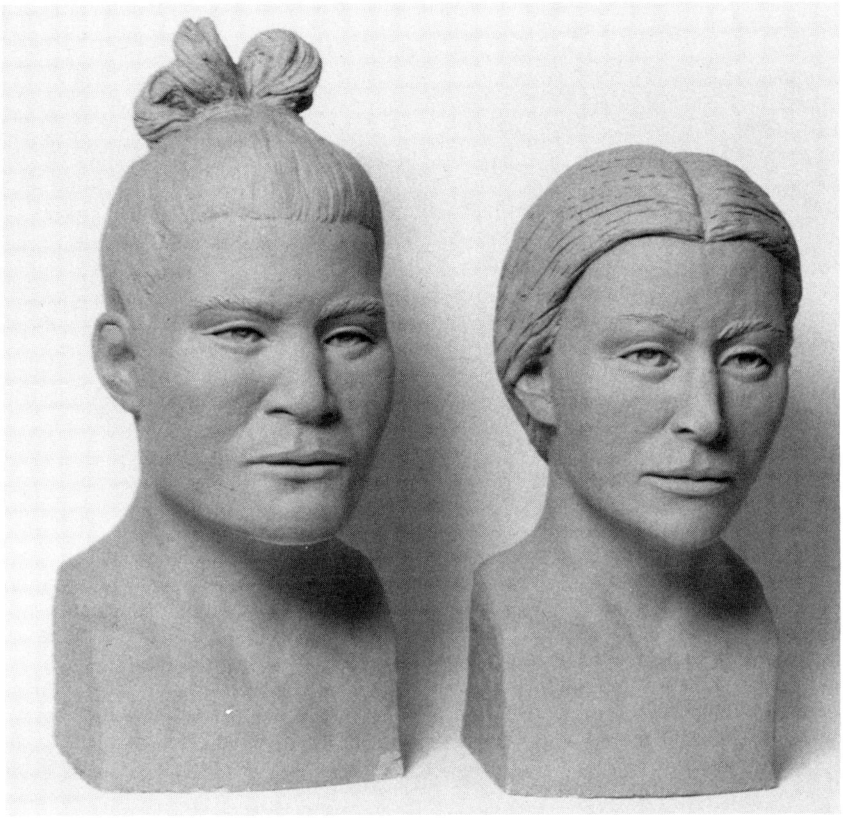

FIGURE 87 Reconstruction of facial features from skulls of 5000–6000-year-old individuals from the Gauthier site. The male was about 25-35 years old when he died; the female was in her late teens or early 20s (features reconstruction by Betty Pat Gatliss; photo credit Stanley Bloneley, Florida Museum of Natural History).

the rim is ca. 7.5 cm (2.4–3.5 inches); it weighs approximately 38 kg (83 pounds). The age of the bowl is 920 ± 65 BP (UM-1544) or AD 1030 ± 65. Mr. Yeats loaned the bowl to the Florida Museum of Natural History for preservation and study (Cat. No. 78-45). The specimen was placed in water in November 1978 and treatment with PEG 540 Blend began in October 1979 and terminated on November 16, 1980. As of spring 1990, the bowl is in excellent condition. (Information on file, Department of Anthropology, University of Florida, Gainesville.)

Duval County—At the Mission Period site of San Juan del Puerto (8-DU-53) on Fort George Island (Fig. 1), 15 pine logs (Fig. 92) were found protruding from the bank of a tide water creek (Jones 1981). The logs were about 71 cm long and 7–10 cm in diameter. All but three of the logs had crudely cut notches near each end and all of them had been subjected to fire. The function of the structure is uncertain. Mission Period ceramics were found in close association. A wood sample submitted for radiocarbon analysis yielded a date of 470 ± 55 BP or AD 1480 (DICARB 2031). Two of the posts were brought to the Florida Museum of Natural History for preservation and analysis; the others were returned to the slough.

A

FIGURE 88 (A,B,C) Three views of a 2400-year-old wooden bowl or mortar from Stricklin Peat Co., Grandin, Florida: (A) upright, (B) upside down to show rounded base that was probably placed in the ground, (C) interior showing circular lines resulting from the motion of a pestle. A small section of the bowl was missing when recovered. This specimen weights about 13 kg (28.6 pounds), is 34 cm (13.34 in) in height, 26.0 cm (6.5 in) in diameter at the bottom and 35.5 cm (13.98 in) in diameter at the top with a wall thickness of 4.5 cm (1.78 in) at the top. (D,E) Two views of a 3000-year-old wooden bowl or mortar from Stricklin Peat Co.: (D) upside down to show flat base, (E) showing shape of the interior. About 1/3 of this specimen was missing when found. The two remaining sections weigh about 9 km (20 pounds). It is 34 cm (13.39 in) in height on one side and 35 cm (13.78 in) in height on the other, 27.0 cm (10.63 in) in diameter across the bottom, 34 cm (13.39 in) in diameter across the top with a wall thickness at the top of 2.5 cm (approximately 1 in). These bowls have been preserved in polyethylene glycol by the author. (All pictures taken by the Office of Instructional Resources, University of Florida.) See also Fig. 125.

A carved wooden club, made of hickory and representing a human face in profile (Fig. 93), also was recovered at mission San Juan del Puerto by William Jones. The club is about 23 inches long and about 5 inches greatest width; the handle is 9.9 inches long. It is thought to have been made by the Guale Indians in the 17th century using iron tools (FSM Cat. 101679, Acc. No. 4513).

Flagler County—In 1971, a small wooden figure (Fig. 94) was recovered from an organic deposit in the Tomoka River, Flagler County (Fig. 1). The wood was identified as *Peltophorum* sp., probably *P. adnatum,* commonly

FIGURE 88B

called Brazilwood by the Wood Products Laboratory, Madison, Wisconsin. There are several species of this genus, all called Brazilwood. *P. adnatum* is the West Indian and Central American species. Brazilwood also grows in Africa. Radiocarbon analysis of the figurine produced a date of 470 ± 90 BP or AD 1480 with a range from AD 1390–1570 (Morrell 1974; Kenner 1974; Purdy 1988). This specimen is now on display at Tomoka State Park, Ormond Beach. It was in good condition when I photographed it in May 1990.

Glades County—See Manatee County.

Lake County—In Cassia, at the extreme west end of Owens Pond (Sec. 36, T18S, R28E, USGS Pine Lakes, Florida 1962; Fig. 1), the remains of what was probably a fish weir were found in 1976 during commercial peat mining activities (Fig. 95). The pieces of the weir were taken to the Florida Museum of Natural History (Cat. No. A-6990, Acc. No. 76-16) and preserved for four years in increasing concentrations of PEG. They were 26–61 cm long (ca. 10–24 inches) and 2–11 cm (ca. 1–4 inches) in diameter.

Lee County—In 1928, J. Walter Fewkes, Chief of the Bureau of American Ethnology, described two wooden "slabs" that had been presented to the U.S. National Museum by Mr. George Kinzie (Fig. 96). The specimens are thought to have been found during dredging operations below a shell mound on the

FIGURE 88C

Caloosahatchee River west of Fort Myers (probably 8-L-17; Fig. 1). They resemble those from the Key Marco Site reported by Cushing. Objects with similar shapes and designs were also made out of stone and metal. Fig. 96 is 21 cm (8.25 inches) long and 7.5 cm (3 inches) wide. It is carved on both faces. Figure 96b is 25 cm (9.75 inches) long and 8.0 cm (3.20 inches) wide. It has a bulbous profile on the nontendoned half. For more detail see Allerton et al. (1984:46). (U.S. National Museum 329,598 and 329,599.) A carved wooden figurehead of a bird, also from Lee County, was found in 1971 near Pineland (see Fig. 97A and B).

Manatee County—In 1980, the curator at the South Florida Museum and Bishop Planetarium in Bradenton reported that the following wooden artifacts were in the possession of that institution: A trough (Fig. 98) 67.5 cm long, 23.5 cm wide, and 11.5 cm high; a double-ended pestle (Fig. 99) 27.5 cm long and 8.0 cm in diameter; a mortar (Fig. 100) 18.0 cm high, 32.0 cm in diameter (found on Observation Island, Glades Co.; Fig. 1); a mortar (Fig. 101) 52.5 cm high, 28.2 cm in diameter; a mortar or bowl (restored; Fig. 102) 43.0 cm high, 39.0 cm in diameter (Myakka, Sarasota County; Fig. 1), and two "post" fragments 29.6 cm long, 12.5 cm wide, and 14.0 cm long and 6.3 cm wide. All of these objects are thought to have been recovered by Montague Tallant and/or J.E. Moore. In May 1990, the museum photographed for me the specimens shown in Figs. 98–102.

FIGURE 88D

Marion County—A burl club was recovered from the bottom of Lake Weir (Fig. 1) in the early 1960s. When Lee A. Newsom and I examined it in 1982 it was very hard and stable. It measured 42 cm long; the handle was 4.2 cm wide and the distal end was 11 cm wide. The specimen was very polished, especially the handle. The club was called to my attention by Mr. John Love of Summerfield, Florida. It is in his possession. (Information on file, Department of Anthropology, University of Florida, Gainesville.)

Okeechobee County—A wooden "idol" (Fig. 103) said to have been carved from lignum vitae (some accounts say black mangrove) was plowed up on the north shore of Lake Okeechobee (Fig. 1) in 1921 by Mr. M.A. Millar, artesian and deep-well constructor. His letter that accompanied the specimen when he donated it to the United States National Museum, where it remains today, states that, "Originally the water came up to this mound where we plowed up the idol at the foot of the mound. The tractor was plowing deep and brought

FIGURE 88E

it up from the bottom of the furrow. The location of this mound is the southeast quarter of the northwest quarter, Section 22, Township 40 South, Range 32 East Tallahassee" (Accession No. 66473, Smithsonian Institution, U.S. National Museum). A newspaper article about the object mentions that "the outer surface had bleached to a brownish gray. Beneath, where a chip had been knocked from the image, the wood showed black. The idol represented a figure in a sitting posture with the hands resting on the bent knees. It is about 8 inches high and is heavy for its size, weighing about 2 pounds. Where Mr. Millar plowed up the idol, Lake Okeechobee's waters formerly stood 6 feet deep. The land is muck that has been reclaimed by drainage. . . . many small crockery articles were found in the same vicinity as the idol. These pieces of pottery are glazed" (*De Soto County News*, March 25, 1921). J. Walter Fewkes, who was once Chief of the Bureau of American Ethnology, makes the following comments about the figure.

> The form and technique of this idol would seem to leave little doubt of its having been made by Indians, although a similar idol found in Cuba is regarded by some ethnologists as African. . . . The two specimens differ in form and in the position of the legs, that from Cuba resembling a sitting figure, with the legs bent at the knee, while in the Florida

FIGURE 89 Wooden club recovered during dredging of Collins Canal in Dade County in 1913. Specimen in 29.21 cm (11.5 in) long and 8.89 cm (3.5 in) wide at top (picture credit: Historical Museum of Southern Florida, Miami).

> specimen the figure is squatting. The hair of the latter is cut in such a way as to remind one of the Muskhogean stone idols. Both seem to have been made of the same kind of wood and they are weathered to the same color (Fewkes 1928).

Palm Beach County—The kneeling effigy pictured in Fig. 104 was found near Pahokee (Fig. 1) in 1928 by Karl Riddle, a civil engineer and the first city manager of West Palm Beach. It is 22.9 cm (9 inches) high; in circumference

FIGURE 90 Wooden implement and club with carved handle, both found while digging a boat slip on the north bank of the Miami River (picture credit: Historical Museum of Southern Florida, Miami).

the head is 21.60 cm (8.5 inches), the bust is 254 cm (10 inches), the waist is 19,7 cm (7.75 inches), and the knees are 28.8 cm (11.25 inches) around. The arms have been broken off. The wood was identified as cypress but the species has not been verified by microscopic examination. The foregoing information was provided by Harriet H. Mahoney, Karl Riddle's daughter (Mahoney, personal communication 1981). In 1990, I wrote to Mrs. Mahoney, asking for quality photographs of this carving to be used in this volume. In her return correspondence, she informed me that she had donated her father's Indian artifact collection to the Henry Flagler Museum in Palm Beach, where it was on display with a plaque honoring her father (Mahoney, personal communication 1990). I contacted the museum and was told that Riddle's collection had been given to the Historical Society of Palm Beach. The individual in charge at that organization informed me that the collection was in storage and refused to retrieve the carving for me so I could photograph it. This situation accounts for the poor quality of the pictures in Fig. 104.

A small wooden bowl (Fig. 105) was found on Kreamer Island in Lake Okeechobee (Fig. 1) in 1916 by W.H. Waggoner. It is 6.35 cm (2.5 inches) high, 16.5 cm (6.5 inches) in diameter at the rim, 7.6 cm (3 inches) in diameter across

FIGURE 91 A 900-year-old wooden bowl from De Soto County, Florida. It weighs about 37.72 kg (83 pounds), is 24.13 cm (9.5 in) high and 60.96 (nearly 2 feet) in diameter at the top with a rim width averaging 7.62 cm (ca. 3 in). (Picture taken by Office of Instructional Resources, University of Florida.)

the flat bottom, and .76 cm (.3 inch) thick. It was made from a burl and has tool marks inside and out (FSM Cat. 16730). In 1976, Mrs. Leslie Strickland, Director of the Belle Glade Municipal Library asked permission to borrow the bowl from the Florida Museum of Natural History to place on exhibit at the library in Belle Glade. Before the loan was made, the bowl was preserved in 50% polyethylene glycol and extensive notes were made about its appearance. Before treatment, the bowl weighed 191.8 g (6.77 ounces); after treatment and at the time of shipment to Belle Glade in January 1977, the weight was 200.2 g (7.06 ounces). The specimen is still in Belle Glade and in good condition. In May 1990, Phyllis J. Lilley, branch manager of the Palm Beach County Public Library in Belle Glade, took the picture shown in Fig. 105.

Pinellas County—Warren (1971:71–72; 1990 personal communication) reports the recovery of a unique bird effigy made of unidentified wood (Fig. 106) found on a road constructed of oyster shell that had been dredged from Tampa Bay (Fig. 1). The figure is incomplete. One side is incised with three straight lines in three locations. The wood is mineralized and Warren suggests that the specimen may be old because of the mineralization and because artifacts belonging to the Paleoindian and Archaic Periods have been recovered in a similar shell matrix. In May 1990, Dr. Warren had the specimen rephotographed for me (Fig. 106).

FIGURE 92 Photograph showing the position of 15 pine logs associated with a structure of unknown function at the Mission Period site of San Juan del Puerto on Fort George Island; Note the nature of the deposit in which they were imbedded (courtesy of W.M. Jones).

A fragment of a wooden vessel with straight sides and an expanded rim, similar to the shape of late Orange Period Incised ceramic containers was recovered from the Canton Street site (8-Pi-55), St. Petersburg in fill dredged from Tampa Bay (FSM Cat. No. 98850, Acc. No. 4408).

Putnam County—In 1982, Michael G. Palmer recovered a large wooden bowl (Fig. 107) made of pine lighter knot when he was excavating a peat deposit to create a small lake on his property in Crescent City (Fig. 1). The bowl was preserved in the Department of Anthropology, University of Florida and latex molds were made by Robert Leavy, Interpretation Department, Florida Museum of Natural History. It measures 42.5 cm (16.75 inches) long, 19 cm (7.5 inches) wide, 18 cm (7.24 inches) deep; the opening is 16 cm (6.25 inches) long, 12 cm (4.75 inches) wide, and ca. 14 cm (5–6 inches) deep; the holes on both ends were each 11.5 cm (4.5 inches) long and about 2.5 cm (1 inch) in diameter; wet weight was 15 pounds. There is charring on the specimen that may have resulted from

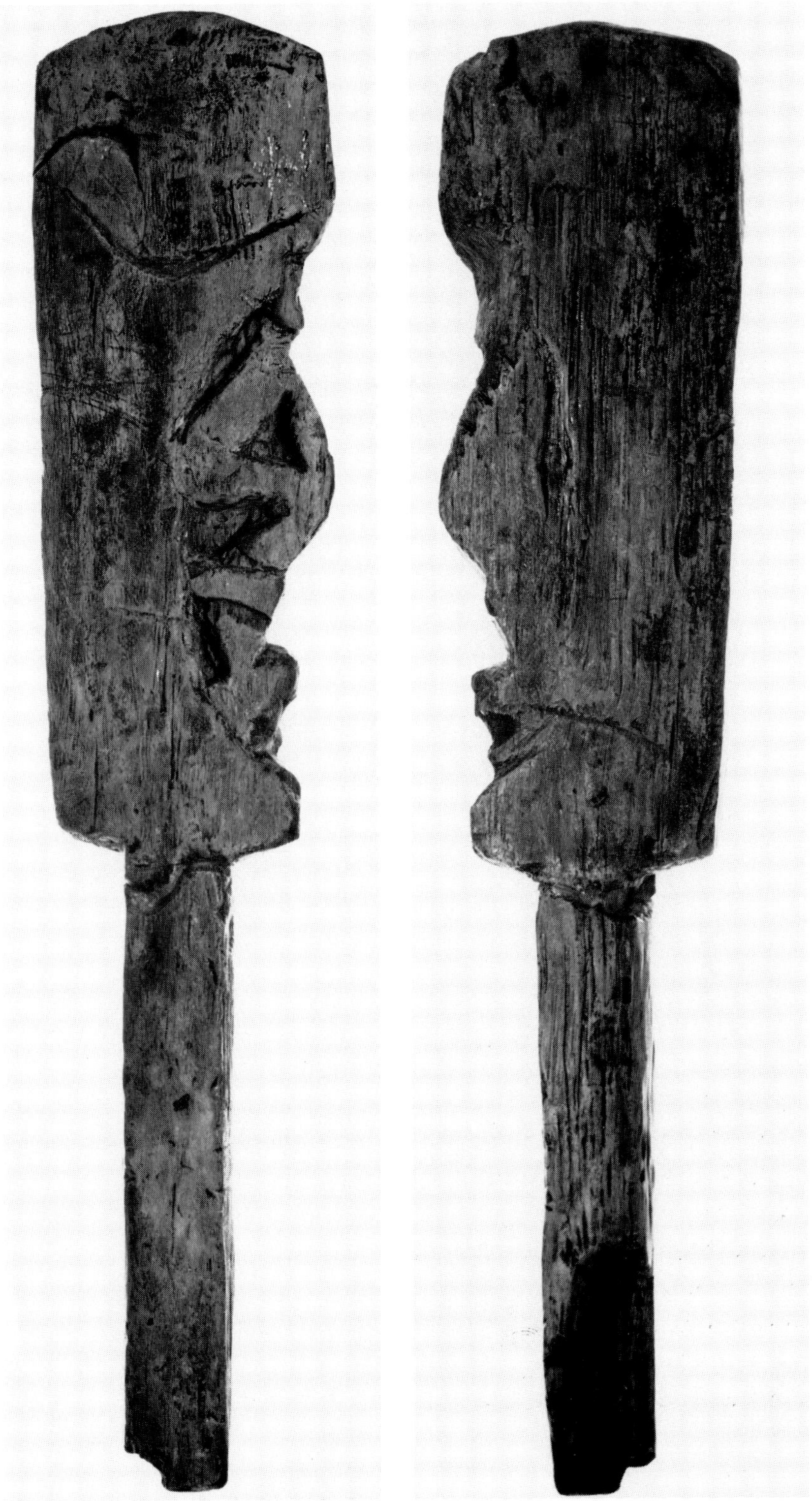

A B

FIGURE 93 (A,B) Both sides of a wooden club recovered at Mission San Juan del Puerto by William M. Jones (FSM Catalog 101679, Acc. No. 4513).

FIGURE 94 (A,B,C) Three views of a small figurine recovered from the Tomoka River in Flagler County. See text for additional information about this specimen (photos by the author).

a manufacturing technique or from a peat fire. The age of the bowl is 3340 ± 70 BP (Beta-5954). When calibrated to dendrochronology, the age is increased by approximately 300 years. The bowl was returned to Mr. Palmer in 1985. (Information on file, Department of Anthropology, University of Florida.)

Sarasota County—See Manatee County.

Walton County—A digging stick was found by Elston Fagen in 1962 in Alaqua Bayou inland from the entrance to Chostawhatchee Bay (Sec. 24, T1S, R20W) below the water table (Fig. 1). Yulee Lazarus, Curator at the Temple Mound Museum in Fort Walton Beach reported in 1980 that the digging stick was in good condition. It is 3.5 cm long and 5.1 cm wide and dates to the Fort Walton Culture (Mississippian/European contact).

Other Wooden Artifacts

A number of canoe paddles and posts have been recovered from time to time throughout the state. I do not have a complete record of these items. One paddle was reported by the Broward County Archaeological Society as coming

FIGURE 94C

from the muck burial mound at the Margate Blount site (8-Bd-41). It was preserved with white glue and was in good condition in 1980 when our records were compiled. Another paddle was donated to the Department of Anthropology by George R. Freeman who found it in Goodson Lake, Putnam County. These paddles have not been dated by radiocarbon analysis.

Comment

The foregoing account of sites at which wooden artifacts have been found demonstrates that the Indians in Florida were skilled woodworkers who created a wide range of utilitarian and artistic items from a great variety of wood species. We will never know how many wooden objects were not reported when they were discovered and were allowed to degrade without proper treatment. Perhaps this book will stimulate people to come forward with knowledge they have about additional wooden artifacts and make them aware of the possibility that others remain entombed in organic wetland deposits.

FIGURE 95 Part of a fish weir found in 1976 during commercial peat mining activities in Lake County (FSM Catalog A-6990, Acc. No. 76-16).

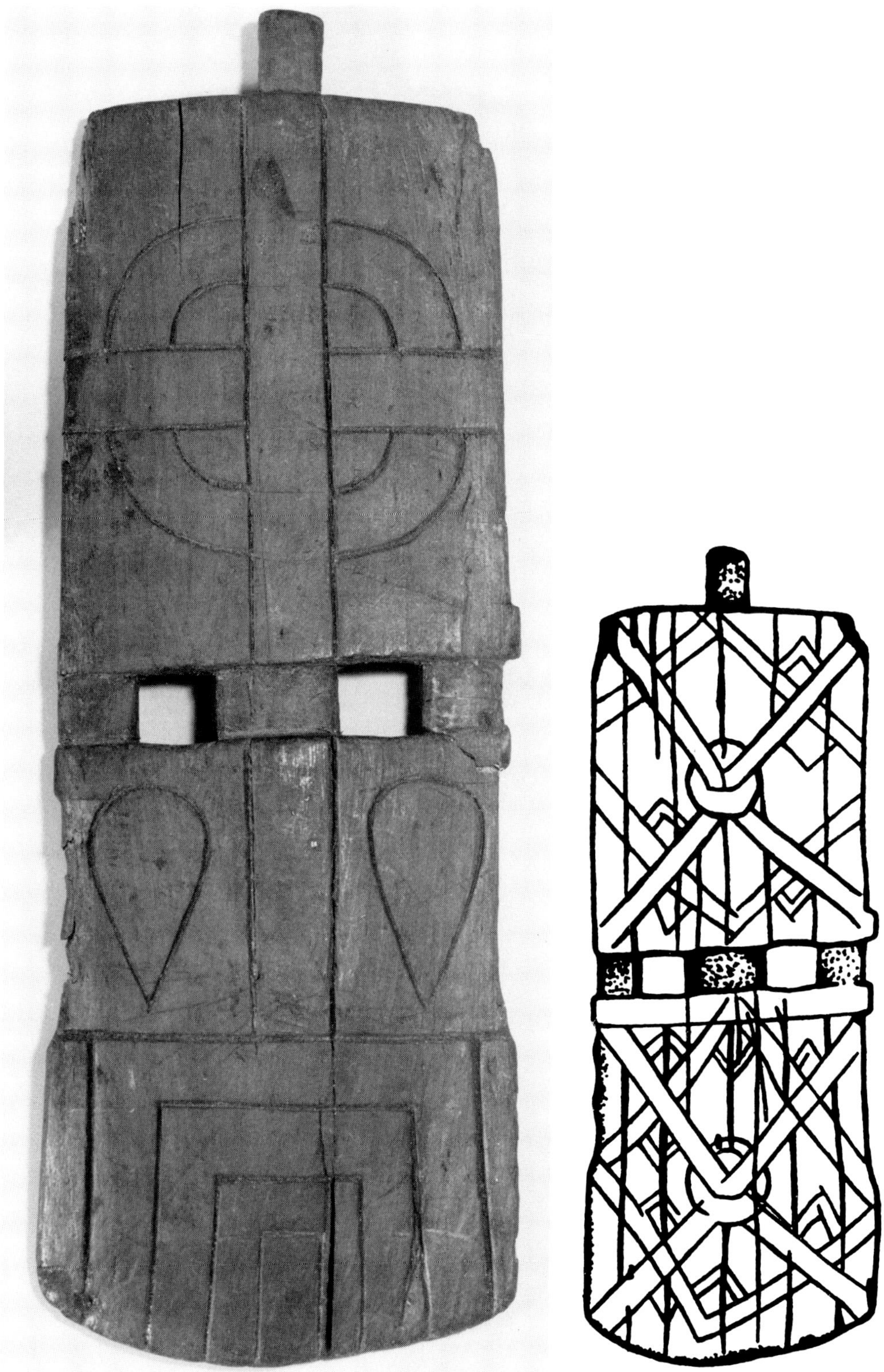

FIGURE 96A Front and back views of wooden ceremonial specimen from Lee County. Front: Smithsonian Institution Catalog No. 329599, Neg. No. 1169-B; Back: from Fewkes 1928).

FIGURE 96B Wooden ceremonial specimen from Lee County. (Smithsonian Institution Catalog No. 329598, Neg. No. MNH 1169-A.)

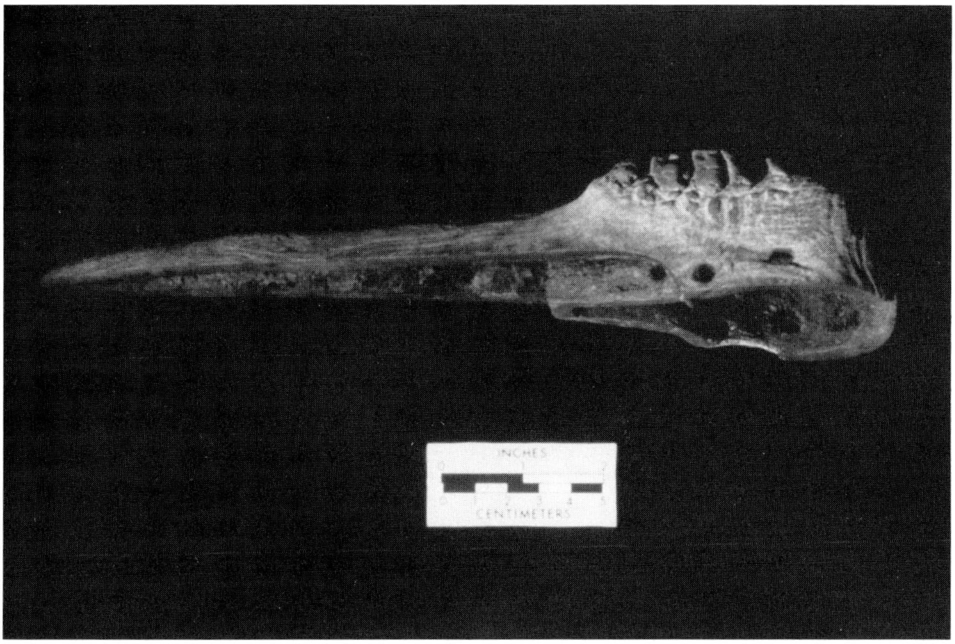

A

B

FIGURE 97 (A) Carved wooden figurehead (upper part of bird beak?), cypress (Taxodium sp.), found in 1971 by Ed and Phyllis Thomasson in muck near Pineland, Lee Co.; donated to Florida Museum of Natural History, 1990 (cat. no. 90-24-1); (B) end view. Photographs courtesy of W. H. Marquardt.

FIGURE 98 A wooden trough from Observation Island, Glades County; made of mulberry(?). (SFM-A6601.) (Picture credit: Bishop Planetarium and South Florida Museum.)

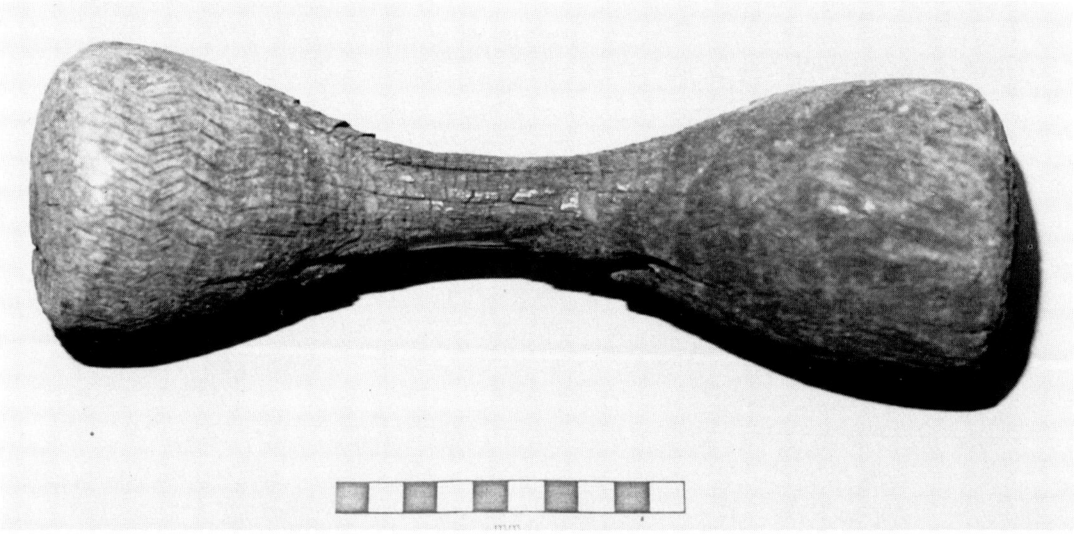

FIGURE 99 A double-ended pestle from Levy County; made of mangrove(?) (SFM-A6568.) (Picture credit: Bishop Planetarium and South Florida Museum.)

FIGURE 100 Wooden mortar from Observation Island, Glades County; made of mangrove(?). (SFM-A6565.) (Picture credit: Bishop Planetarium and South Florida Museum.)

FIGURE 101 Bell-shaped wooden mortar from Myakka, Sarasota County; made of pine (SFM-A6566.) (Picture credit: Bishop Planetarium and South Florida Museum.)

FIGURE 102 Wooden mortar in possession of Bishop Planetarium and South Florida Museum (no documentation or SFM Number, probably modern); made of cypress. (Picture credit: Bishop Planetarium and South Florida Museum.)

A B

FIGURE 103 (A,B,C) Three views of a wooden "idol" found on the north shore of Lake Okeechobee in 1921. See text for additional information. (Smithsonian Institution Catalog 316254, Acc. 66473, Neg. 86-3440.)

Figure 103C

A

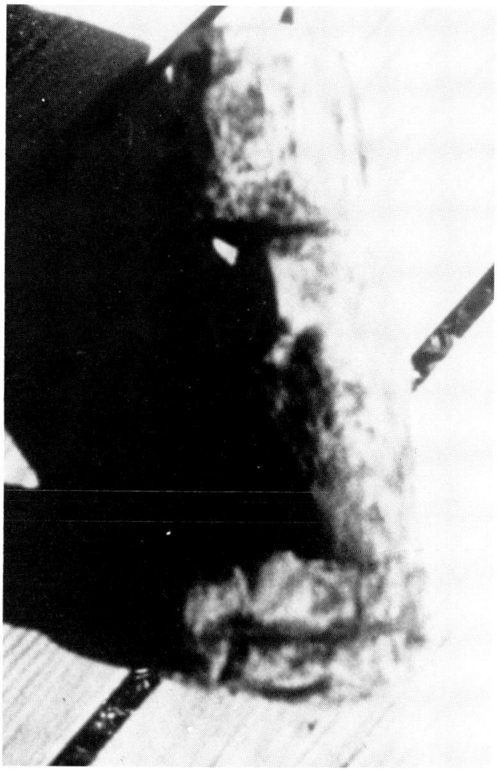

B

FIGURE 104 (A,B) Wooden carving of a human figure recovered in 1928 near Pahokee, Palm Beach County. This specimen is interesting because it is kneeling in a manner similar to the famous feline figure from Key Marco excavated by Cushing in 1896 (see Figure 9d). See text for additional information (photos courtesy of Harriett Mahoney).

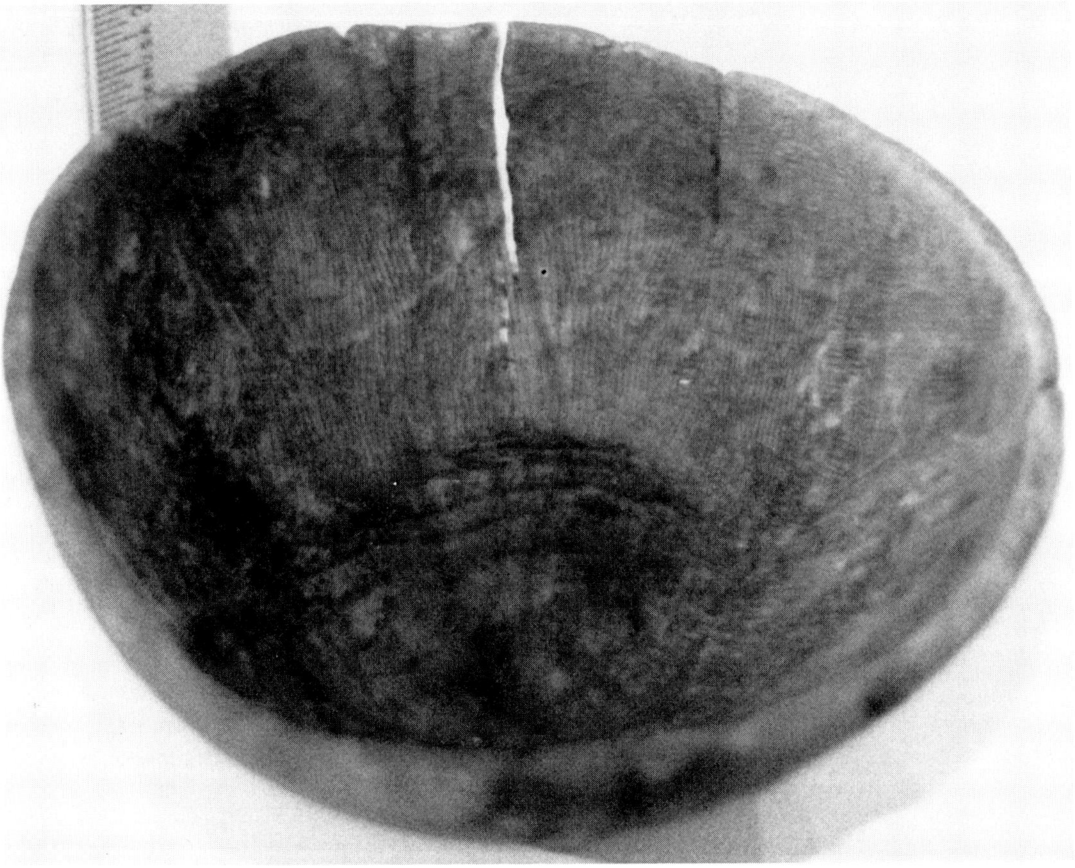

FIGURE 105 Small wooden bowl found in Kreamer Island in Lake Okeechobee in 1916 (photo credit: Palm Beach County Public Library).

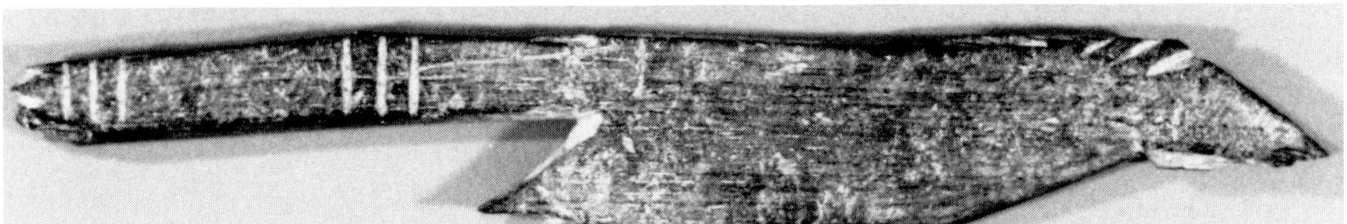

FIGURE 106 Bird effigy recovered in Pinellas County (photo courtesy of Lyman O. Warren).

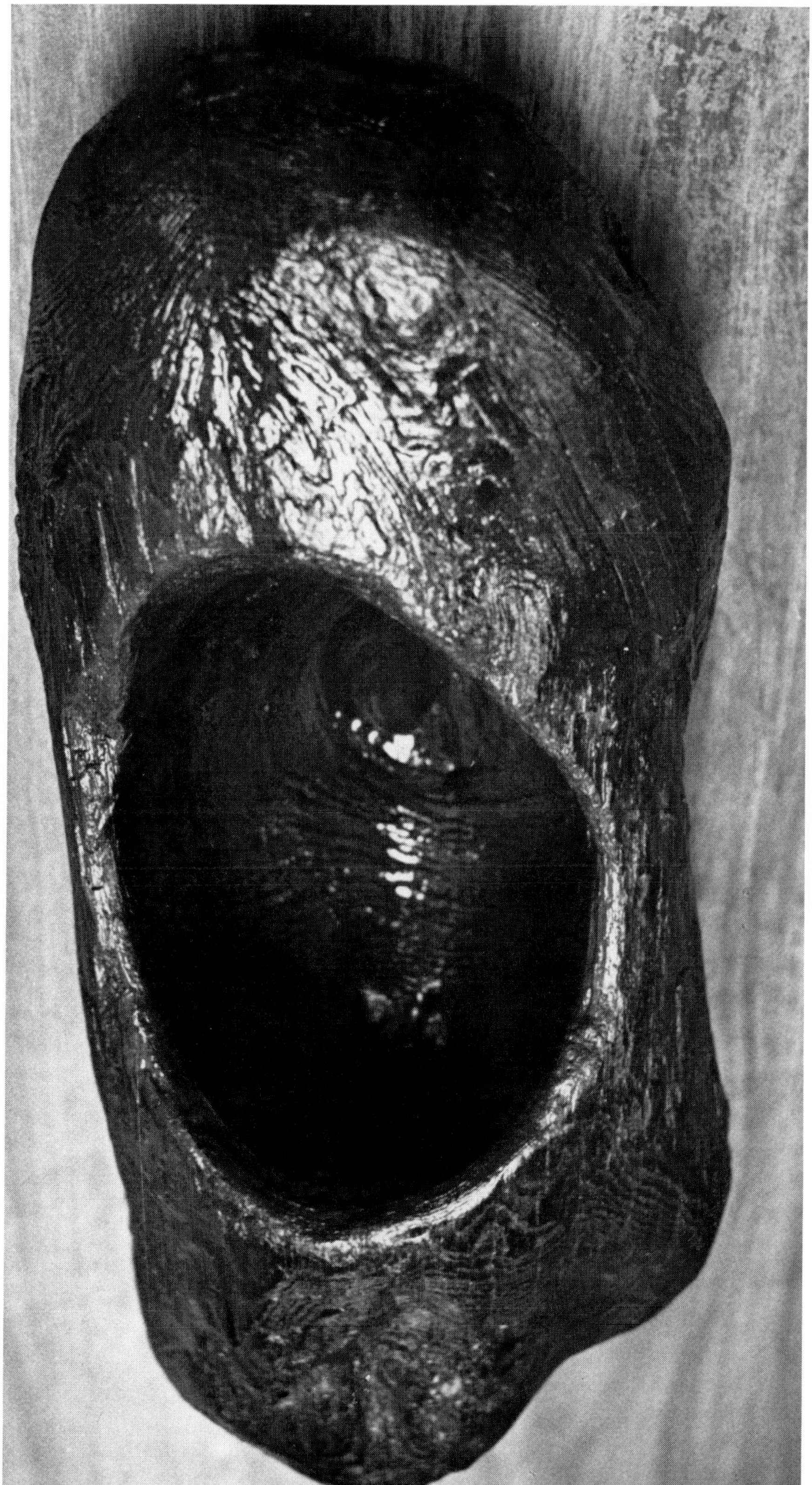

FIGURE 107 Large wooden bowl from Putnam County recovered from a peat deposit near Crescent City. (Photo property of the author.)

References

Allerton, David, George M. Luer, and Robert S. Carr 1984 Ceremonial Tablets and Related Objects from Florida. *The Florida Anthropologist* 37(1):5–54.

Benson, Carl A. 1967 A Unique Wood Carving from Tick Island. *The Florida Anthropologist* 20(3–4):178–179.

Bushnell, Francis F. 1960 The Harris Creek Site, Tick Island, Volusia County. *The Florida Anthropologist* 13(1):25–31.

Carr, Robert S. 1986a Preliminary Report on Excavations at the Cutler Fossil Site (8Da-2001) in Southern Florida. *The Florida Anthropologist* 39(3):231–232.

1986b Sinkhole Yields Rare View of Dade's Past. *The Miami News*, Friday, March 28:13A.

Clausen, Carl J., H.K. Brooks, and Al B. Wesolowsky 1975 The Early Man Site at Warm Mineral Springs, Florida. *Journal of Field Archaeology* 2(3):191–213.

DeSoto County News 1921 Wooden Idol Plowed Out of Muck on the Shore of Okeechobee. *DeSoto County News*, March 25.

Fewkes, J. Walter 1928 Aboriginal Wooden Objects from Southern Florida. *Smithsonian Miscellaneous Collections* 80(9):1–5.

The Florida Anthropologist 1981 Florida Anthropologist Interview with Calvin Jones: Excavations of an Archaic Cemetery in Cocoa Beach, Florida. *The Florida Anthropologist* 34(21):81–89.

Goggin, John M. 1942 Prehistoric Wooden Club from Southern Florida. *American Anthropologist* 44:327–328.

Işçan, M. Yasar. 1983 Skeletal Biology of the Margate-Blount Population. *The Florida Anthropologist* 36(3-4): 154–166.

Jahn, Otto L. and Ripley P. Bullen 1978 The Tick Island Site, St. Johns River, Florida. *The Florida Anthropologist* 31(4):Part 2, 25 pages plus 60 figures.

Jones, William M. 1981 Wooden Objects Recovered at the Site of San Juan del Puerto (8-DU-53) Fort George Island. Manuscript on file, Department of Anthropology, University of Florida, Gainesville.

Kenner, Joe 1974 Rare Art from Florida's Past. *Florida Conservation News* (October):8.

Mahoney, Harriett H. 1981 Letter dated February 20, 1981 on file with Barbara A. Purdy, Department of Anthropology, University of Florida, Gainesville.

Martin, Robert A. and S. David Webb 1974 Late Pleistocene Mammals from The Devils' Den Fauna, Levy County. In *Pleistocene Mammals of Florida*, pp. 114–145. The University Presses of Florida, Gainesville.

Miller, James J. and John W. Griffin 1978 Cultural Resource Reconnaissance of Lake Woodruff National Wildlife Refuge. Contract No. PX5880-8-0205, Cultural Resource Management, Inc., Tallahassee.

Moore, Clarence B. 1892a A Burial Mound of Florida. *American Naturalist* 26(February):129–143.

1892b Supplementary Investigations at Tick Island. *American Naturalist* 26(July):568–579.

1893 Certain Shell Heaps of the St. Johns River, Florida, Hitherto Unexplored. *American Naturalist* 27:8–13, 113–117, 605–624, 708–723.

Morrell, L. Ross 1974 Letter on file, Division of Recreation and Parks, Department of Natural Resources, Tallahassee.

Purdy, Barbara A. 1988 American Indians After AD 1492: A Case Study of Forced Culture Change. *American Anthropologist* 90(3):640–655.

Russo, Michael, Lee A. Newsom, Ray M. McGee, and Barbara A. Purdy n.d. Groves Orange Midden Site (8-VO-260l). In press *Southeastern Archaeology*

Schwehm, Alice Gates 1983 The Carved Wood Effigies of Fort Center: A Glimpse of South Florida's Prehistoric Art. Master of Arts thesis, Department of Fine Arts, University of Florida, Gainesville.

Stackhouse, Edgar J. 1966 A Unique Wood Carving from Florida. *Pennsylvania Archaeologist* 36(1–2).

Stackhouse, John R. 1990 Personal communication, May. Mr. Stackhouse found the wooden carving from Tick Island. Mr. Stackhouse provided me with photographs of the specimen.

Warren, Lyman O. 1971 Unique Bird Effigy Dredged from Tampa Bay. *The Florida Anthropologist* 24(2):71–72.

1990 Personal communication. Dr. Warren's son found the bird effigy dredged from Tampa Bay. Dr. Warren provided me with photographs of the specimen.

Williams, Wilma B. 1983 Bridge to the Past: Excavations at the Margate-Blount Site. *The Florida Anthropologist* 36(3-4): 142–153.

FLORIDA CANOES: A MARITIME HERITAGE FROM THE PAST*

Introduction

A MARITIME HERITAGE exists in Florida that deserves to be investigated and preserved. The largest number of prehistoric and early historic canoes in the world have survived in the state's numerous waterways. Records exist for 185 canoes that date from 5140 years BP to the 19th century; thus, they span the Archaic through the Seminole Indian periods. Hundreds must have been found that were not reported or were destroyed during peat mining operations, development projects, or installation of drainage canals for agriculture. Countless more are probably still entombed. Destruction of canoes is not a thing of the past as is illustrated by the following story.

In the summer of 1985, we received a call from the public relations people of a large corporation who told us that a number of canoes, wooden structures, and other artifacts had been uncovered by their earth-moving equipment. The people were helpful when we visited the area and permitted us to take a sample of one of the canoes for radiocarbon analysis for which they paid. The water table was very high at the time and we were unable to conduct a thorough investigation. The public relations people promised to call us when the water table dropped. When they did not call by mid-winter, we contacted them and in May 1986, we visited the area again. As before, we were treated well and were taken to the location where the canoes had been found. To our dismay and, seemingly, to the dismay of the corporate representatives, the entire area had been destroyed. We do not believe that this was done maliciously but, perhaps worse, they did not understand the importance of these remains. The canoe was 3300 years old. At that time, it was the oldest canoe in the Western Hemisphere, but since then a canoe from DeLeon Springs, Florida has been

*See also: Newsom, Lee A. and Barbara A. Purdy, The Florida Anthropologist 43: 164–180.

dated at 5140 BP and another from Florahome is ca. 4000 years old. With the rapid development occurring in Florida it is probable that the above incident will be repeated over and over again.

We need to salvage as much information as possible about the existing canoes and attempt to increase the public's awareness of their significance. This chapter summarizes what is known and what needs to be done in order to make the records more complete. Investigations are necessary to organize and expand the existing canoe records in a four-phase project. Phase 1: contact canoe owners (including institutions such as museums) to determine if the canoes still exist and to request permission to measure, photograph, and sketch them, acquire wood samples for radiocarbon dating and microscopic identification of the species, and assess their present condition and preserveration requirements. Phase 2: obtain radiocarbon dates and identify the wood species. Phase 3: conduct a literature search to bring together historic descriptions of canoe manufacture and use, and produce a monograph that includes all of the data about the canoes and examines the importance of watercraft to the early inhabitants of Florida. It is possible to speculate about universal behaviors in maritime environments that crosscut time, space, and culture. Phase 4 of the canoe project includes the preservation of the canoes and their future exhibit either in a centralized location or in the various areas where each was found.

Prehistoric and early historic canoes form one segment of our investigations of all wooden artifacts that have survived in Florida (MacDonald and Purdy 1982; Purdy and Newsom 1985; Purdy 1988; Chapter 3). Canoes are reported at the rate of one or more a month. Twenty canoes, for example, were reported in 1985 (Newsom and Purdy 1986) and 19 canoes were retrieved and examined when Florida experie nced drought conditions in 1977. Emergency funds for the 1977 project were provided by the National Science Foundation and the National Endowment for the Arts. In that year also, two wooden totems were found in the St. Johns River near Hontoon Island in the same vicinity as the owl totem recovered in 1955 (Bullen 1955; see Chapter 3).

Present State of Knowledge

Fig. 108 shows the geographic distribution and density of the canoe finds. The majority of the finds have come from the northcentral highlands where the many lakes and streams in the region may have been interconnected prior to the advent of ambitious drainage projects in the late 19th and early 20th centuries. Even under natural conditions these lakes, located on relic dune ridges in an area of karst topography, are frequently affected by drought. We believe the concentration of canoes here may be more a function of geology and hydrology than a reflection of the greater cultural importance of the dugout in the Northcentral Highlands. The proximity of this area to the University of Florida may also be a factor.

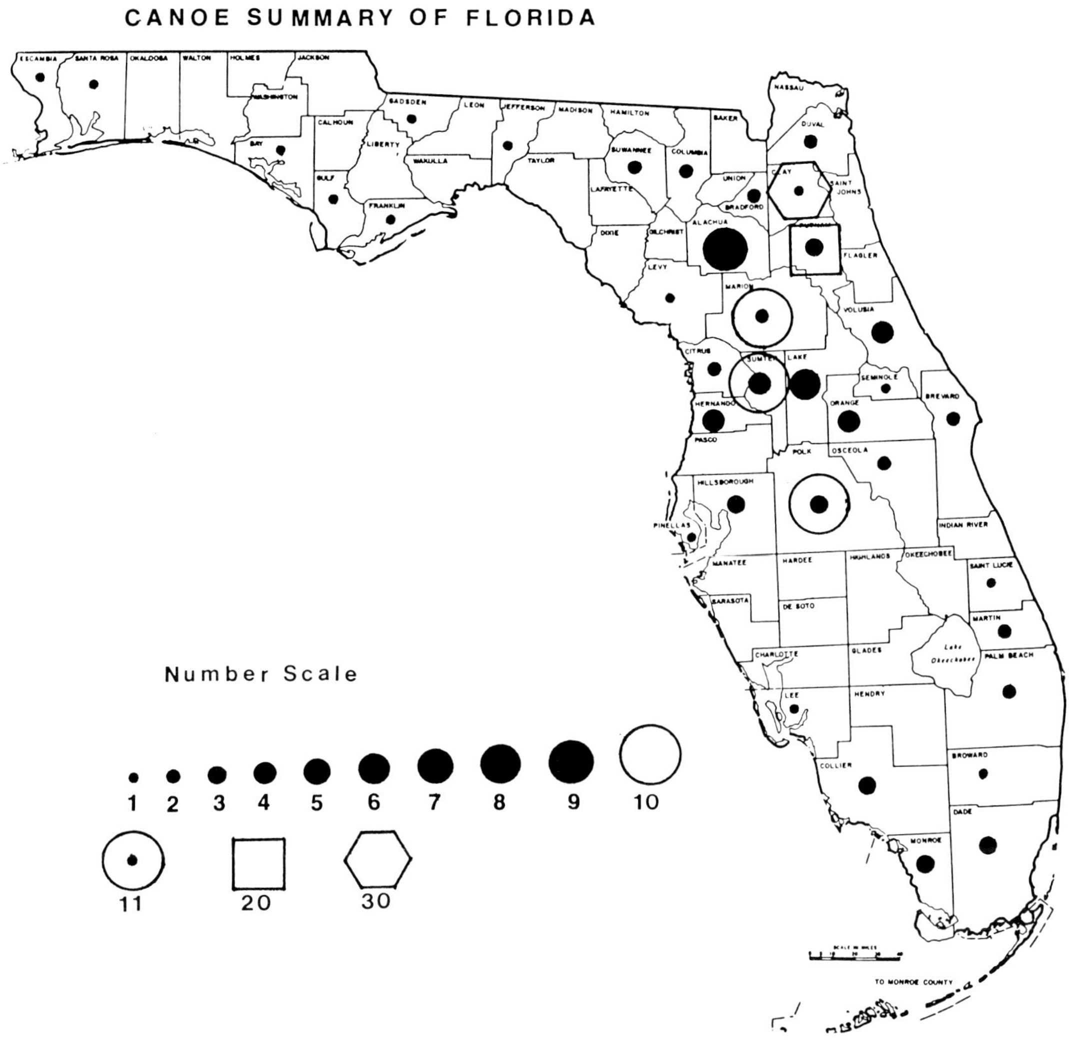

FIGURE 108 Geographic distribution and density of canoe finds in Florida. (compiled by R.M. McGee, 1990).

We do not have records of the matrices in which many of the canoes were found. Some of the canoes were reported to have been buried in sand but the great majority were found entombed in organic deposits of varying thicknesses. Their integrity was maintained for the same reason that the twigs, seeds, leaves, etc. comprising the deposit remained intact and can be identified after thousands of years; that is, because of the anaerobic environment. Florida has the second largest peat resources in the nation. It is not surprising, therefore, that cultural items were incorporated and preserved in these organic deposits in areas where people exploited the resources of swamp or marsh (Fig. 109).

Most of the canoes have not been dated by radiocarbon analysis. There is as yet no evidence of canoes from the Paleoindian Period. The oldest canoe in Florida thus far is from DeLeon Springs in Volusia County near the St. Johns

FIGURE 109 Illustrates the nature of the deposit in which many canoes are entombed (picture is of Ollie Stricklin, Stricklin Peat Co., Clay County). See also Fig. 124.

River (Fig. 110). Its Archaic date of 5120 years BP makes it more than 1000 years older than its closest four rivals in the Western Hemisphere, three from Florida and one from Ohio, all of which date from ca. 3100–4000 years BP. A majority of the Florida canoes, for which radiocarbon analysis has been conducted, date between AD 700–950 (see Table 1). At least three styles are recognized for the prehistoric canoes. From the Historic Period, we have

FIGURE 110 The oldest canoe in the Western Hemisphere being removed from DeLeon Springs. Its age is 5140 ± 60 BP (photo courtesy of the Florida State Park Service).

records of 37 canoes. While only three of these have been dated by radiocarbon, other evidence suggests they are recent. Of these, at least two types were probably made by the native Florida Indians and a more recent type is characteristic of those made by the Seminoles.

Florida canoes can be divided into several stylistic groups based on shape, size, and fineness of finishing. The first group (Fig. 111) includes the crudest and the earliest canoes. These are roughly hewn with blunt, largely unmodified ends and outer surfaces. The bow and stern are indistinguishable. These canoes were manufactured by the fire-hollowing technique described in numerous ethnohistoric accounts (see, for example, Ribaut 1964). Often, a large portion of the charred surface is not scraped away and, in a few examples, the bark of the tree is partially intact. Canoes of this type were all made of hard (or yellow) pine. They average 15 feet (3.6 m) long by 16 inches (40.6 cm) wide.

A second style appears to coexist with and later replaces the first type. It is the most common form found. It was also manufactured by the fire-hollowing technique but the finishing process was carried beyond simply hollowing out the inside and roughly finishing the outer surfaces as was done in the first type. There was an attempt to shape the bow and stern so that both ends are usually beveled slightly upward and smoothed on the upper surface. The finishing of the gunwales and inside was also carried out to a much greater degree (Fig.

Table 1
Uncorrected ages of Florida canoes listed chronologically

Name	County	C-14 Age BP	C-14 Age
Stricklin #4	Clay	320 ± 80	AD 1630 ± 80
Lake Hancock	Polk	380 ± 50	AD 1570 ± 50
Crooked Lake #3	Polk	440 ± 50	AD 1510 ± 50
Suwannee River #2	Suwannee	470 ± 80	AD 1480 ± 80
Hall Lake	Clay	575 ± 50	AD 1375 ± 50
Stricklin #7	Clay	630 ± 100	AD 1320 ± 100
Florida Rock	Putnam	800 ± 60	AD 1150 ± 60
Crooked Lake #1	Polk	800 ± 70	AD 1150 ± 70
Lake Galilee	Putnam	990 ± 55	AD 960 ± 55
Stricklin #5	Clay	1000 ± 60	AD 950 ± 60
Magnolia Lake	Clay	1060 ± 50	AD 890 ± 50
Pine Lakes	Lake	1120 ± 50	AD 830 ± 50
Cowpen Lake	Putnam	1160 ± 65	AD 790 ± 65
Wildcat Lake	Marion	1160 ± 65	AD 790 ± 65
Loch Low Lake	Seminole	1160 ± 80	AD 790 ± 80
Hampton Lake	Bradford	1180 ± 50	AD 770 ± 50
Zellwood	Orange	1185 ± 75	AD 765 ± 75
Dog Island	Franklin	1240 ± 80	AD 710 ± 80
Swisher Lake	Putnam	1635 ± 60	AD 315 ± 60
Lakeland	Polk	3040 ± 115	1090 ± 115 BC
IMC #1	Polk	3350 ± 60	1400 ± 60 BC
Davis #4	Putnam	3880 ± 80	1930 ± 80 BC
Harney Flats	Hillsborough	3950 ± 70	2000 ± 70 BC
DeLeon Springs	Volusia	5140 ± 100	3190 ± 100 BC

112). These canoes average 18.4 feet (5.6 m) long by 16 inches (40.6 cm) wide. They range from 11 feet (3.35 m) to 30 feet (9+ m) long. The narrowness of the canoes supports Dickinson's statement made in 1696 that "the canoe was just wide enough for us to sit down in" (Andrews and Andrews 1945). Canoe types one and two are both commonly found in the inland lakes and streams.

A third type appears to coexist also with the two types described above. It is distinguished by a distinctive bow which is possibly a modification for larger and rougher bodies of water. These canoes have been discovered primarily on the Atlantic coast, St. Johns River, and some of the bigger lakes. The style is comparable to type two in overall dimensions, manufacturing technique, and finishing except that the bow extends upward and outward into a marked "platform" (Fig. 113). This extended, overhanging bow is believed to have been designed to override large waves while negotiating rougher bodies of water. In some of these the undersurface beneath the bow has been slightly shaped into a "V"-form which would also facilitate travel through such waters. Aft of the bow, the undersurface is smooth and flat.

A

B

FIGURE 111 Style 1 canoe (A) canoes from Florahome (third from left is style 1); (B) style 1 canoe from Stricklin Peat Co., Clay County. (Pictures property of the author.)

At least four canoes represent a fourth stylistic class which seems to have appeared first in the early historic period. In overall shape they resemble most closely canoes of the second type but the use of metal tools (Fig. 114) is clearly evident. All lines and surfaces are very sharp and cleanly cut, and the square,

A

B

FIGURE 112 Style 2 canoe (A) in storage, and (B) recovered in Putnam County during drought of 1977-78. (Pictures property of the author.)

FIGURE 113 Style 3 canoe showing overhanging bow. (Picture property of the author). A Style 3 canoe can be seen also in Fig. 112.

broad-bladed tool marks of a metal axe are numerous. In contrast, the tooling marks found on prehistoric canoes, when present, generally indicate a tool with a rounded bit. In shaping some of the type four canoes, the fire-hollowing technique was still used in conjunction with the metal adze but others exhibit no sign of the use of fire. It could, of course, have been adzed away during the final finishing process. It is possible that type four was made by 16th and 17th century Europeans or by more recent white settlers, but a radiocarbon date of AD 1500 on one canoe supports the probability that they were manufactured by aborigines who had come into possession of iron tools.

The four canoe types described thus far were manufactured predominantly of pine. Exclusive of Seminole canoes (described below) only 5% of the canoes have been positively identified by thin section analysis as manufactured of cypress. Those made of cypress came from river basin areas where cypress is more readily available than pine.

"Composite" craft make up the fifth type. The catamaran as observed in use by both Pedro Menendez and Jonathon Dickinson in the 16th and 17th centuries falls into this category. Menendez tells us that the Calusa chief Carlos "came with as many as 12 canoes and two of them fastened one to the other, with decks covered with awnings of hoops and matting" (Connor 1964). We have records of two canoes that may have been catamarans. Both of these canoes have carved or drilled holes along the length of the gunwales which may have been used for the purpose of lashing the center platform or connecting ties in place. One of the canoes exhibits the use of fire hollowing and metal tools and

A

B

FIGURE 114 Style 4 canoe of the early historic period shwoing metal tool marks (picture property of the author; see also Figure 118).

is, reportedly, made of cedar although this has not been verified by us. Both canoes have tapered, uplifted bows and are 16 feet (5.08 m) and 19 feet (5.93 m) long. Also, a toy catamaran was found at the Key Marco site (Cushing 1897:365; Gilliland 1975).

Four other "composite" craft have been located. Three of these can be

FIGURE 114C

compared to a dugout that has been sawn in half and the sides reattached. In both cases hollowed logs of hard pine were spliced together with a series of cleats or cross pieces and wooden pegs. The halves apparently were hewn without the use of fire but metal tool marks are abundant (Fig. 115). Another type of composite canoe is from Key West and was fashioned of three planks or boards. The bow is pointed and the stern is squared. It is very similar to a modern rowboat except it is very narrow. It would have been a good craft for navigating the shallow shoals, inlets, and estuaries of the keys. Swanton (1946) made brief mention of a similar craft citing Garcia (1902) which, according to Garcia, "had been made by the Indians in the Spanish style."

The sixth canoe type is that of the historic period Seminole Indians. We have records of 20 Seminole canoes (Fig. 116). They average 19–20 feet (5.7–6 m) long and 23 inches (58 cm) wide. They tend to be wider at the bow and taper toward the back into a slightly elevated and rounded stern. The bow is even more elevated and the forward portion of the hull is usually shaped into a "V" cross section with a projecting anterior keel. A Seminole informant (Neill 1953) stated that this design aided travel through the sawgrass marsh of the Everglades. In some cases, the bow was shaped upward and forward to project over the water as in the third native canoe type described earlier. All of the Seminole canoes examined by us have been made of cypress. The Seminole commonly employed an interesting manufacturing technique. This involved cutting a series of holes into the hull along the centerline while the canoe was being hollowed out. These holes were used to gauge the thickness of the hull during construction (Fig. 117). They were later plugged with wooden dowels. At least five of the canoes we have recorded sport gauging holes, two with plugs still

FIGURE 115 Style 5 composite variety (photo property of the author).

intact. Some of the Seminole canoes are fitted with thwarts to support seats or decking. The Seminoles were also known to have constructed sailing canoes (Neill 1953) and three of the canoes in our sample have a mast support near the center bow. One of the canoes, found in the Everglades in 1979, has two keels, one anterior and the other astern.

Four canoes do not fit comfortably into our typology. Two of these are exceptionally large, 50 feet and 46 feet long, and the other two are made of tropical woods that do not occur in Florida. All of these probably belong to the historic period but they have not been radiocarbon dated.

The present condition of the canoes is a topic that cannot be neglected. All of the canoes are in stages of preservation ranging from fair to extremely poor. The most intact canoes are those with both ends remaining. All of the canoes seem to be missing portions of the gunwales. Many are fragmentary and often this occurred in antiquity rather than from present-day recovery methods. There are some amazing stories about canoe acquisition and subsequent

FIGURE 116 Style 6, Seminole Indian canoe.

"curation" procedures. Since 1977–1978, when we received emergency funds to investigate canoes reported during that drought period, we have had growing cooperation from people who wish to be informed of the right way to retrieve a canoe.

New problems arise after canoes are removed from the water. Often they fall apart because they are left on the bank to dry out. In some cases, canoes will not disintegrate under these conditions but usually within a day or two they will crack and split as moisture is liberated from the interstitial spaces of the wood. The large size of canoes presents a major problem in keeping them wet until preservation can proceed. We have brought some canoes together in holding ponds or have returned them to the lakes in which they were found awaiting the construction of vats. There are many disadvantages to the holding pond "solution" to the canoe problem; damages that occur during transport or in the new environment are two major concerns.

Some landowners have opted to try their hand at preserving canoes. In one case, since the price of building a vat and purchasing polyethylene glycol was prohibitive, a gentleman was advised to preserve the canoe with turpentine and linseed oil. This procedure may have worked fairly well but after the canoe was treated, he placed it in a specially constructed glass enclosure that caught a good portion of Florida's daily sun. After four years in this display case, the canoe had disintegrated (Fig. 118).

Major difficulties pertain to the size and quantity of canoes found in

FIGURE 117 Manufacturing a Seminole Indian canoe.

Florida and the cost of preservative chemicals. An ideal situation would be to have a preservation laboratory with enough vats to treat several canoes at one time and space to study and exhibit them after treatment. If space is not available to exhibit the canoes after they are preserved, an alternative solution would be to return them to the area in which they were found and place them on display somewhere, like a local historic museum. The Division of Historical Resources in Tallahassee has been cooperating with us in an effort to preserve the canoes.

FIGURE 118 Disintegration of the canoe shown in Fig. 114 four years after it was recovered. (Photo taken by the author). See also Fig. 124.

Significance

Since the use of watercraft presumably predates the peopling of the Western Hemisphere, it is possible that people were using canoes 10,000 or more years ago. The probability of finding a canoe that old in Florida is slim, however, since the widespread development of organic deposits, in which most canoes have survived, did not occur in Florida until about 6000 years ago. Surface water in rivers and lakes, in fact, may not have been very plentiful until around 6000 years ago and, therefore, canoes would not have been too functional. In the beginning, the rate of deposition of peats was only a few centimeters a century but deposition increased exponentially with time. Between AD 1200 and the late 19th and 20th centuries, when drainage projects began, deposition is estimated at 16 cm per century. The oldest canoe in Florida was found at DeLeon Springs. It was resting slightly above the contact between the bottom of the peat deposit and the underlying limestone bedrock. The date of 5140 BP on this canoe, therefore, supports data from geologic and climatic studies about the deposition of organic deposits and about the growth of aquatic resources and their use by aboriginal Floridians.

Most large prehistoric settlements in Florida are found along the coasts and inland waterways. Canoes were used for transportation of people, goods, trade, warfare, exchange of ideas, and in the exploitation of aquatic and flanking terrestrial resources. The Middle Preceramic Archaic Period throughout the southeast (approximately 7000–4000 years BP) marks the beginning of a continuing trend toward expansion and diversification of the subsistence base including a heavy reliance on aquatic resources. The dugout must have played an integral role in this expansion making widely distributed resources more readily available and cost effective. That the exploitation of Florida's aquatic resources increased through time may be reflected in the greater numbers and stylistic diversity of the canoes. Evidence of prehistoric economies demonstrates that aquatic resources provided a large percentage of the diet even after the cultivation of crops began. Insights into the value of the canoe to the Indians of Florida can be gained by reviewing the early historic literature. Nearly all accounts of Indian travel mention movement by water, which the Europeans soon emulated.

It is interesting to contemplate why watercraft in Florida and most of North America did not evolve beyond canoes hollowed from a single log into composite vessels. (Those described in this paper were probably influenced by Europeans.) Possible explanations are there was no need (or perceived need) or the technological skills had not been developed. This latter reason hardly seems tenable given the fact that woodworking was advanced and all manner of lashings and resins were known. We are left then with lack of perceived need or diffusion as incentive or stimulus for innovation. Even in the Old World, complex watercraft did not appear very early. Its occurrence is contemporaneous with expansion or trade into areas where perilous seas had to be traversed. In Florida, the Indians traveled long distances but apparently did not cross oceans, although the short passage between Cuba and the United States may have been traversed. In the Historic Period, travel by Arawaks and "Spanish" Indians was common between Florida and Cuba (Smith 1854). Canoes varied in size and design somewhat depending on the body of water to be navigated, the purpose they served, and the size of available trees. The use of the catamaran in Florida probably predated the historic period as suggested by the toy catamaran from the Key Marco site (Gilliland 1975) and may have solved the need, at least partially, for larger vessels.

We are examining the use of watercraft in Florida from a holistic point of view:

1. We are studying the technology of canoe manufacture beginning with selection of the tree, properties of the wood, the methods of felling, splitting, and shaping the wood, and the implements used to perform these tasks. It was important to socialize the young to learn how to make canoes. We have come across one canoe that appears to represent a blunder as one might expect from a novice. It is crude and was left unfinished, apparently because a large hole burned completely through the deck during the fire-hollowing process.

2. The integration of canoes into society as a whole was important to the smooth functioning of the culture. For instance, compare the things that could

A

FIGURE 119 (A) Canoe from Magnolia Lake, Clay County that has a circular charred area probably resulting from carrying fire; (B) close-up of charred area (pictures property of the author).

not or would not be done without a canoe with those things that could or would be done if a canoe were available. It was important to teach young people how to operate canoes; this was also part of the socialization process and part of the value system of the culture.

3. The implication exists that the use of waterways determined energy capture, capital bulding, and culture evolution. By using canoes to transport people, supplies, and raw materials, thousands of calories were saved that could be put to other tasks (MacDonald and Purdy 1982). Some of the Natchez canoes, for example, were 40 feet long and could carry 12 tons. The capacity of such large canoes was recognized by one Henry Hudnel who, prior to the Civil War, commissioned the Seminole Indians to construct a canoe for the purpose of hauling crates of oranges from his groves on the St. Johns River. This same canoe was purchased in 1880 by a Mr. Bartley who used it to haul wood to Jacksonville. One load consisted of 1 cord of oak, or 1-1/4 cords of pine, or 18 persons. (A cord is 128 cubic feet—8 feet long, 4 feet high, and 4 feet wide.) Canoes were pack animals but they did not have to be fed. The output of canoes was more complex than their input, but, even so, much energy and expertise were expended in their manufacture and, thus, it can be assumed

FIGURE 119B

that they were cared for. Although there is no mention of it for Florida, there are accounts from other areas that Indians deliberately sunk their canoes if they were going to be gone for awhile in order to prevent their destruction from natural or enemy forces (Bakeless 1964:249,359). This practice may explain why we have found many canoes near shorelines in several small lakes. Canoes may have been needed to exploit the resources of a lake but the population may have been seasonally nomadic and for some reason never returned to retrieve their sunken canoes. There have been at least five instances of multiple canoes found together. In one recent find, a cluster of nine canoes was discovered in a peat area.

We have verified another practice that is reported in many ethnohistoric accounts, that of carrying fire in a canoe. ". . . they killed him with a hatchet while he fanned a fire trying to cook a fish in the middle of the canoe" (Bennett 1968). While all descriptions state that fire was used to facilitate canoe manufacture, we believe that some burned areas represent the cultural practice of carrying fire in a canoe. One canoe with a circular charred area is 1100 years old (Fig. 119). Eight additional canoes have similar burned areas, including one with six and another with eight. The Indians of Virginia and Tierra del Fuego are pictured in early historic documents carrying fire in their canoes.

A final comment about the significance of the canoes is in order. These objects are made of wood, an organic material. Their preservation in Florida's waterways is unique in the Western Hemisphere since very few have been found in other regions. The canoes will eventually disintegrate without treatment with polyethylene glycol and continuing curation no matter how well preserved they appear at present. Perhaps some of the objects will last for 50 years or so but if they are to last for 1000 years or more, they must be preserved and protected, at least by written records.

Summary

The development of the technology to produce and use canoes is an analyzable part of the infrastructure of Florida's prehistoric and early historic Indians. The behavior associated with the manufacture and use of canoes involves the capture of large amounts of energy and, thus, this behavior is of significant systemic importance. Since the records we have compiled about Florida's dugouts are still fragmentary, the ideas presented in the previous sections and summarized below should be considered hypotheses to be tested when additional data are available.

1. The growing availability and utilization of aquatic resources about 6000 years ago in Florida led to the development of canoe technology as a cost-efficient means to exploit both aquatic and flanking terrestrial resources.
2. Canoe manufacture became more refined through time as is demonstrated by better construction and stylistic variability (including size) depending upon the waterway to be traversed and the tasks to be performed.
3. Pine was chosen over cypress for canoe manufacture in the prehistoric periods because of (1) tradition, (2) preference, (3) availability, and/or (4) because of the resin canals that aided in the fire-hollowing process.
4. Cypress was chosen over pine in the historic period (1) because of the availability of iron tools that negated the need to use the fire-hollowing technique, (2) because the idea of using cypress was introduced in the 16th to 18th centuries by non-Florida Indians and Europeans, and/or (3) because of the abundance of cypress in the Everglades where the Seminole Indians reside.
5. Indians from areas to the south traveled to Florida by canoe at least by the early historic period in an attempt to flee Spanish enslavement and disease. Such journeys probably took place prehistorically also.

In 1978 emergency funds from the National Science Foundation and the National Endowment for the Arts, and a generous supply of polyethylene glycol (CARBOWAX) from Union Carbide Corporation were provided to retrieve and care for the canoes. In 1987, Union Carbide supplied additional polyethylene glycol and a small grant was received from the state of Florida. Occasionally, a land owner has paid for radiocarbon analysis. The canoes have generated a great deal of publicity, cooperation, and interest but no monetary support has been attained other than that mentioned above. The rapid growth that Florida is experiencing impacts the canoes as well as other aspects of Florida's maritime heritage.

A study of Florida's canoes has numerous ramifications because canoes functioned in many realms of Indian culture including economic, social, and ritual. Watercraft, in general, provide a way to look at continuities through time and space. In many places in the world today, all major activities occur along waterways. The development of modern Florida, for example, depended on the use of water routes. Until the advent of cars and trains, steamboats were the lifeline that linked Florida's people and resources to the rest of the world (Mueller and Purdy 1984). In the prehistoric and early historic periods, canoes provided the same link.

All available information about Florida canoes has been computerized by Ray M. McGee. These files, plus additional records, photographs, slides, illustrations, and wood samples are located in the Department of Anthropology, University of Florida, Gainesville.

References

Andrews, E.W. and C.M. Andrews, Editors 1945 Jonathan Dickinson's Journal or, God's Protecting Providence (published originally by Dickinson in 1699). Yale University Press, New Haven.

Bakeless, John 1964 *The Journals of Lewis and Clark*. Mentor Press.

Bennett, Charles E. 1968 *Settlement of Florida*. University Presses of Florida, Gainesville.

Bullen, Ripley P. 1955 Carved Owl Totem, Deland, Florida. *Florida Anthropologist* 8:61–73.

Connor, Jeannette Thurber, Translator 1964 *Pedro Menendez de Aviles: Memorial by Gonzalo Solis de Meras*. University Presses of Florida, Gainesville.

Cushing, Frank Hamilton 1897 Exploration of Ancient Key Dweller Remains on the Gulf Coast of Florida. *Proceedings of the American Philosophical Society* 35(153):329–448.

Gilliland, Marion Spjut 1975 *The material culture of Key Marco, Florida*. University Presses of Florida, Gainesville.

MacDonald, George F. and Barbara A. Purdy 1982 Florida's Wet Sites: Where the Fragile Past Survives. *Early Man* 4:6–12.

Mueller, Edward A. and Barbara A. Purdy 1984 *The Steamboat Era of Florida*. Proceedings of a conference funded by the Florida Endowment for the Humanities.

Neill, Wilfred T. 1953 Dugouts of the Mikasuki Seminole. *Florida Anthropologist* 6(3).

Newsom, Lee Ann and Barbara A. Purdy 1986 Florida's canoes. Paper presented in a symposium about canoes and canoeists at the annual meeting of the Society for American Archaeology New Orleans, April 1986.

1990: Florida Canoes: A Maritime Heritage from the Past. *The Florida Anthropologist* 43: 164–180.

Purdy, Barbara A. 1988 Archaeological Wet Sites: Untapped Archives of Prehistoric Documents. In *Wet Site Archaeology*, pp. 325–335. Edited by Barbara A. Purdy. The Telford Press, West Caldwell, New Jersey.

Purdy, Barbara A. and Lee Ann Newsom 1985 Significance of Archaeological Wet Sites: a Florida Example. *National Geographic Research* 1(4):564–69.

Ribaut, Jean 1964 *The Whole & True Discouerye of Terra Florida*. A facsimile reprint of the London edition of 1563. University Presses of Florida, Gainesville.

Smith, Buckingham 1854 *Memoir of Do d'Escalente Fontaneda Respecting Florida*. (Reprinted with revisions in 1944 by the University of Miami and the Historical Association of Southern Florida, edited by David O. True.)

Swanton, John R. 1946 The Indians of the Southeastern United States. *Bureau of American Ethnology Bulletin* No. 137. Smithsonian Institution, Washington, D.C.

Wyman, Jeffries 1875 *St. Johns' River, Florida*. Peabody Academy of Science, Salem, Mass.

PRESERVATION OF ORGANIC MATERIALS FROM ARCHAEOLOGICAL WET SITES

Introduction

THE ITEMS SHOWN in Figs. 120–122 illustrate, as no words can describe, why it is important to be aware of the degradation that occurs when organic materials, particularly botanical specimens, are removed from deposits that have entombed and preserved them. The figures represent entirely different sets of circumstances but the outcome is the same: irreversible deterioration.

Fig. 120 is a small wooden carving from the Ozette Village site on the Olympic Peninsula, Washington. The figure lay face down and partially excavated with only the head projecting out from the moist clay deposits for nearly 2 years when further excavation exposed the rest of it. The body portion that had remained buried in the moist clay deposits was in good shape when discovered but the exposed, untreated head had become badly degraded as the wood dried (Daugherty 1988:26).

Fig. 121A is a cucurbit stem recovered at Hontoon Island, Florida in 1980. It was placed in formic acidic acid (FAA), a fixative used to preserve plant parts, but the vial accidentally tipped over spilling the contents with consequent results as pictured in Fig. 121B.

Fig. 122 shows how a "healthy" specimen of archaeological wood is altered within hours if it is not kept waterlogged prior to preservation. This specimen was unaccountably removed from its waterbath and left on a counter for 12 hours. It is a significant object because it is one of the few wooden artifacts recovered from the 6500-year-old Bay West site (see Chapter 3) and is thought to be a fragment of an atlatl or spear shaft. The great interest in it was probably what caused its "death."

The primary objective of this chapter is to attempt to explain why degradation of floral materials occurs even when they are embedded in a saturated

FIGURE 120 Small wooden carving from the Ozette Village site showing degradation of the head (see text for explanation; picture credit Ruth and Louis Kirk).

organic soil deposit, and to describe what procedures should be implemented to preserve wood and other plant materials removed from those deposits in order to prevent further degradation. Faunal remains also require some treatment when removed from a waterlogged matrix. We use a 1:3 mixture of polyvinyl acetate emulsion (such as Elmer's Glue, Bull Dog grip, Lepages White glue, etc.) and water for this purpose. Composite implements, such as those made of wood/bone, wood/metal, bone/metal, or metal/metal pose special problems of preservation. Investigations concerning the proper course of treatment for composite implements are still in their infancy.

The world of wet site archaeology is fortunate that The International Council of Museums (ICOM) has an active and productive Working Group on Wet Organic Archaeological Materials. The roster includes chemists, conserva-

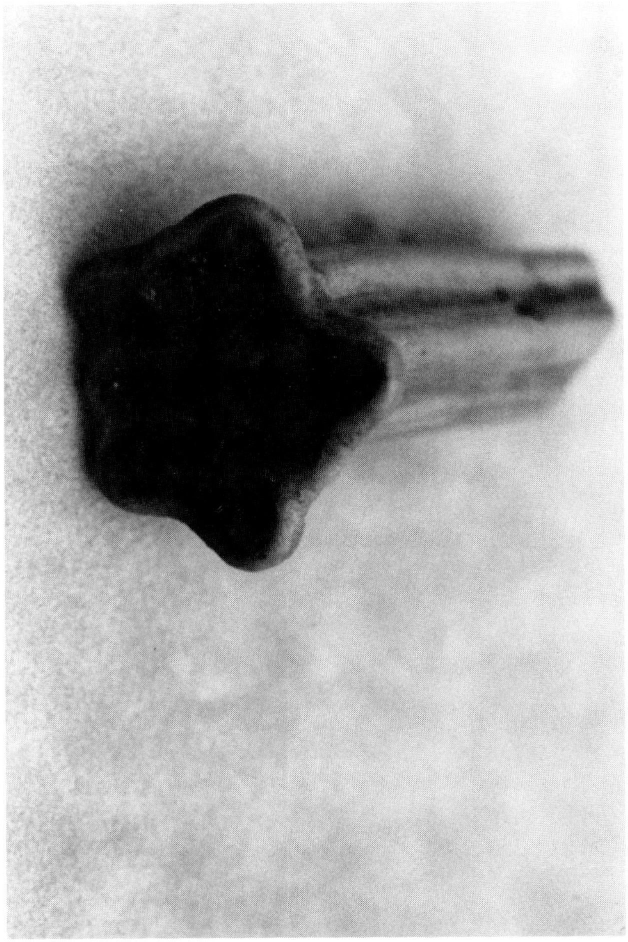

A

FIGURE 121 (A) Curcurbit stem from Hontoon Island; (B) same specimen after degradation (pictures by the author).

tors, wood anatomists, wood products specialists, foresters, historians, artists, archaeologists, and museum curators and exhibitors. The expertise of some members of this group when applied to the needs of other members has resulted in truly significant contributions. Anyone who intends to become involved in the preservation of materials from archaeological wet sites *must* consult the important publications emanating from the labors of this organization in order to take advantage of the knowledge that has accumulated and to avoid making unnecessary mistakes.

Waterlogged Woods and Their Environments

The breakdown of various constituents of plants in the process of peat formation was discussed in Chapter 2. Wooden artifacts that become incorporated into developing organic soils undergo similar kinds of transformations.

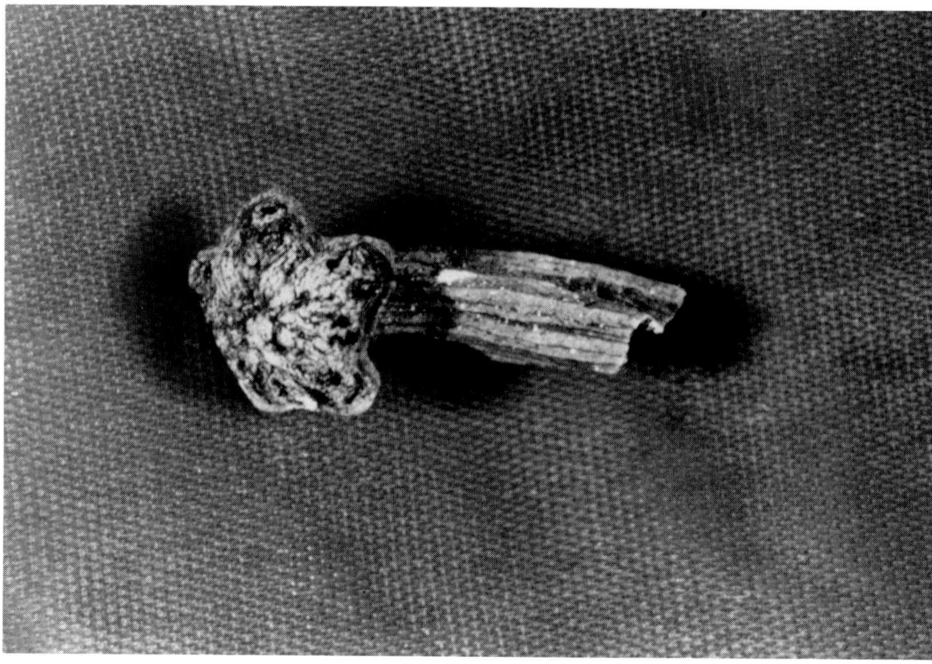

FIGURE 121B

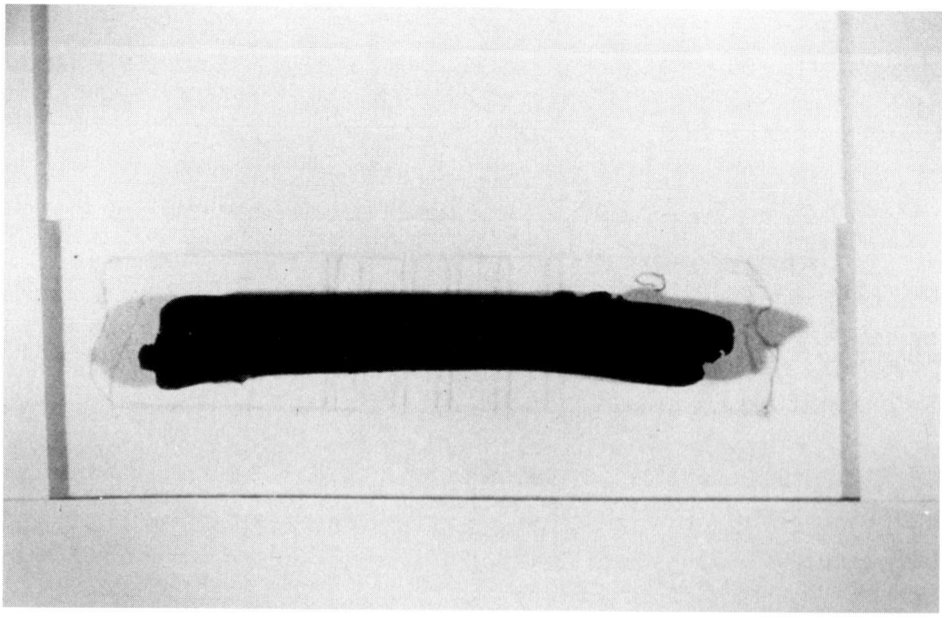

FIGURE 122 Degraded specimen from the Bay West site, Collier County (see text for additional information). (Picture property of the author.) See also Chapter 3, Fig. 16.

A number of other circumstances exist with artifactual woods, however, and must be taken into account, such as whether or not an object was charred or painted, how it was used, and if it was already worn or had suffered insect, fungal, or bacterial damage before it was deposited. All of these factors and more have an affect on the survival of wood in the soil and on its subsequent treatment after removal.

The amount of reduction/oxidation that occurs at the interface between totally waterlogged deposits (reduced) and those that are occasionally oxidized is another condition that plays an important part in the state of preservation of wood in an organic soil matrix. The moisture content dictates reduction/oxidation (the redox potential) and controls aerobic or anaerobic conditions. Within a range of moisture content from fully waterlogged down to about 30%, the lower the moisture, the higher is the level of biological and chemical activity and this is where maximum degradation can be expected (Florian 1982:63–64, 194–199; Goode 1982:26,48). "The only problem which remains for me is that I do not know the level at which wood degradation *is* at a maximum, so I do not know if the value of 30% is accurate. I do know, however, that as moisture level decreases below the fibre saturation point, biological activity decreases until at about 13–15% moisture content it ceases" (Grattan 1989b, personal communication). It is impossible to have a situation where wetting/drying occurs and expect to have preservation of botanical materials. The reason that botanical materials are preserved so superbly in dry caves and tombs is because the moisture content is constantly *less* than that necessary to sustain microbial decay (Ember 1988:14).

In Florida, environmental factors have not been measured for many sites where wooden artifacts have been found. Exceptions include Little Salt Spring, Page-Ladson, Warm Mineral Springs, and Windover. It is critical to know the ecological situation and its variations with depth and location in order to predict if floral materials may have survived, and to determine proper storage conditions and preservation regimes if excavations are conducted. The best natural preservation conditions for waterlogged woods are anaerobic without the influence of light, low temperature, steady pressure from soil or water, and mostly steady pH value (Jesperson 1984). Florida's undisturbed organic deposits satisfy all of these requirements, including temperature, and this explains why spectacular wooden artifacts have survived at numerous sites in Florida. About temperature, Davis says: "Peat bogs and marshes frequently occur in depressions where the air temperature becomes colder than the surround territory, and peat soils are cold soils because peat soils do not give up heat to the atmosphere as rapidly as mineral soils" (Davis 1946:55).

Studies of waterlogged woods have demonstrated, as with organic soils, that 95% of the celluloses and hemicelluloses are gone, lignin remains, and the mineral components are increased 10–60 times over fresh wood because they are not extracted over time and because more are absorbed from the water and soil. In general, cellulose fibrils are surrounded by hemicellulose which is embedded in lignin in fresh wood. For woods that are not very old a larger percentage of celluloses may still be present but one would expect that 500-year-old woods are just as degraded as 5000-year-old woods (Coles 1982:140; Hoffmann 1982:105). "This is an interesting point to me, because I have measured cellulose and lignin in 5000-year-old Somerset Levels wood and also in some 30–40 million-year-old fossil wood preserved in the permafrost in the Arctic. Both had rather similar analyses!" (Grattan 1989b personal communication). From both chemical and microbiological standpoints it is clear why lignin is the most stable component of wood. Its aromatic ring structure and bonding patterns make lignin resistant to hydrolysis. Lignin is also not appetizing to microorganisms since only a few organisms can digest it and then only a certain percentage; none is known to utilize it as a sole nutrient source (Ember 1988:15). The degradation of the woods, therefore, is mainly a degradation of the carbohydrates cellulose and hemicellulose removed in approximately the same ratio as they appeared in the living wood (Hoffmann 1982:76). When carbohydrates are removed from the fibrils of the cell wall further cavities are formed leaving a residual lignin-carbohydrate skeleton that will eventually break and collapse if proper treatment is not applied.

When excavations are underway in a waterlogged deposit, it is extremely difficult to keep fragile organic materials waterlogged while, at the same time, removing enough moisture from the deposits so that excavation can proceed efficiently. If the soil matrix is dried beyond a critical percentage, artifacts and other floral and faunal remains embedded in it will be endangered. Individuals who intend to excavate waterlogged sites need to be aware of this problem (which will probably vary for every site) and prepare for it. A similar problem needs to be addressed if large items like canoes or timbers are placed in temporary storage, like holding ponds, awaiting preservation. The new environment will introduce radical changes of temperature, oxygen, light, pressure, algal growth, and bacteria to these objects.

Preservation of Waterlogged Wood

Polyethylene Glycol

There are probably hundreds of agents of deterioration complicated by the number of wood species present in the environment that may have been selected by people in the past to manufacture implements and artistic items. It seems hopeless to expect that a cookbook type of preservation recipe might be available that can be applied to all wooden objects. Waterlogged woods, however, independent of age and species, share several common characteristics.

Waterlogged means that wood is at or near its maximum water content (Barbour 1982:209); that is, the complete filling with water of all of the pore spaces—the capillaries and the microcapillaries (Grattan 1989a:55). Our concern in this section is whether water in the cell wall can be extracted from the wood by replacing it with another substance (a bulking agent) that will prevent dimensional changes when treatment is completed so that objects can be studied, stored, or exhibited without fear of further deterioration.

Because lignin and other components of the cell wall, such as tannins, fats, or waxes, do not fill the cell wall completely, a manifold system of open capillaries and cavities (namely, the secondary cell space) is left between and within the fibrils and microfibrils. This capillary system makes up about 40% of the cell wall (Hoffmann 1982:75) but this percentage is increased radically when degradation occurs. The capillary system of the wood cell wall is the structure that needs to be treated to stabilize waterlogged wood. Dimensional change, called capillary tension collapse, occurs when "free" water evaporates from capillaries that form the permanent void structure of the wood. Desorption occurs when "bound" water is removed from intermolecular association with the cell wall substance. As bound water is removed from the cell walls, the molecules making up the cell walls draw together. Technically, shrinkage is the dimensional change that results from desorption of bound water (Barbour 1982:210). If waterlogged archaeological woods are air dried, a 70% volumetric loss of dimension can occur because of capillary tension collapse and some cell wall shrinkage (Barbour 1984:23).

In his review of wood preservation techniques, Grattan (1988a) examines the pros and cons of various methods that have been utilized since the 1850s. We can take advantage of Grattan's critical analysis and eliminate from consideration the use of methods that are ineffective, dangerous, irreversible, or yield undesirable attributes of appearance and texture. Other factors include cost, time, and special requirements of space, facilities, or training.

Extensive experimentation over a 35-year period has shown that the chemical of choice at present is a synthetic polymer known as polyethylene glycol (PEG). Although typical applications of PEG include everything from agriculture to printing (Union Carbide Corporation 1981), its antishrink effect on wood was discovered almost immediately after its manufacture began around 1950 (Grattan 1988a:239). PEG becomes part of the molecular structure of the cell walls by replacing bound water and holding the walls in a swollen or semiswollen state. As PEG is not covalently bonded to hemicellulose or lignin in the cell wall, treatment is, in theory, reversible (Ember 1988:13).

When experiments with PEG began in the 1950s and 1960s, little was known about its various molecular weights and their ability to penetrate wood depending upon its degree of degradation. PEG with a molecular weight of 4000 (now PEG 3350) resembles paraffin wax and thus gives strength and cohesion to very deteriorated wood. Early treatments proceeded as follows. Wood was placed in a heated solution of PEG 4000 and the temperature was increased gradually causing evaporation to take place so that the submerged wood was exposed to increasing concentrations of PEG. Eventually the wood was surrounded by pure molten PEG 4000. All of the water in the wood was extracted so that the cell cavities supposedly were entirely full of the wax. But there were drawbacks. The PEG solution and contents tended to turn inky black, wood became unnaturally heavy with a waxy feel and appearance, and fine surface detail was often obliterated. The principal problem, however, was connected with penetration (Grattan 1988a:240).

Over the years many of these problems have been addressed and solved. PEG 3350 is still recommended for highly deteriorated woods but lower molecular weights are often used satisfactorily in conjunction with long treatment time involving a gradual increase in concentration of the chemical. If PEG, especially PEG 3350, is applied initially in high concentrations to the surface of waterlogged wood, dehydration of the interior will occur. Low molecular grades and concentrations of PEG will penetrate rapidly and counterbalance the outward flow of water (Grattan 1982:65).

It is now known that PEG should not be subjected to very high temperatures. Heating causes considerable darkening of the wood especially in the presence of oxygen (Brownstein 1982:106) and at temperatures above 100°C, PEGs become insoluble in water as a consequence of rupturing the hydrogen bonds (Brownstein 1982:279). Most waterlogged woods contain tannins and iron salts (Watson 1982:238) that are leached out and cause discoloration of the wood if a hot PEG solution is used (Brownstein 1982:283). Iron compounds depolymerize PEG at high temperatures and the "conserved" wood may become a sticky mess (Hoffmann 1982:76).

High concentrations of iron salts in waterlogged wood may also impede the penetration of PEG solution but they cannot be removed chemically without causing considerable damage to the wood. The reason for this is that the iron salts are in the form of unstable sulfides that oxidize readily. Oxidation is accompanied by the formation of gypsum precipitated from solution as long needle-shaped crystals. This crystallization of gypsum can cause distortion of the wood cell walls. An attempt should be made to stabilize iron salts in situ (Watson 1982:238; 1984:213–218,357). "The disodium salt of EDTA, 5% in water, has been used to remove iron. Wooden artefacts from the Mary Rose were soaked for 24–36 hours; longer immersion periods softened the wood. . . . An important point about iron contamination of wood is that chances of removal are much higher if the storage before treatment is kept anaerobic and slightly acidic" (Grattan and Clarke 1989:195; Grattan, personal communication 1989). Some success with partial removal of iron compounds using dilute hydrochloric acid has been reported. It has been noted that iron deposits encourage the colonization of certain bacteria on ancient timbers causing severe surface damage (Watson 1984:218). Barbour describes possible detrimental effects if pyrites are present in degraded woods and recommends speed in conservation because the reactions stop once the wood is dry (Barbour 1984:218). High ash values, especially iron, also contribute to the inability to measure specific gravity accurately (Seifert and Jagels 1984:277–278).

Another characteristic of PEG is its hygroscopicity but it will not "weep" unless the relative humidity (RH) is above 75–80% (Grattan 1982:246; Hoffmann 1984:103) although some workers report that relatively undegraded wood treated by low molecular weight PEGs (400 or 540) Blend should not be exposed to RH above 60% (Watson 1982:250). Woods treated with small amounts of PEG 400 and then freeze-dried, however, are not very hygroscopic at all (Hoffmann 1984:95). (Freeze-drying of degraded woods is discussed below.) Humidity is not high in most buildings or climates but in places like Florida and the tropics, a problem will arise if PEG-treated wood is exhibited or stored out-of-doors or in a facility lacking a climate control system. (It is interesting to note, however, that in 1984 a number of samples of wood that had been preserved with 50% PEG 540 Blend in 1980 were taken to Hontoon Island so that the public could see them. For 2 months the wood was in a room lacking air conditioning. The relative humidty was about 90% and the temperature was about 85°F. The PEG migrated to the surface of the wood and it was moist all of the time. When it was returned to the laboratory in Gainesville, the wood apparently returned to its previous state and, in 1990, is still in good condition.)

When the proper molecular weight and correct conditions of application are used, PEG has proven to be a superior technique for the preservation of degraded waterlogged woods. Experiments are still underway and much is yet to be learned about how PEG works, the most satisfactory treatment regimes, and the long-term survival expectancy of PEG-impregnated woods. The success of any treatment will depend on the extent the consolidating material replaces the water and where replacement occurs in the wood. PEG can be expected to bulk the cell lumina (the cavity that the cell wall encloses) and infiltrate the cell walls. The final location of the PEG depends on its molecular weight. The higher molecular weights may bulk the cell lumina but their solvation sphere size is thought to be too large to infiltrate the cell walls (Young and Wainwright 1982:107).

In a series of rigorous experiments designed to gain more insight into the complex relations between the extent of wood degradation, the size of PEG molecules, the amount of PEG taken up, the stabilization achieved, and the hygroscopicity of the wood after treatment, Hoffmann (1984:95–115) demonstrated, as already suspected, that the rate of penetration of PEG is a function of both the degree of degradation of the wood and the molecular weight of the PEG. Very degraded woods required PEG 3000 to PEG 3350 for best stabilization results while only slightly degraded woods were treated successfully with PEG 200 to PEG 400. Even slightly degraded wood should be treated in order to prevent shrinkage and distortion upon drying. Young and Wainwright (1982:107) and Hoffmann (1984:95–115) showed that PEG stained with cobalt thiocyanate suppresses the natural fluorescence of the cell wall. Using this technique they followed the movement of PEG into the wood and demonstrated that higher moelcular weight PEGs entered the lumina only, while PEGs with low molecular weights bulked the cell wall. Few tests have been conducted to determine how much PEG is actually taken up by the wood. The percent bulking is taken as the ratio of the volume of the impregnant compared to the original volume occupied by the waterlogging water. In other words, it tells you how much of the void space in the wood is filled with impregnant. It can be calculated if you know the wet weight and final weight of the object, the water content of the wood in question, and some idea of the hygroscopicity of the treated object and the density of the impregnant (Grattan 1982:251).

In order to determine the extent of wood degradation so that the correct treatment can be undertaken, it is important to measure how much of the wood substance remains. However, there is some question about the best way to obtain this value. Specific gravity, which is normally a good indicator of the strength of fresh wood, is a very poor evaluator of strength for waterlogged woods (Seifert and Jagels 1984:277). Maximum moisture content (Mmax) is considered to be an indirect measure of solid wood loss, the assumption being that, for a given species, a high Mmax indicates greater deterioration of wood substance (Seifert and Jagels 1984:276). If, for instance, the maximum water content is 400%, the void volume is about 80% and this tells something about the wood which is missing. The formula is

$$\frac{\text{Wet Weight - Dry Weight}}{\text{Dry Weight}} \times 100$$

To carry out the method a sample is submerged in distilled water in a suction flask. The flask is closed and evacuated for 30 minutes. Then the vacuum is released for 15 minutes. This cycle is repeated two times. The sample is then weighed, oven dried at 105°C and weighed again. The maximum water content is calculated as 1% of the dry sample (Hoffmann 1982:83). The procedure is simple but interpretation of values obtained is less straightforward because of numerous variables that are difficult to measure (Seifert and Jagels 1984:276). Cook and Grattan (1984:236) suggest that water contents of any sort are better avoided altogether. This is because a water content is by itself a rather meaningless quantity depending on a whole range of factors other than the degree of deterioration. These factors, mentioned also by Seifert and Jagels (1984:276) include the degree of waterlogging of the wood, and all of the factors which affect density such as the species, the type of tissue, the location in the tree, the extent of deterioration, and the presence of extraneous materials in the wood. Thus, it is quite possible to have several different samples of the same water content which have widely differing states of deterioration and condition. Furthermore, even if one knows the species and something else about the condition, water content is not much help since all of the published information concerning wood uses the density. Maximum water content can easily be converted to density using the formula below if a simple assumption is made:

$$1/R = W/100 + 2/3$$

where R is the density of wood (equivalent to a green volume, oven dry weight density), and W is the maximum water content in percent calculated on the basis of the oven dry weight of wood. The assumption on which this depends is that the cell wall has a density of 1.5 g/cc. There is much support in standard texts of wood technology for this value for all species of wood (Cook and Grattan 1984:236). Mechanical tests or a simple needle test also provide indications of the degree of wood degradation but they too should be used with caution.

Freeze-Drying

Experiments with freeze-drying of degraded archaeological woods have nearly as long a history as experiments with PEG. We can benefit from these efforts by presenting the most recently recommended procedures.

Attempts to freeze-dry wood in the 1950s met with little success until the water of the wood was exchanged with tertiary butanol (a nonaqueous solvent) and then impregnated with PEG 4000 (now PEG 3350) after which objects were frozen and then freeze-dried. This process gave good results but investigators sought an approach that did not entail the use of a toxic nonaqueous solvent. In the 1970s, a solution to the problem was found by placing wood in a cold bath of dilute PEG 400 for 3 months or so and then freeze-drying (Ambrose 1975; Grattan 1988a:245). The theory behind the freeze-drying was that the PEG would remain solid and prevent deformation as the moisture passed from the ice to vapor phase and then out of the wood. With very deteriorated wood, however, not all collapse or shrinkage was prevented and the resulting wood was very fragile, spongy, and moist to the touch (Cook and Grattan 1984:219; Grattan 1988a:242,245). Nevertheless, at the present time, successful freeze-drying of archaeological woods usually employs a modification of the Ambrose technique.

The purpose of freeze-drying (or lyophilization) is to convert the water to ice and to effect the removal of the ice by direct sublimation so that the surface tension acting in the surface of a liquid layer is eliminated during the drying process, i.e., dehydration without the contraction of cells (MacKenzie 1984:190). In reality, freeze-drying is not by itself a conservation method but is one phase of a multistep conservation procedure. During the normal drying process of organic material, cell walls and fibers cannot resist the capillary attraction and surface tension. The cell walls approach one another and then become permanently bound so the material shrinks, cracks, and becomes brittle and unsuitable for further treatment or exhibit. The task of the conservation process is to avoid these damages in the drying process by means of excluding the consequences of capillary attraction following the evaporation of water. Freeze-drying is a physical process that prevents capillary attraction. Sublimation is the transition from the solid to the gaseous state, bypassing the liquid state, i.e., a direct evaporation of ice from the frozen object (Nielsen 1984:307). It is interesting to place the technique of freeze-drying in an historic perspective. Freeze-drying by sublimation at atmospheric pressure was accomplished with potatoes by Andean Indians in South America, with salmon by the Eskimo/Aleuts, and with clothes hung out to dry in winter by Americans in the midwest (MacKenzie 1984:191–192).

There is no question that water freezes in the lumina with freeze-drying but evidently water does not freeze-in small capillaries (the bound water). The most important aspect of freeze-drying from a PEG solution is that the water is removed and migration of the impregnant to the wood surface is prevented. Freeze-drying enables the PEG to be evenly distributed throughout the wood. It allows shrinkage but prevents collapse. The reasons for this are explained as follows: a piece of wood which has been impregnated with a solution of 25% PEG 400 in water at –20°C will consist of finely divided crystals of pure ice in a matrix of a super cooled water/PEG solution containing about 50% of each component. As freeze-drying takes place, the crystals of ice will gradually recede until the overall concentration of PEG is about 50%, at which point only the super-cooled PEG/water mixture will be present and drying will not take place by freeze-drying. As more water is removed solid PEG 400 separates out until it alone remains. As for the bound water, this will certianly not be frozen, because it is in some respects in a condition which resembles the gaseous state. Presumably, as freeze-drying proceeds, bound water is lost and is replaced by the mobile super-cooled water/PEG solution. One reason why the PEG freeze-drying process works to avoid shrinkage is that the movement of the super-cooled solution of PEG enables PEG to replace the bound water continuously as drying proceeds leading to complete replacement when the wood is dry. Thus, throughout the freeze-drying process, the cell walls are kept in an expanded condition (McCawley et al. 1982:260–261).

As noted above, there was occasionally some dissatisfaction with the results of using the Ambrose method which employs a 20–25% v/v solution of PEG 400 as pretreatment before prefreezing and then freeze-drying at –20°C. The method functions adequately in most instances but there are problems when waterlogged wood is extremely deteriorated or contains pockets of extreme deterioration. Although most of the dimensional change on going from the wet to the dry state is prevented, such wood may become dark in appearance, soft, wet, and spongy to the touch after treatment. Furthermore, capillary tension type collapse may not be entirely eliminated so that checking or cracking on the surface may occur. Wood in this condition is a problem to handle because it retains dirt, is fragile, extra sensitive to high relative humidity, and not very durable (Cook and Grattan 1984:219). Successful results were obtained by Watson (1982:238) using a consolidating solution made up of 10% PEG 400 and 15% PEG 3350. This procedure, now called the "double PEG" method, has gained support for moderately deteriorated wood and has been carried out at various laboratories (Grattan 1988a:248).

An international comparative study was conducted to determine the most beneficial treatment for waterlogged woods yielded the following conclusions: (1) low molecular weight PEG is most effective in controlling shrinkage of undeteriorated wood, (2) very deteriorated woods respond best when freeze-drying is used in combination with higher molecular weights of PEG, and (3) the use of mixed PEG 400 and 3350 as a pretreatment for freeze-drying gives wood a superior appearance and is most acceptable for a majority of degraded woods. When mixed PEG solutions are used, impregnation with the lower grade should start first building up to the final mixture in 5% or 10% stages. Nearly all of the treatments in the study project involved the use of PEG and freeze-drying (rather than PEG and air drying) which substantiates the widespread acceptance and good performance of the technique (Grattan 1988a:247, 1988b, 1988c:11–14).

Experiments with freeze-drying wood without PEG pretreatment resulted in samples collapsing and splitting in a day or two. Later examination by scanning electron microscope showed the presence of crystals which had been precipitated out of solution during freezing and had ruptured vessel walls. The crystals contained calcium and sulfur and it was concluded that the crystals were probably composed of gypsum. The formation of these crystals has not been observed in wood treated with PEG and it may be that the PEG either complexes the inorganic salts preventing crystallization or the PEG may obscure the presence of crystals (Watson 1982:241).

The biggest drawback with freeze-drying is the cost of equipment. Acquiring and maintaining a freeze-drying apparatus large enough to be useful for all but the smallest artifacts can be expected to cost more than the budgets of most archaeological projects can accommodate (Barbour 1985:ms.).

Wood Preservation in Florida

When Frank Hamilton Cushing excavated the Court of the Pile Dwellers at Key Marco, he made several observations that subsequently have been noted by all of us who have struggled with the problems of wet site archaeology: the great abundance of botanical remains over all other materials, the deceptively well-preserved condition of wooden artifacts, and the frustration of saving these fragile specimens from deterioration.

> Some of the things thus recovered could be preserved by very slow drying, but it soon became evident that by far the greater number of them could not be kept intact. No matter how perfect they were at first they warped, shrunk, split, and even checked across the grain, like old charcoal, or else were utterly disintegrated on being exposed to the light and air if only for a few hours. . . . fully twenty-five per cent of these ancient articles in wood and other vegetal material were destroyed in the search; and again, of those found and removed, not more than one-half retained their original forms unaltered for more than a few days (Cushing 1897:359).

Cushing sought by every means at his disposal to remedy these difficulties, but found that "ordinary glue, shellac, and silicate of soda proved to be comparatively ineffective" (1897:359).

At the time of Cushing's report, experiments with preserving wood had been underway in Europe for about 50 years but, as discussed earlier, they were not very satisfactory. It is unfortunate that the Key Marco site was excavated in the late 19th century. In 1972, I obtained permission to preserve a wooden stool or head rest from the Key Marco site that is part of a collection housed at the Florida Museum of Natural History (Purdy 1974:105–109; Gilliland 1975). This specimen was a desirable choice because there is a drawing and a detailed description of it, including its dimensions when it was recovered in 1896 (Cushing 1897:363). I was interested in the Key Marco remains and other wooden artifacts found in Florida primarily as a result of my involvement with the famous Ozette Village Site on the Olympic Peninsula, Washington. The hundreds of masks, figurines, and utilitarian items from the Ozette site were being preserved with PEG and I chose this product also. After recording the dimensions of the stool, I placed it in simmering water. It floated for 5 hours before it absorbed enough water to sink. After 4 days, during which I changed the water each day which was the color of coffee or very strong tea, I assumed that the stool was completely waterlogged. I preserved the artifact by soaking it in a 10% solution of heated PEG 1000 for 1 month, 25% for 2 weeks, and a 50% solution for 1 month (Purdy 1974:108). In August 1989, I measured the stool and compared the results to the 1972 posttreatment figures. They were nearly identical. Based on information that has been compiled in the last decade, I would proceed somewhat differently today in preserving the stool, but it is still in better condition than it was before treatment in 1972. We no longer heat the PEG and we use PEG 540 Blend (a blend of PEG 300 and 1450) which was not available in the early 1970s.

PEG 540 Blend was used to preserve large collections of wood from two archaeological wet sites in Florida, the Bay West Site (about 500 objects) and Hontoon Island (about 3000 objects). See Chapter 3 for a history of these sites. Both collections were treated in the early 1980s and both are still stable (Purdy 1982:159–169). This is important because a cookbook type of recipe was used. Specimens were kept waterlogged after excavation until treatment began. They were weighed, measured, and placed in increasing concentrations of PEG 540 Blend up to 50%. After treatment, the wood was allowed to dry slowly and was monitored daily at first and then occasionally. While the overwhelming majority of the wood from these two sites is in excellent condition, there are exceptions that should be mentioned. Some of the hardwoods did not preserve as well as the softwoods and it is probable that the PEG should have been taken to a higher concentration for hardwoods, a different molecular weight used, or the wood freeze-dried. In addition, a broken canoe paddle made of red cedar (*Juniperus* sp.), a softwood, remained stable for 8 months after treatment, then suddenly began to shrink. It was re-waterlogged immediately and treated again but this time the concentration of PEG 540 Blend was elevated ultimately to more than 90%. It is stable at this time. Specimens, or portions of specimens, that were charred did not preserve well at all. This is a topic that needs to be addressed by the ICOM Working Group on Wet Organic Archaeological Materials because many archaeological woods have been subjected to fire.

Twenty samples preserved in 1980 from the Hontoon Island site were chosen for special monitoring. Some interesting observations are as follows. It took approximately 2 months for the weight of the specimens to stabilize after they were removed from a 50% solution of PEG; that weight has been maintained. During the summer months the weight increases as much as 3.9%. This gain is undoubtedly due to Florida's high summer humidity. (See earlier discussion of hygroscopicity.) The wood is in uncovered containers and, although the laboratory is air conditioned, it lacks a special climate control. The woods in this sample consisted of bald cypress (*Taxodium distichum*), pine (Pinus sp. southern hard pine), American elm (*Ulmus americana*), and an unidentified hardwood. All of the identifications were made by Lee A. Newsom, a qualified wood anatomist. The only problems noted were with the unidentified hardwood piece which developed a longitudinal split and remained "spongy", and with the woods that were charred. Small pieces of charcoal broke off of one specimen from time to time and the fragments were placed in a plastic bag to keep them together. No dimensional changes were recorded with the flat pieces of wood but one of the round branch pieces shrunk slightly in the middle. In August 1989, I measured and weighed 17 of these samples. They had not changed since March 1985. No additional shrinkage or deterioration was noted on the hardwood or the specimen that had undergone differential shrinkage in the middle. The burned piece was not remeasured but it is very fragile and cannot be handled without additional fragments breaking off.

In 1983, another experiment was conducted to record the changes that occur when various species of waterlogged archaeological wood are allowed to air dry without preservation. The specimens included three hardwoods (*Ulmus americana* [elm], *Salix* spp. [willow], and *Diospyros virginiana* [persimmon] and two softwoods (*Taxodium* sp. [cypress] and *Pinus* sp. [pine]). For all species, the weight loss was between 87 and 93%. Dimensional changes, however, are quite different. The average of all values for each species was

	Longitudinal (%)	Radial (%)	Tangential (%)
Willow	30	35	50
Elm	15	40	60
Persimmon	13	50	53
Pine	20	24	24
Cypress	10	11	15

Pine and cypress underwent less overall shrinkage than the hardwoods and they did not undergo as radical changes in the three dimensions. As a result, it may be possible to suggest that one reason myself and others have experienced greater success in preserving softwoods is because even without preservation, the softwoods do not distort as drastically as the hardwoods. Another interesting but very preliminary result of this experiment was the finding that when one of the samples of persimmon was rewetted after having lost 52% of its original wet weight, it did not regain its original wet weight and dimensions completely (i.e., it had already undergone some permanent shrinkage), but when a sample of pine was rewetted after it had undergone a 52% loss, it regained all but 5% of its original wet weight and there was no measurable change in the original dimensions. It is interesting to compare these experimental results with the information available about the Key Marco stool which Cushing thought was manufactured of buttonwood (*Conocarpus erecta*), a hardwood. From the measurements provided by Cushing (1897:363), I was able to calculate that the stool had shrunk 15% longitudinally, 40–50% radially, and 50–60% tangentially (Purdy 1974:107). These values compare well with the dimensional changes observed in the hardwoods mentioned above, especially elm and persimmon.

Preservation of canoes has been underway since the mid-1970s. Chapter 4 is devoted to a description of Florida's canoes and their preservation needs.

Preliminary results of experiments at the University of Florida with freeze-drying of degraded archaeological wood using a modified Ambrose technique were not too successful. This suggests that further experimentation with solution concentrations and soaking times is needed (LeCompte 1987:86). At the present time, conservation facilities at the University of Florida do not include freeze-drying equipment.

I have never used a product other than PEG to preserve wood in Florida but because of the widespread belief that polyvinyl acetates (PVA or, for example, Elmer's Glue) can be used successfully for this purpose, I directed an experiment to demonstrate that this product usually is not desirable. The experiment was conducted by Craig C. Dochniak in the summer and fall of 1981 as a Florida Foundation for Future Scientists project. Dochniak's procedures and observations were excellent. In 1985, the specimens were reexamined and the results of his experiment summarized as follows:

> A study carried out at the University of Florida indicates that the use of PVA solutions is an undesirable technique. The experiment involved soaking pieces of wood, approximately 3 cm in diameter and i cm thick cut from the same branch, in 10%, 25%, and 50% solutions of PEG 540 Blend and PVA over a month's time. Several specimens remained in water as controls. Upon drying, it was found that all pieces soaked in PVA lost about 85% of their original weight and 65% of their original volume. The loss in weight and volume of the pieces soaked in PEG decreased with the increasing concentrations of PEG. . . . the sample from a 50% solution decreased only 38% in weight and 7% in volume . . . One of the control samples which was removed from water and allowed to air dry lost 91% of its original wet weight and 71% of its volume.
>
> Based on these preliminary results, a second experiment was performed in which one of the control specimens was submitted to increasing solution of PEG 540 Blend (10–80%) over a 1-month period. Upon drying, this piece retained its original wet weight and volume, thus showing 100% efficiency for preservation using a final solution of 80%

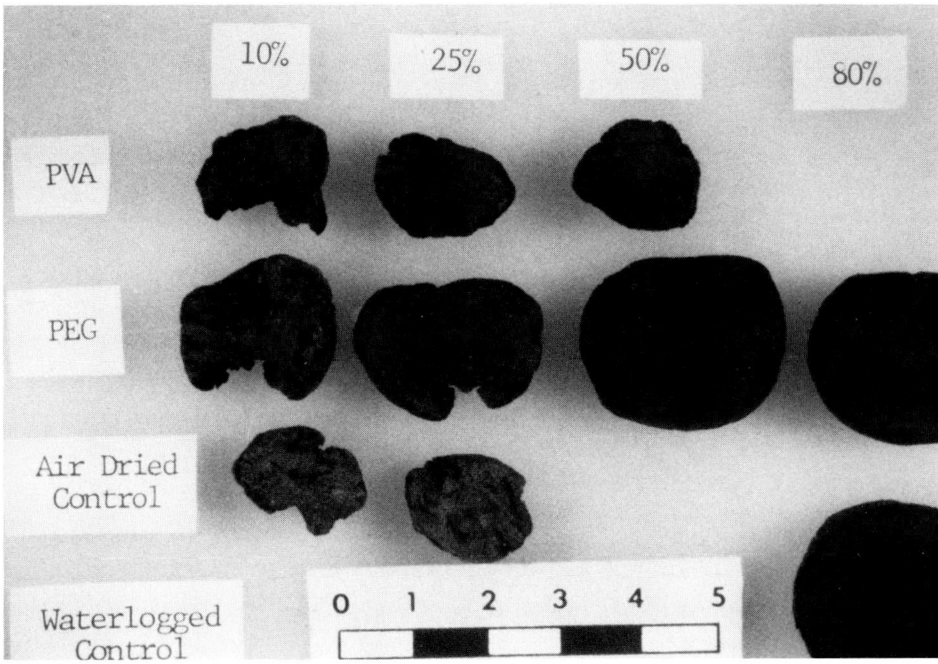

FIGURE 123 Results of experiment designed to test the differences between preserving wood with polyethylene glycol versus a polyvinyl acetate. (Photo property of the author).

> PEG. The specimens preserved in PEG have remained stable thus providing confidence in PEG as a long-term preservation method. In summary, preservative solutions of PEG 540 Blend were more effective than solutions of PVA or air drying, and higher concentrations of PEG worked better than lower concentrations (LeCompte 1987:85–86).

In August 1989, I reweighed and rephotographed the samples used in this experiment. The weights varied only slightly from the 1985 figures and were well within acceptable instrumental and human error. A visual comparison of the specimens is provided in Fig. 123.

The wood carvings and other wooden objects recovered at the Fort Center site in the 1970s (see Chapter 3) were soaked in a commercial grade of white glue (PVA). Despite the findings of the experiment reported above, the Fort Center woods appear to be stable at the present time. Perhaps a plausible explanation is that the objects were so badly degraded when retrieved from their burial matrix that the large PVA molecules were able to penetrate the structure and fill the voids left when the water was drawn out. Another possibility may be that degradation had advanced so far that the specimens were impenetrable and the PVA is merely serving as a sealant protecting the surface. Grattan makes a further suggestion that "the wood might not have been very degraded so that capillary tension collapse might not have occurred. This is a good example of why analytical information is so useful. Without the data on this wood we are really floundering to explain the observations!" (Grattan, personal communication, 1989b). It would be interesting to conduct an experiment to determine if the white glue has permeated the interior of these unique objects.

A preservation method employing sucrose shows promise and has gained some support recently (Cook and Grattan 1984; Grattan and Clarke 1989; Grosso 1981; Parrent 1985). It has been considered by the Florida Division of Historical Resources in Tallahassee as a product to preserve canoes. According to Grattan (1988a:249; Grattan and Clarke 1989:184), however, the evidence suggests that sucrose does not exercise as much control over wood shrinkage as PEG and there is greater danger of invasion by microorganisms and insects. Additional reserch is needed.

Additional Comments about Preservation of Archaeological Specimens from Wet Sites

Preservation of materials from water-saturated sites is a complex subject because many variables need to be considered. It is precisely for this reason that a divided camp exists between conservators who advocate a cookbook type of treatment and those who believe that each object should be treated individually. In Florida, I have been quite successful in subjecting large quantities of wood to the same treatment regime. The majority of these woods, however, have come from freshwater peat deposits and have not been associated with metals of any kind. The rare discovery of objects made of copper offer the only examples of metal found at pre-European sites in the Southeastern United States. Sometimes wood adhering to the copper has survived because of the copper salts, but both wood and copper are usually in poor condition when found. Organic remains from marine environments often pose special problems, particularly those from shipwreck sites where metals form part of the artifacts or have lain in close proximity for hundreds of years. Conservation of metals includes extracting chlorides or other salts, drying the metal, and creating a stable surface. For wood, as already discussed, preservation includes limiting the dimensional changes when going from waterlogged to a dry state, and producing a stable, durable material. PEG cannot be used to preserve metal/wood composite objects because metals corrode when placed in a PEG solution unless PEG is used in conjunction with corrosion inhibitors. On the other hand, some products are effective in removing chlorides but destroy the archaeological wood of a composite object. Substitutes for PEG have been tested, some of which are fairly effective in preserving the wood as well as metal. Apparently, each metal (i.e., iron, copper, bronze) requires a different product. The problems involved in preserving composite objects are being studied (Cook et al. 1984:147). Currently, work is being continued by metal corrosion specialists led by Dr. Lyndsie Selwyn. "Absolute corrosion rates have been determined for various metals in the presence and absence of wood. The combination of Hostacor KS1 and PEG still seems to be useful for ferrous metal/wood, indeed corrosion rates of zero have been determined.

Work at present is devoted to the effect of corrosion inhibitors in sea-water" (Grattan personal communication 1989b).

Other problems pertaining to the preservation of materials recovered from shipwrecks are discussed in various articles published in *Problems of the Conservation of Waterlogged Wood* (Oddy 1975), *Pacific Northwest Wet Site Wood Conservation Conference* (Grosso 1976), the *Proceedings of the Ninth Conference on Underwater Archaeology* (Arnold 1978; and other publications of the Council of Underwater Archaeology), *Conservation of Waterlogged Wood* (1979), *Canadian Conservation Institute Journal* (1980), the *Proceedings of the ICOM Waterlogged Wood Working Conference* (1982), *Waterlogged Wood: Study and Conservation* (1984), *From Shipwreck to Museum Exhibit* (VandenHazel 1987), and *Conservation of Marine Archaeological Objects* (Pearson 1989). In Florida, preservation of shipwrecks and their artifacts is conducted by the State Division of Historical Resources, Tallahassee.

Conservators agree that the simplest answer to a preservation problem should be sought. There may be times when "no treatment" is the procedure of choice. Such a situation exists when objects have been impregnated with oils as at a Basque whaling site in Labrador (Tuck 1982:171) and at a site in Yukon Territory (Titus 1982:153).

> There are examples when wood is found to be undeteriorated and requires [no treatment]. This is more true the further north you go! From an ethical point of view it is always desirable not to have to add anything to the wood. Thus, I believe that if wood is found to be little deteriorated, slow drying should always be tried first. This becomes particularly important for big timbers where impregnation and all the paraphernalia that goes with it, comes very expensive (Grattan, personal communication 1989b).

Most studies of waterlogged wood, however, show that no drying, no matter how slow, will save the wood without replacement of water with a bulking agent (Kazanskaya and Nikitina 1984:142). Cushing recounts his frustration in attempting to save the Key Marco artifacts by slow drying without a preservative (1897:359). For additional examples, see Figs. 124 and 125. Cushing would have had problems with some of the wooden artifacts from Key Marco if he had used PEG, however, because many of the masks and figurines were painted and PEG may remove paint. In such cases, other methods must be employed, if possible.

The use of biocides in controlling microorganisms on degraded waterlogged materials has not been discussed up to this point. In Florida, we have observed the growth of organisms in the preservation vats. We used to use a biocide called Cytox and also skimmed off the growth as it occurred. No growth has appeared on any woods after they have been removed from the preservation vat and dried. The conclusions drawn by Dawson (1982:269–276) and other members of the ICOM Working Group on Wet Organic Materials are that it is better to avoid biocides because some of them are not effective, all of them

FIGURE 124 The canoe pictured in Fig. 109 was placed in a holding pond that went dry during drought conditions. This photograph was taken two years later in 1990. (Photo credit Cindy Shraeder, Orlando Sentinel.)

A

B

FIGURE 125 (A, B) Two views of a 2300-year-old wooden mortar made of pine found in the Wekiva River, Orange County. It is nearly identical in size and age to those shown in Fig. 88 (photo courtesy of Don Baty). The bowl was not kept waterlogged or preserved. When it began to disintegrate two years later, it was coated with white glue. (C) In December 1990, the specimen was cored to obtain a sample for radiocarbon dating at which time it was learned that the wood had undergone extreme degradation.

FIGURE 125C

are toxic, and skimming the growth periodically is probably just as beneficial. In addition, although PEGs are not effective biocides, they do not support bacterial growth and, in high concentrations, PEGs may even dehydrate or rupture bacteria cells (Brownstein 1982:282–283). In some cases, the use of snails or fish to keep the vats clean of growth has been successful (Titus 1982:158, discussion section).

It is important to realize that fresh wood is not the same as degraded wood. The difference in anatomy and chemistry between deciduous and nondeciduous wood is present in fresh wood such that single species have different characteristics. The characteristics of fresh wood cannot be carried over to archaeological wood because the factors of deterioration enter into the picture (Sæterhaug 1984:196–198). This is apparent when considering that average shrinkage in fresh woods when dried is 0.2% longitudinally, 5.0% radially, and 10.0% tangentially (Brown, Panshin, and Forsaith 1952:40). As noted earlier, these figures are much higher in degraded woods. Most workers have found that for very degraded wood, the species has little influence on the success of conservation (Schweingruber 1982:105) and this has generally been the case in Florida. There appears to be a simple relationship between the extent of collapse and the state of deterioration of the wood overriding the effects of source, species, and condition (Cook and Grattan 1984:224). In Florida, the large batch of wood from the 6500-year-old Bay West site on the lower Gulf coast preserved the same as the collection retrieved from the much younger Hontoon Island site along the St. Johns River. One wonders, then, if measurements of water content, specific gravity, density, or mechanical strength have any real value for degraded woods. Some archaeological woods are very impregnable probably because they have undergone as much degradation as they are going to for a while. If tested, they would probably have a low water content and fairly high mechanical strength. I believe the important factors to consider are the proper molecular weight(s) needs to preserve an object and a way to determine if the PEG has penetrated the wood structure. It is also helpful to know how much PEG has been incorporated into the wood, but this may not be easy to evaluate (Grattan 1982:251).

At the present time, no one knows what will happen to wood preserved with PEG after 50 years or 1000 years. Because of this, some people would advocate no course of treatment until there is positive proof of a product's longterm success. Although we have not had problems with keeping specimens waterlogged over a long period of time (for exceptions see Chapter 4), many workers report detrimental changes in wood if it is not preserved quickly. The fact that PEG can be removed from an object, in theory at least, pleases conservators who may find a more effective product in the future. The manufacturers of PEG caution that when PEG is incorporated into wood, it forms a new composite material and the properties of pure PEG are no longer applicable. PEG very readily forms complexes with many materials and this complexing distorts the properties of PEG making it difficult to relate the properties of pure PEG to those found in treated wood (Brownstein 1982:62; Sawada 1982:117). This may be a problem when, for the sake of economy, a conservator wishes to reuse PEG after treatment of other specimens is completed. If the PEG has broken down or changed into another product, it may have lost its effectiveness for preserving additional objects (1982:283).

Although many unknown variables exist because ideal systems, such as pure water in intact cells (1984:192), do not prevail when preservation products are being considered, the bottom line is that organic objects are more useful for study and exhibit if they are preserved. Experience demonstrates, in fact, that most waterlogged materials, if not preserved, will deteriorate so badly that they are useless for any purpose. Discussions at the ICOM Working Group on Wet Organic Materials conferences have gone a long way in resolving conflicts concerning the requirements of conservators who preserve art objects for exhibit that will not be handled further, and those who wish to use preserved items to study wood species for what they can offer about environment, species preference, and woodworking technology at archaeological sites. The "standards" for these two objectives can be very different. The exhibitor wants the specimen to look as natural as possible while the archaeologist may wish to coat the surface with a sealant so that it will withstand examination. He may even want to use an archaeological wood collection for classroom purposes.

There are other problems related to waterlogged materials: who decides what should be saved and preserved when large quantities of perishable organic materials are recovered? Is it the archaeologist or the conservator? If the archaeologist is in charge, there is a big danger that he will not know how to provide even initial protection for the materials recovered from archaeological wet sites. If the conservator makes the decision, there is a big danger of not knowing what may be considered important now or in the future to understanding the past of a particular human group. It is obvious that interdisciplinary dialogue and action are needed. Conservation is not something that should be carried out in isolation. Perhaps decisions should be made by a committee representing various factions.

An alternative choice to preservation may be reburial after casts, replicas, models, photographs, or scale drawings are made.

References

Ambrose, W.R. 1975 Stabilizing Degraded Swamp Wood by Freeze-Drying. *International Council of Museums (ICOM) Committee for Conservation.* Fourth Triennial Meeting. International Council of Museums, Paris.

Arnold, J. Barto III 1978 Beneath the Waters of Time. *Texas Antiquities Committee Publication* No. 6. Austin.

Barbour, R.J. 1982 Shrinkage and Collapse in Waterlogged Archaeological Wood: Contribution III, Hoko River Series. *Proceedings of the ICOM Waterlogged Wood Working Group Conference*, pp. 209–225. Edited by D.W. Grattan and J.C. McCawley. International Council of Museums (ICOM), Committee for Conservation, Waterlogged Wood Working Group, Ottawa.

1984 The Condition and Dimensional Stabilization of Highly Deteriorated Waterlogged Hardwoods. *Waterlogged Wood: Study and Conservation*, pp. 23–37. Grenoble, France.

Brown, H.P., A.J. Panshin, and C.C. Forsaith 1952 *Textbook of Wood Technology*, Vol. 2. McGraw-Hill, New York.

Brownstein, Allen 1982 The Chemistry of Polyethylene Glycol. *Proceedings of the ICOM Waterlogged Wood Working Group Conference*, pp. 279–285. Edited by David W. Grattan and J.C. McCawley. International Council of Museums (ICOM), Committee for Conservation, Waterlogged Wood Working Group. Ottawa.

Coles, John M. 1982 The Somerset Levels: A Waterlogged Landscape. *Proceedings of the ICOM Waterlogged Wood Working Group Conference*, pp. 129-141. Edited by D.W. Grattan and J.C. McCawley. International Council of Museums (ICOM), Committee for Conservation, Waterlogged Wood Working Group. Ottawa.

CCI 1980 *The Journal of the Canadian Conservation Institute* 4, 52 pages. National Museums of Canada.

Conservation of Waterlogged Wood 1979 *International Symposium on the Conservation of Large Objects of Waterlogged Wood.* Government Printing and Publishing Office, The Hague.

Cook, Clifford and David W. Grattan 1984 A Practical Comparative Study of Treatments for Waterlogged Wood: Part III: Pretreatment Solutions for Freeze-Drying. *Waterlogged Wood: Study and Conservation*, pp. 291–239. Grenoble, France.

Cook, Clifford, A. Dietrich, David W. Grattan, and N. Adair 1984 Experiments with Aqueous Treatments for Waterlogged Wood-Metal Objects. *Waterlogged Wood: Study and Conservation*, pp. 147–159. Grenoble, France.

Cushing, Frank Hamilton 1897 Exploration of Ancient Key Dweller Remains on the Gulf Coast of Florida. *American Philosophical Society Proceedings* 35, pp. 329–448.

Daugherty, Richard D. 1988 Problems and Responsibilities in the Excavation of Wet Sites. In *Wet Site Archaeology*, pp. 15–29. Edited by Barbara A. Purdy. The Telford Press, West Caldwell, New Jersey.

Davis, John H. 1946 The Peat Deposits of Florida: Their Occurrence, Development, and Uses. *State of Florida Geological Bulletin*, No. 30.

Dawson, John 1982 Some Considerations in Choosing a Biocide. *Proceedings of the ICOM Waterlogged Wood Working Group Conference*, pp. 269–277. Edited by D.W. Grattan and J.C. McCawley. International Council of Museums (ICOM) Committee for Conservation, Waterlogged Wood Working Group, Ottawa.

Ember, Lois R. 1988 Preserving the Past. *Chemical and Engineering News*, November 18: 10–19.

Florian, Mary-Lou 1982 Anomalous Wood Structure: A Reason for Failure of PEG in Freeze Drying Treatments of Some Waterlogged Wood from the Ozette Site. *Proceedings of the Waterlogged Wood Working Group Conference*, pp. 85–106. Edited by D.W. Grattan and J.C. McCawley. International Council of Museums (ICOM), Committee for Conservation, Waterlogged Wood Working Group. Ottawa.

Gilliland, Marion Spjut 1975 *The Material Culture of Key Marco, Florida*. University Presses of Florida, Gaineville.

Goode, Richard W., III 1982 Vertical Distribution of Sulfide in the Organic Soils of the Northern Everglades, Palm Beach County, Florida. M.S. thesis, Department of Geology, University of Florida, Gainesville.

Grattan, David W. 1982 A Practical Comparative Study of Treatments for Waterlogged Wood, Part II: The Effect of Humidity on Treated Wood. *Proceedings of the ICOM Waterlogged Wood Working Group Conference*, pp. 243–252. Edited by D.W. Grattan and J.C. McCawley. International Council of Museums (ICOM) Committee for Conservation, Waterlogged Wood Working Group. Ottawa.

1984 Round Table Conference on Freeze-Drying. *Waterlogged Wood: Study and Conservation*, pp. 241–258. Grenoble, France.

1988a Treatment of Waterlogged Wood. In *Wet Site Archaeology*, pp. 237–254. Edited by Barbara A. Purdy. The Telford Press, Caldwell, New Jersey.

1988b The International Comparative Waterlogged Wood Treatment Project. *ICOM Committee for Conservation, Waterlogged Wood Working Group Newsletter* No. 16. Ottawa.

1988c International Comparative Study on the Treatment of Waterlogged Wood. *NewsWarp*, No. 4, pp. 11–14.

1989a Waterlogged Wood. In *Conservation of Marine Archaeological Objects*, pp. 55–67. Butterworths, London.

1989b Personal communication, September. Dr. Grattan reviewed this chapter for me. He is presently Senior Conservation Scientist at the Canadian Conservation Institute, Ottawa.

Grattan, D.W. and R.W. Clarke 1989 Conservation of Waterlogged Wood. In *Conservation of Marine Archaeological Objects*, pp. 164–206. Butterworths, London.

Grosso, G.H. (ed.) 1976 *Pacific Northwest Wet Site Wood Conservation Conference*, Vol. 1. Washington State University, Pullman.

1981 Experiments with Sugar in Conserving Waterlogged Wood. Paper presented at the International Council of Museums (ICOM), Committee for conservation, Sixth Triennial Meeting, Ottawa.

Hoffmann, Per 1982 Chemical Wood Analysis as a Means of Characterizing Archaeological Wood. *Proceedings of the ICOM Waterlogged Wood Working Group Conference*, pp. 73–83. Edited by D.W. Grattan and J.C. McCawley. International Council of Museums (ICOM), Committee for Conservation, Waterlogged Wood Working Group, Ottawa.

1984 On the Stabilization of Waterlogged Oakwood with PEG: Molecular Size Versus Degree of Degradation. *Waterlogged Wood: Study and Conservation*, pp. 95–115. Grenoble, France.

Jespersen, Kirsten 1984 Extended Storage of Waterlogged Wood in Nature. *Waterlogged Wood: Study and Conservation*, pp. 39–54. Grenoble, France.

Kazanskaya, Svetlana Yu. and Klara F. Nikitina 1984 On the Conservation of Waterlogged Degraded Wood by a Method Worked Out in Minsk. *Waterlogged Wood: Study and Conservation*, pp. 139–146. Grenoble, France.

LeCompte, Elise V. 1987 The Conservation of Wooden Remains from Archaeological Wetsites. *The Florida Anthropologist* 40(1):85–89.

MacKenzie, Alan 1984 Round Table Conference on Freeze-Drying (led by David Grattan). *Waterlogged Wood: Study and Conservation*, pp. 241–258. Grenoble, France.

Matsuda, Takatsugu 1984 On Preproducing the Shape of the Tissues of the Dried and Shrunk Waterlogged Wood for the Identification of Its Species. *Waterlogged Wood: Study and Conservation*, pp. 55–62. Grenoble, France.

McCawley, J.C., D.W. Grattan, and Clifford Cook 1982 Some Experiments in Freeze-Drying: Design and Testing of a Non-Vacuum Freeze-Dryer. *Proceedings of the ICOM Waterlogged Wood Working Group Conference*, pp. 253–262. Edited by D.W. Grattan and J.C. McCawley. International Council of Museums (ICOM), Committee for Conservation, Waterlogged Wood Working Group, Ottawa.

Nielsen, Hans-Otto 1984 The Treatment of Waterlogged Wood from the Excavation of the Haithabu Viking Ship. *Waterlogged Wood: Study and Conservation*, pp. 299–312. Grenoble, France.

Oddy, W.A. 1975 Problems in the Conservation of Waterlogged Wood. *Maritime Monographs and Reports*, No. 16. National Maritime Museum, Greenwich, London.

Parrent, J.M. 1985 The Conservation of Waterlogged Wood Using Sucrose. *Studies in Conservation* 30:63–72.

Pearson, Colin (Ed.) 1989 *Conservation of Marine Archaeological Objects*. Butterworths, London.

Purdy, Barbara A. 1974 The Key Marco Florida Collection: Experiment and Reflection. *American Antiquity* 39(1):105–109.

1982 Survey, Recovery and Treatment of Wooden Artifacts in Florida. *Proceedings of the ICOM Waterlogged Wood Working Group Conference*, pp. 159–169. Edited by D.W. Grattan and J.C. McCawley. International Council of Museums (ICOM), Committee for Conservation, Waterlogged Wood Working Group, Ottawa.

Sæterhaug, Roar 1984 Investigations Concerning the Freeze-Drying of Waterlogged Wood Conducted at the University of Trondheim. *Waterlogged Wood: Study and Conservation*, pp. 195–206. Grenoble, France.

Sawada, Masaaki 1984 Some Problems of Setting of PEG 4000 Impregnated in Wood. *Waterlogged Wood: Study and Conservation*, pp. 117–124. Grenoble, France.

Schweingruber, Fritz H. 1982 Conservation of Waterlogged Wood in Switzerland and Savoy. *Proceedings of the ICOM Waterlogged Wood Working Group Conference*, pp. 99–106. Edited by D.W. Grattan and J.C. McCawley. International Council of Museums (ICOM), Committee for Conservation, Waterlogged Wood Working Group, Ottawa.

Seifert, Betty and Richard Jagels 1984 Conservation of the Ronson Ship's Bow. *Waterlogged Wood: Study and Conservation*, pp. 269–297. Grenoble, France.

Titus, Larry 1982 Conservation of Wooden Artefacts. *Proceedings of the ICOM Waterlogged Wood Working Group Conference*, pp. 153–158. Edited by D.W. Grattan and J.C. McCawley. International Council of Museums (ICOM), Committee for Conservation, Waterlogged Wood Working Group, Ottawa.

Tuck, James A. 1982 Conservation of Waterlogged Wood at a 16th Century Whaling Station. *Proceedings of the ICOM Waterlogged Wood Working Group Conference*, pp. 171–175. Edited by D.W. Grattan and J.C. McCawley. International Council of Museums (ICOM), Committee for Conservation, Waterlogged Wood Working Group, Ottawa.

Union Carbide Corporation 1981 *Carbowax: Polyethylene Glycols*. Danbury, Connecticut.

VandenHazel, Bessel J. 1987 From Shipwreck to Museum Exhibit. *Underwater Archaeology Report* No. 3. Nipissing University College, Ontario, Canada.

Watson, Jacqui 1982 The Application of Freeze-Drying on British Hardwoods from Archaeological Excavations. *Proceedings of the ICOM Waterlogged Wood Working Group Conference*, pp. 237-242. Edited by D.W. Grattan and J.C. McCawley. International Council of Museums (ICOM), Committee for Conservation, Waterlogged Wood Working Group, Ottawa.

1984 Research into Aspects of Freeze-Drying Hardwoods Between 1982 and 1984. *Waterlogged Wood: Study and Conservation*, pp. 213–218. Grenoble, France.

Young, Gregory S. and Ian N.M. Wainwright 1982 Polyethylene Glycol Treatments for Waterlogged Wood at the Cell Level. *Proceedings of the ICOM Waterlogged Wood Working Group Conference*, pp. 107–116. Edited by D.W. Grattan and J.C. McCawley. International Council of Museums (ICOM), Committee for Conservation, Waterlogged Wood Working Group, Ottawa.

INDEX